S0-ABV-538

DISPERSION OF POWDERS IN LIQUIDS

With Special Reference to Pigments

SECOND EDITION

Edited by

G. D. PARFITT

Tioxide International Limited
Billingham, Teesside

A HALSTED PRESS BOOK

JOHN WILEY & SONS
New York—Toronto

PUBLISHED IN THE U.S.A. AND CANADA BY
HALSTED PRESS
A DIVISION OF JOHN WILEY & SONS, INC., NEW YORK

First edition 1969
Second edition 1973

Library of Congress Cataloging in Publication Data

Main entry under title:
Dispersion of powders in liquids.

"A Halsted Press book."
Includes bibliographical references.
1. Pigments. 2. Dispersion. 3. Colloids.
I. Parfitt, G. D., ed.
TP936.D57 1973 667'.29 72-11560
ISBN 0-470-65900-9

WITH 28 TABLES AND 100 ILLUSTRATIONS

Printed in Great Britain by Galliard Limited, Great Yarmouth, Norfolk, England.

CONTENTS

List of Contributors xi

Introduction xiii

Chapter 1

FUNDAMENTAL ASPECTS OF DISPERSION — *G. D. PARFITT*

The Dispersion Process 1
The three stages of the dispersion process 1
Dispersibility 3
Wetting of the external and internal surfaces 3
Mechanical breakdown of clusters 8
Stability 10
Maximum rate of flocculation 10
Forces of interaction between particles 13
London–van der Waals attractive forces 14
The coulombic repulsive force 17
The total interaction energy 19
Rate of slow flocculation 22
Application of the DLVO theory 23
Stability arising from adsorbed layers 31
Concentrated dispersions 39

Chapter 2

PROPERTIES OF THE SOLID–LIQUID INTERFACE OF RELEVANCE TO DISPERSION — *M. J. JAYCOCK*

Forces Between Atoms, Ions and Molecules 44
Attractive and repulsive forces between molecules 44
Coulombic forces between ions 45
Interaction between a permanent dipole and an ion 46
Interaction between two permanent dipoles 46
Polarisability 47

Description of an Interface . . . 47
Surface tension . . . 48
Total surface energy . . . 49
Surface entropy . . . 49
Interfacial tension . . . 49
The Solid Surface . . . 50
Surface mobility . . . 51
Conditioning of solid surfaces . . . 52
Surface tension and surface free energy . . . 53
Calculated values of surface tension and surface energies . . . 54
Energies associated with edges and corners . . . 57
Specific surface area . . . 58
The surface of real solids . . . 63
Wetting and Contact Angles . . . 68
Magnitude of contact angles . . . 68
Spreading coefficients . . . 70
The adhesion of liquids to solids . . . 71
Flotation . . . 71
Wetting and non-wetting . . . 73
Adhesion . . . 74
Adhesive forces . . . 74
The stickiness of particles . . . 75
Sliding friction and lubrication . . . 76
Rolling friction . . . 77
Adsorption from Solution . . . 77
The composite isotherm . . . 78
Adsorption from completely miscible liquids . . . 78
Adsorption from partially miscible liquids . . . 79
Adsorption of dissolved solids from solution . . . 79
The adsorption of polymers . . . 80
The adsorption of electrolytes . . . 80

Chapter 3

ELECTRICAL PHENOMENA ASSOCIATED WITH THE SOLID–LIQUID INTERFACE — *A. L. SMITH*

Introduction . . . 86
The Electric Double Layer . . . 86
Origin of the electric double layer . . . 86
The nature and definition of electric potentials in the double layer 87
Potential determining ions and the Nernst equation . . . 90
Applicability of the Nernst equation to surfaces . . . 91
Adsorption of non-potential determining ions and the inner part of the double layer . . . 94
Adsorption of dipolar molecules . . . 98
The Stern adsorption isotherm and the discreteness of charge effect 99
The diffuse region of the double layer in solution . . . 102
Capacity of the diffuse region of the double layer . . . 104
Double layer on the solid side of the interface . . . 107
Free energy of the diffuse double layer . . . 109
Electrokinetic Properties of Dispersions . . . 111
Classification . . . 111
The zeta potential . . . 112

Experimental determination of electrophoretic mobilities . . 114
Stationary levels in microelectrophoresis cells 115
Choice of microelectrophoresis cell 117
Experimental detail 119
Choice of electrodes 122
Optical corrections 123
Accuracy of electrophoretic mobilities 124
Conversion of mobilities to zeta potentials 125

Chapter 4

SURFACE-ACTIVE COMPOUNDS AND THEIR ROLE IN PIGMENT DISPERSION — *W. BLACK*

Introduction 132
Surface Activity 133
Properties of Surface-active Agents 136
Selection of Surface-active Agents 141
Dispersion of Pigments in Aqueous Media 145
Foaming 152
Water-based stoving paints 154
Miscellaneous aqueous dispersions 156
Dispersion of Pigments in Non-aqueous Media 156
Easily dispersed pigments 166
Universal Tinters 167
Pigment Flushing 168
Pigment-resin Printing of Textiles 169
Problems and Prospects 171

Chapter 5

PRINCIPLES OF PRECIPITATION OF FINE PARTICLES — *A. G. WALTON*

Introduction 175
The Metastable Limit 176
Formation and Stoichiometry of Clusters and Metastable Phases . 177
Nucleation Theory 181
Homogeneous nucleation 181
Heterogeneous nucleation 183
Criticisms of nucleation theory 184
Nucleation of Polymers 184
Secondary or Ancillary Nucleation 185
The Role of the Impurity Substrate—Epitaxy 185
Experimental Tests of Nucleation Theory 188
Heterogeneous nucleation 188
Homogeneous nucleation 190
Solution Phase Nucleators 194
Effect of Nucleation Mechanism on Precipitate Characteristics . 195
Precipitation Kinetics 197
Precipitate Morphology 203
Ageing and recrystallisation 211
Morphology and supersaturation 214
Nucleation and precipitation in electric fields 216
Summary 218

Chapter 6

TECHNICAL ASPECTS OF DISPERSION AND DISPERSION EQUIPMENT — *I. R. SHEPPARD*

Introduction 221
Definitions and Objectives 222
Classification of Type of System Involved 226
Range of Equipment Available 228
Low shear rate equipment 230
High shear rate mixers 234
Ball mills 237
Roll mills 242
Range of Industries and Types of Product 245
Adhesives 245
Ceramics and refractories 246
Chemicals 246
Paint 247
Paper industry 248
Pharmaceuticals 249
Pigments and dyestuffs 249
Plastics 250
Printing inks 251
Rubber 252
Economics of Dispersion Operations 253
Dispersion Stage Optimisation and Formulation 255
Optimisation experiments 255
Use of rheology 261
Extraction from final formulation 263

Chapter 7

ASSESSMENT OF THE STATE OF DISPERSION — *S. H. BELL and V. T. CROWL*

Introduction 267
The Nature of Powders 268
The Process of Dispersion 272
Technological Properties of Dispersions 273
Sedimentation behaviour 275
Rheological behaviour 277
Methods of Assessment of the Degree of Dispersion and Dispersion Stability 282
Control tests for degree of dispersion 282
Particle size analysis 283
Sedimentation behaviour 291
Rheological properties 295
Optical properties 297
Miscellaneous methods 301
The Nature of Practical Dispersions 303

Chapter 8

DISPERSION OF INORGANIC PIGMENTS — *H. D. JEFFERIES*

Introduction 308
Classification 308
Pigments in paints and inks 309
Pigments: manufacturing processes 312
Particle size, opacity and colour of inorganic pigments . . 313
Dispersion media 318
Characterisation of Inorganic Pigments: Physical and Chemical Aspects 327
Particle shape. 327
Particle size and size distribution 328
Rugosity, roughness or smoothness factors 329
The effect of milling (micronising) of pigments 330
Structured pigments; carbon blacks 333
The composition of the pigment surface 335
Dispersion of Inorganic Pigments 339
Rheology of pigment suspensions 340
Oil absorption value of pigments 341
The relative viscosity of pigment suspensions 345
Pigment/dispersant/liquid ratio. 347
Rheological aspects of dispersion 350
The Dispersion Process in Practice 353
The breakdown of aggregates and agglomerates 353
Pigment wetting in practice 355
Stabilisation of pigment suspensions 360
Pigment Dispersion in the Final Product 367
Paint film structure 367
Paint film gloss 373
Flooding and floating 376
Pigment dispersion and hiding power 377

Chapter 9

DISPERSION OF ORGANIC PIGMENTS — *J. BERESFORD and F. M. SMITH*

Introduction 383
Chemical Classification of Organic Pigments 384
The Physical Properties of Pigments 387
Pigments as powders 387
The crystallography of organic pigments 390
The optical properties of pigments 393
Organic pigments in dispersion. 394
The Application of Pigments 399
Extremely high viscosity systems 399
Very high viscosity systems 400
High viscosity systems 400
Medium viscosity systems 401
Low viscosity systems 403
The Dispersion of Organic Pigments 405
Conclusions 407

Index 411

LIST OF CONTRIBUTORS

S. H. Bell
Paint Research Station, Waldegrave Road, Teddington, Middlesex

J. Beresford
Ciba–Geigy (UK) Limited, Hawkhead Road, Paisley, Scotland

W. Black
Imperial Chemical Industries Limited, Organics Division, P.O. Box 42, *Hexagon House, Blackley, Manchester* 9

V. T. Crowl
Paint Research Station, Waldegrave Road, Teddington, Middlesex

M. J. Jaycock
*Chemistry Department, Loughborough University of Technology, Loughborough, Leicestershire LE*11 3*TU*

H. D. Jefferies
4 *Tintern Avenue, Billingham, Teesside*

G. D. Parfitt
*Tioxide International Limited, Central Laboratories, Portrack Lane, Stockton-on-Tees, Teesside TS*18 2*NQ*

I. R. Sheppard
41 *Twiss Green Lane, Culcheth, Warrington, Lancashire*

A. L. Smith
*Chemistry Department, Liverpool Polytechnic, Byrom Street, Liverpool L*3 3*AF*

F. M. Smith
Ciba–Geigy (UK) Limited, Pigments Division, Roundhorn Estate, Wythenshawe, Manchester 22

A. G. Walton
Division of Macromolecular Science, Case Western Reserve University, Cleveland, Ohio 44106, *USA*

INTRODUCTION

One clear objective of the first edition of this book was to combine in a single volume the fundamental principles of colloid and interface science involved in the dispersion of powders in liquids with critical accounts of those technical aspects of the dispersion process that particularly concern pigment producers and users. Perhaps the first edition served to demonstrate the wide gap that apparently exists between theory and practice—indeed a number of reviewers felt the need to say so. But as time passes the gap narrows, the principles are being applied to real systems, and there is increasing recognition of the fact that the basic rules of wetting, flocculation, etc. can be applied with success.

The chapters in the second edition have been updated. They are also in a different order from those in the first, the hope being that the new approach provides a more logical development from principles to practice.

The fundamentals and terminology of the dispersion process are defined in Chapter 1. It is useful to divide the overall process into three distinct stages. The first involves the wetting of the powder surface, and Chapter 2 considers those aspects of the surfaces of solids and their interactions with liquids that are relevant to this stage. Once the powder is wetted out a milling stage is usually necessary but for pigments this area of mechanical breakdown of aggregated structures is not well defined, and other than recognising the problem there is little to be said at present. The third stage concerns the stability of the dispersion to flocculation, and in Chapter 1 the various theories of stability are described for systems containing either charged or uncharged colloidal particles. The relevant theoretical and practical aspects of electrical phenomena associated with charged solid/solution interfaces, including electrokinetic phenomena and the calculation and significance of zeta potential, are discussed in Chapter 3. These first three chapters contain a number of references to scientific studies of the various aspects of the dispersion process: to the casual reader they may seem too remote from the practical situation to merit further study, but succeeding chapters focus attention on the fundamentals and how they find application in practice, and endeavour to pinpoint the problem areas.

In all three stages it might be said that 'surface activity' predominates, and it is not surprising that in a variety of ways surface active agents play an important role in the process. In the large range of dispersion processes in which pigments are used, the surface-active compounds employed are legion. Their types, properties and applications are reviewed in Chapter 4. The alternative technique for the formation of a dispersion of fine particles in a liquid medium, that is by precipitation from supersaturated systems, is discussed in Chapter 5. This is, of course, an important aspect of the manufacture of many inorganic and organic pigments, but the principles involved in the development of precipitate morphology are also relevant to the ageing processes that can occur in certain pigmented systems. The techniques used for dispersion in industrial processing are surveyed in Chapter 6, and a wide range of products are covered to illustrate how the choice of machine is related to the nature of the system involved, and how optimisation can be achieved. The degree of dispersion achieved may be difficult to assess—the ultimate experiment would clearly define the particle size distribution in the dispersed system, but in so many practical cases this experiment is still somewhat remote. Several quite reliable techniques are in regular use; these are based on the principles discussed earlier in the book and something of a link between these principles and the practical problem of assessing the state of dispersion is created in Chapter 7. The final two chapters are concerned with the current position in the dispersion of inorganic (Chapter 8) and organic (Chapter 9) pigments in a variety of media. Where feasible the authors demonstrate the application of fundamentals while focusing attention on the areas where knowledge is limited and empiricism prevails.

It is hoped that this book will help to bridge the gap between the science and technology of the dispersion process, and in doing so make both sides increasingly useful to each other.

G. D. Parfitt,
Tioxide International Ltd

CHAPTER 1

FUNDAMENTAL ASPECTS OF DISPERSION

G. D. PARFITT

THE DISPERSION PROCESS

The term *dispersion* is used here to refer to the complete process of incorporating a powder into a liquid medium such that the final product consists of fine particles distributed throughout the medium. The dispersion of fine particles is normally termed *colloidal* if at least one dimension of the particles lies between 10 Å (1 mμ) and 10^4 Å (1 μ), and the term *sol* is used for *any* colloidal system in which the dispersion medium is liquid (mostly when the dispersed phase is solid).

Sols are classified, in terms of the affinity of the colloidal particle for the medium, as *lyophobic* (possessing aversion to liquid) or *lyophilic* (possessing affinity for liquid); *hydrophobic* and *hydrophilic* respectively for aqueous media. Some difficulties may arise with the use of these terms. In the sense of the definition lyophobic implies no affinity between the colloid and the medium, *e.g.* an insoluble powder, but taken to its logical extreme this would suggest no wetting of the powder by the liquid and hence no dispersion could be formed. But although a metal oxide powder may be insoluble in water it is nevertheless normally wetted by water and hence the powder surface is hydrophilic, but the dispersion is classed as hydrophobic. We might also wish to consider the degree of hydrophobicity of a surface in terms of, say, the heat of wetting in water; there is no hard and fast rule when these terms are used for interfaces, but the application to dispersion is clearly defined. Solutions of macromolecules and association colloids are of the lyophilic type and form spontaneously when the components are brought into contact. They are true solutions and are stable in the thermodynamic sense whereas lyophobic sols do not form spontaneously and hence in principle are thermodynamically unstable. The division between solutions and dispersions is without ambiguity.

The three stages of the dispersion process

It is useful to consider the overall process of dispersion as consisting of three stages, and take each stage separately in order to assess the importance of the various factors involved. These three stages are quite distinct in their nature, but in practice they overlap. The principles involved in each stage are fairly well established but because of the overlap it is often not easy to recognise any particular aspect in a practical dispersion system. The three stages to be considered are the following.

Wetting of the powder

In many practical uses of powders the primary particle size is sufficiently small for further sub-division to be unnecessary. But in the dry state the powder usually contains some aggregates of primary particles and these may be attached to other aggregates and/or primary particles forming agglomerates (using Gerstner's terminology[1]). Not only is it necessary for the liquid to wet the external surfaces but it must also displace air from the internal surfaces between the particles in the clusters. Hence this aspect will involve a knowledge of the wetting characteristics of the system and some assessment of the dimensions of the internal surfaces.

Breaking up the clusters to form colloidal particles

Aggregates may require considerable mechanical energy to break them down completely to the point when the surface of each primary particle is available to the wetting liquid. Presumably agglomerates would normally require less energy than aggregates. Many factors are involved in the powder achieving its particular state so that a knowledge of its manufacture and/or storage conditions is necessary for the problem of breaking the bonds between particles in the clusters to be completely understood.

Flocculation of the dispersion

Having wetted the surfaces and broken down the clusters into fine particles, these are then dispersed throughout the medium. The problem is then to maintain the dispersed state since the particles have a natural tendency (as a result of Brownian motion) to reduce in number with time due to irreversible collisions. An attractive force exists between the particles as they approach, the magnitude of which increases significantly as the distance of approach decreases. The reduction in particle number is here to be termed *flocculation* whatever the mechanism involved. The other term, *coagulation*, which is often used to represent this process, has various meanings in technology so is to be avoided in this book; but in the fundamental literature it is usually used either in place of or together with (indiscriminately) the term flocculation, and the casual observer may look for a hidden difference between the two terms. (La Mer[2] has suggested a means of differentiating between them but this is not universally accepted). To resist flocculation some sort of repulsive force is necessary and this is usually achieved through the particles being charged or containing adsorbed layers which protect them, or both.

Dispersibility may be defined as the ease with which a dry powder may be dispersed in a liquid, and this term can be used to express the effectiveness of the first two stages. The *stability* of a system is the resistance to flocculation. These are two separate ideas and will be treated as such in the following discussion. The relationships between the factors and principles involved and the practical aspects of dispersibility have received relatively little attention in the past, in contrast to the extensive investigation of flocculation behaviour. One difficulty in practice is that it is difficult to differentiate between a system that has been dispersed but is unstable and one which cannot readily be dispersed; ill-defined experimental conditions may well lead to confusion in interpretation. Surface active agents often play a leading role in all three aspects of the dispersion process, although they might easily be useful in one but antagonistic in another. There are no simple rules; each case has to be considered in detail.

DISPERSIBILITY

Wetting of the external and internal surfaces (see Chapter 2)

Various terms are used, often loosely, to describe the process that occurs when a solid phase and a liquid phase come into contact so that the solid–air interface is replaced by the solid–liquid interface. In general we refer to this as *wetting* but for practical purposes it is more useful to postulate that more than one kind of wetting may be associated with the formation of a solid–liquid interface, and to describe each type in terms of the changes in free surface energy involved. There are three distinct types of wetting,[3] designated as *adhesional* wetting, *spreading* wetting and *immersional* wetting, according to the mechanical process taking place.

(*i*) Adhesional wetting. When 1 cm^2 of a plane solid (S) surface is brought into contact with 1 cm^2 of a plane liquid (L) surface, unit surface area of each phase disappears to form unit area of the new solid–liquid interface. The work W_a involved in this process carried out under isothermal conditions (equals the decrease in free surface energy) was given by Dupré[4] as

$$W_a = \gamma_{S/L} - (\gamma_S + \gamma_{L/V}) \tag{1.1}$$

where $\gamma_{S/L}$, γ_S and $\gamma_{L/V}$ are the interfacial free energies at the solid–liquid, solid–vapour and liquid–vapour interfaces respectively, the vapour here being that of the separate phases. ($-W_a$ is the work of adhesion, the reversible work required to restore the initial condition.)

(*ii*) Spreading wetting. When a drop of liquid spreads over a plane solid surface, for unit area of solid surface that disappears equivalent areas of liquid surface and solid–liquid interface are formed

The work involved is

$$W_s = (\gamma_{S/L} + \gamma_{L/V}) - \gamma_{S/V} \quad (1.2)$$

where $\gamma_{S/V}$ is the free energy of the surface at equilibrium with the vapour of the spreading liquid. $\gamma_{S/V}$ is related to γ_S by

$$\gamma_{S/V} = \gamma_S - \pi_e \quad (1.3)$$

where π_e is the equilibrium spreading pressure which takes into account the extra energy associated with the adsorption of liquid vapour on the solid surface.

(*iii*) Immersional wetting. The total immersion in a liquid of 1 cm^2 of solid surface involves exchange of solid–vapour for solid–liquid interfaces without any change in the extent of the liquid surface, and the immersional work is given by

$$W_i = \gamma_{S/L} - \gamma \quad (1.4)$$

The three types of wetting differ in that in the first there is unit decrease in liquid–vapour interface, in the second a unit increase in this interface and in the third there is no change.

Values of γ_S, $\gamma_{S/V}$ and $\gamma_{S/L}$ are not readily accessible by experiment, but they are related by the Dupré equation for contact angle equilibrium (often referred to as Young's equation)

$$\gamma_{S/V} = \gamma_{S/L} + \gamma_{L/V} \cos\theta \quad (1.5)$$

which provides a means of evaluating the wetting functions in terms of the measurable quantities $\gamma_{L/V}$ and θ, where θ is the contact angle between the solid and liquid phases. In this equation the vapour refers to that of the liquid, *i.e.* the system is at equilibrium with the vapour at its saturated vapour pressure. It is important to remember that eqn. (1.5) only applies to a system at equilibrium and for which $\gamma_{L/V}$ and θ have their equilibrium values.[5] For the wetting of a solid surface which is initially free of liquid molecules but in equilibrium with its own vapour, the π_e term should be included.[6] In principle π_e may be obtained from an adsorption isotherm for the vapour on the solid, but there is a general uncertainty on values of π_e and often the term is neglected leading to the (inaccurate) assumption that $\gamma_S = \gamma_{S/V}$.

The simplest way to describe the wetting phenomena that occur when a solid particle is immersed in a liquid is to consider a cube of side length 1 cm. The three stages, adhesion, immersion and spreading, involved in the complete wetting of the cube are shown in Fig. 1.1 and are represented by the changes (a) to (b), (b) to (c) and (c) to (d) respectively. The energy changes that take place are given by

$$W_a = \gamma_{S/L} - (\gamma_{L/V} + \gamma_{S/V}) = -\gamma_{L/V}(\cos\theta + 1) \quad (1.6)$$

$$W_i = 4\gamma_{S/L} - 4\gamma_{S/V} = -4\gamma_{L/V}\cos\theta \quad (1.7)$$

$$W_s = (\gamma_{S/L} + \gamma_{L/V}) - \gamma_{S/V} = -\gamma_{L/V}(\cos\theta - 1) \quad (1.8)$$

and for simplicity we assume that the solid surface before wetting is in equilibrium with the vapour of the liquid. From these relations we may predict whether or not any particular stage of the process is spontaneous, *i.e.* when the appropriate W is negative. When W is positive then work must be expended on the system for the process to take place.

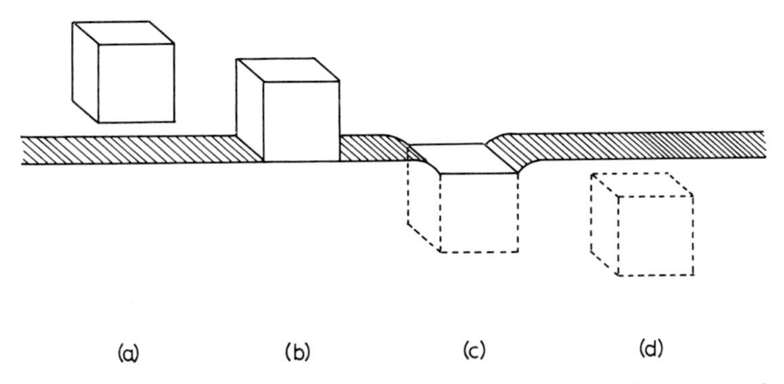

Fig. 1.1. *The three stages involved in the complete wetting of a solid cube by a liquid*: (*a*) *to* (*b*) *adhesional wetting*, (*b*) *to* (*c*) *immersional wetting*, (*c*) *to* (*d*) *spreading wetting*.

The conclusions are that

(*i*) adhesional wetting is positive (W_a negative) when the angle of contact is less than 180°; the process is invariably spontaneous,
(*ii*) immersional wetting is positive and immersion is spontaneous only when θ is less than 90°, and
(*iii*) spreading wetting is positive only when $\theta = 0$, and work must be done to achieve spreading at all larger values of this angle.

The total work W_d for the overall dispersion process is given by the sum of those for the three separate stages

$$W_d = W_a + W_i + W_s = 6\gamma_{S/L} - 6\gamma_{S/V} = -6\gamma_{L/V}\cos\theta \quad (1.9)$$

a result that could have been predicted in terms of the overall energy expenditure. By splitting the operation down into the three stages it is possible to predict whether some intermediate state of the dispersion process may not be spontaneous and therefore retard the overall effect. The requirement that $\theta = 0$ for spontaneous spreading explains why certain solids are only partially submerged and still remain at the surface; work must be done to bring about complete submersion. Perhaps the material might be so dense that gravity could furnish sufficient energy for the liquid to spread over the solid surface. Gans has discussed the relevance of wetting phenomena to the paint technologist and his two articles[7,8] are recommended for further reading.

In practice the addition of surface active agents ensure that θ is often close to zero and hence spontaneous dispersion is common. When θ is zero or so close to zero that spontaneous spreading occurs it is normal to say that the liquid wets the surface; non-wetting means that $\theta > 90°$ so that the liquid tends to form a drop on the surface which runs off easily. From eqn. (1.5)

$$\cos \theta = (\gamma_{S/V} - \gamma_{S/L})/\gamma_{L/V} \tag{1.10}$$

from which it is apparent that if $\theta < 90°$ a decrease in $\gamma_{L/V}$ will reduce θ and hence improve wetting. The addition of a surface active agent usually causes a reduction in $\gamma_{L/V}$ and if adsorbed a decrease in $\gamma_{S/L}$. Both effects lead to better wetting. The change in $\gamma_{S/V}$ is probably negligible in most cases so that the dominating factor in wetting is normally $\gamma_{L/V}$, the surface tension of the liquid phase.

Contact angle measurements on powders present many problems; in particular hysteresis is often observed where the angle varies according to whether the interface at which the measurement is made has advanced from or receded to its equilibrium position. This phenomenon has been variously attributed to surface roughness, surface heterogeneity, penetration of liquid into the solid, friction in the vicinity of the three phase line of contact, and the presence of traces of surface active material.[9] Some observations of the measurement of θ for powders have recently been reviewed,[10] but published data are relatively rare.

The penetration of liquid into the channels between and inside the agglomerates is more difficult to define precisely. The pressure P required to force a liquid into a tube of radius r is

$$P = -2\gamma_{L/V} \cos \theta / r \tag{1.11}$$

and hence penetration is only spontaneous (P negative) when $\theta < 90°$. If θ is not zero then, using eqn. (1.5),

$$P = -2(\gamma_{S/V} - \gamma_{S/L})/r \tag{1.12}$$

so that the important requirement is that $\gamma_{S/L}$ should be made as small as possible since $\gamma_{S/V}$ is virtually constant. However, if θ is zero then

$$P = -2\gamma_{L/V}/r \tag{1.13}$$

so that now $\gamma_{L/V}$ should be large. Changes in θ accompany an increase in $\gamma_{L/V}$, and assuming the only variable tension to be $\gamma_{L/V}$, P will ultimately become constant (from eqns. (1.10) and (1.11)), and the limiting value increases proportionally with $(\gamma_{S/V} - \gamma_{S/L})$. Hence it is desirable to maximise $(\gamma_{S/V} - \gamma_{S/L})$ and keep $\gamma_{L/V}$ as small as possible. Since any surface active agent affects both $\gamma_{L/V}$ and $\gamma_{S/L}$, the best agent for promoting such opposing effects may be difficult to find and will vary from one system to another.

The relevance of eqns. (1.11)–(1.13) to the practical situation is somewhat tenuous, since the channels between particles in agglomerates are not readily defined in geometric terms. Equation (1.11) indicates that in the absence of air a liquid will enter the channel spontaneously if $\theta < 90°$, but agglomerates are normally filled with air and once the liquid has penetrated the channel the air pressure will increase and complete wetting would therefore appear to be impossible. Heertjes and Witvoet[11] have extended the simple treatment to the more complex situation found in practice, and besides demonstrating the relevance to the wetting behaviour of the structure of the agglomerate have shown that when filled with air complete wetting can only occur when $\theta = 0$.

Another important factor in the wetting process is the rate of penetration of the liquid into the capillaries, and it is desirable that this be as large as possible. For the case of horizontal capillaries or in general where gravity may be neglected, the rate of entry of liquid into the tube is given by the Washburn equation[12]

$$l = \left(\frac{rt\gamma_{L/V}\cos\theta}{2\eta}\right)^{\frac{1}{2}} \tag{1.14}$$

where l is the depth of penetration in time t and η is the viscosity of the liquid. For a packed bed of solid particles the radius may be replaced by a factor K, which contains an effective radius for the bed and a tortuosity factor which takes into account the complex path formed by the channels between the particles and aggregates. Thus

$$l^2 = \frac{Kt\gamma_{L/V}\cos\theta}{2\eta} \tag{1.15}$$

and for any particular (stable) powder bed a plot of l^2 versus t should be linear. For a convenient measurement of l for a powder see *e.g.* reference 13. To achieve rapid penetration high $\gamma_{L/V}\cos\theta$, low θ and low η are desirable, with K as large as possible *e.g.* a loosely packed powder. High $\gamma_{L/V}$ and low θ are normally incompatible; a low θ is an important requisite, but when $\theta = 0$ further lowering of $\gamma_{L/V}$ will reduce the penetration rate.

The Washburn equation describes a system in which the surface of the capillary is covered with an adsorbed duplex film in equilibrium with the saturated vapour of the wetting liquid (N.B. a duplex film is one that is sufficiently thick for the tensions at the solid–liquid and liquid–vapour interfaces to act independently). Good[14] has extended the argument to the case where the surface is initially free of adsorbed vapour when the initial driving force for the liquid flow will be larger, and hence the 'dry' capillary will wet faster than one already equilibrated with the vapour of the penetrating liquid. He shows that the maximum rate of penetration is given by

$$l^2 = \frac{Kt[\gamma_{L/V} \cos\theta + \pi_e - \pi(t = 0)]}{\eta} \qquad (1.16)$$

where $\pi(t = 0)$ is the spreading pressure for the film that exists at zero time, but because molecules of the liquid will through diffusion and evaporation processes reach the solid surface before the advancing body of liquid, the actual rate lies between the rates predicted by eqns. (1.15) and (1.16).[15]

Hence the wetting phenomena are directly related to $\gamma_{L/V}$ and θ. In general the overall process is likely to be more spontaneous the lower θ and the higher $\gamma_{L/V}$ although these two factors tend to operate in opposite senses.

An account[16] of Zisman's work on contact angles of liquids on low and high energy solid surfaces in relation to liquid and solid constitution is recommended for further reading. Zisman introduced the concept of critical surface tension for the wetting of a solid, being the intercept at $\cos\theta = 1$ of a plot of surface tension versus $\cos\theta$ for a homologous series of organic liquids in contact with a particular surface, and Gans[17] has used the information to predict the spreading behaviour of liquids on solids as relevant to pigment dispersion. Reports on studies of dispersion in relation to wetting behaviour are rare. Work from the author's laboratory[18,19] has shown that with a carbon black (Graphon) and rutile in aqueous surfactant solutions there is, as expected from the theory, a marked increase in the degree of dispersion at the surfactant concentration for which θ becomes less than 90°; in these experiments the mixtures were subjected to slow end-over-end action so that under the circumstances spreading wetting was possible. With most other published data separation of the effects associated with wetting and surfactant adsorption from flocculation behaviour has not been possible or attempted. Association of dispersion phenomena specifically with surfactant adsorption rather than electrokinetic potential (related to stability) has been assumed for carbon blacks[20–22] and oxides.[23,24]

Mechanical breakdown of clusters

The most difficult part of the dispersion process to define is that concerned with the breaking down of the aggregates and agglomerates into finer particles after all the available surface has been wetted. Particles held together by weak forces in the dry agglomerate state would presumably require little energy to disperse, once wetted, and charge and surface tension effects would be important. Penetration of liquid into the channels between particles in agglomerates may provide sufficient excess air pressure to bring about disintegration.[11]

But mechanical energy is required to destroy aggregates or break down single crystals into smaller units, and in practice the relation between grinding efficiency and the adjustable parameters in any particular dispersion process has usually been established empirically. A great deal of consideration has been given to this problem by Rehbinder and his colleagues in Moscow.[25–27] They have established that the fine grinding of solids to create new interfaces is facilitated considerably by the adsorption of surface active agents at structural defects in the surface. These defects are normally present in the natural state but they might also appear as microcracks during the milling process. The decrease in surface energy associated with the adsorption may even be so large that the colloidal state becomes more stable than the condensed state, and then the material would break down spontaneously, in the absence of external forces, into smaller units. DiBenedetto[28] has discussed how the presence of submicroscopic defects in the solid can lead to fracture occurring at only a small fraction of the theoretical strength, and the fracture energy is further reduced by a wetting liquid causing embrittlement of the solid. A theory for such spontaneous dispersion has been proposed by Rehbinder[29]; the dispersions formed in this way should, perhaps, be classed as *lyophilic*.

Of particular interest to the present discussion is a study of the dispersion of various oxides (quartz, rutile, alumina, etc.) in a laboratory eccentric vibratory mill, in dry air and in various liquids (polar and non-polar and solutions of surface active agents) and their vapours.[30] Dispersion efficiency is shown to be related to the degree of lowering of the interfacial tension and is independent of the polarity of the liquid, supporting the theory that the dominating effect is the lowering of surface energy by adsorption. Adsorbed vapours may also facilitate dispersion, the effectiveness depending greatly on surface coverage. For example, in the grinding of quartz sand it was shown that the degree of dispersion is considerably greater with 0·04% moisture than in dry air, and rises sharply up to about 1%. On further increase in water content from 2 to 30% the degree of dispersion falls considerably, with 30–40% it rises sharply again and reaches optimum level (about the same as with 1%) in wet grinding with 50–80% water. The water contents correspond to the following surface coverages: 0·04%, 0·1 monolayer; 1%, 1 monolayer; 30%, 5–10 monolayers. The decrease in efficiency from 2 to 30% is explained by the formation of aggregates of quartz particles held together by thin intermediate layers of water, reaching a maximum in dispersion retardation at 5–10 monolayers.

Increasing the water layer thickness weakens the structure until at about 40% further increases have no more effect. Rehbinder remarks that the formation of these structures, the breakdown of which consumes a considerable proportion of the mill energy, is an important factor in dispersion studies with addition of small amounts of surface active materials to dry powders. Similar effects were observed with acetone, ethanol and benzene vapours. The weakening of solids by adsorbed material is generally known as the Rehbinder effect. How much it applies to particles of pigmentary size is still an open question.

STABILITY

A colloidal dispersion may be considered as stable if there is no change in the total number of particles with time, although this is to some extent arbitrary since it depends on the time scale over which the observations are made. There are various ways (principally three) by which the number might decrease. One is a consequence of the large interfacial free energy associated with the small particles, which would tend to decrease by crystallisation to a value for which the surface area is minimal, *i.e.* to one large crystal; the rate at which that occurs in lyophobic systems is insignificant. Sedimentation under gravity would also lead to a reduction in number with time but this is only important for large dense particles since thermal agitation is normally sufficient to keep small colloidal particles dispersed. For dispersions of solid particles in liquid media the primary cause of instability is flocculation in which particles under the influence of random Brownian motion come into close contact and form clusters (flocculates) of aggregates and/or primary particles. In the clusters the particles retain their original identity. The frequency of collisions is determined by the concentration and physical properties of the particles, the viscosity of the medium and the temperature, *i.e.* on those parameters which determine the rate of diffusion. The factors involved in flocculation will now be considered in some detail.

Maximum rate of flocculation

The continuous, haphazard motion of colloidal particles suspended in a liquid medium was first observed by Robert Brown in 1827. This Brownian motion is the mechanical result of innumerable collisions which take place between the particles and the molecules of the liquid. Einstein showed theoretically that the average square of the displacement $\overline{x^2}$ along any given axis in time t of a particle of radius a in a medium of viscosity η is given by

$$\overline{x^2} = RTt/3\pi N_0\eta a \qquad \text{or} \qquad \overline{x^2} = 2Dt \tag{1.17}$$

where R is the gas constant, T the absolute temperature, N_0 the Avogadro number and D the diffusion coefficient. Using Einstein's law rates of Brownian motion may be evaluated. For spheres in water at 20°C, $D \approx 2{\cdot}15 \times 10^{-13}/a$ from which the following data were calculated.[31]

(*i*) For $a = 10$ Å, $D = 2{\cdot}15 \times 10^{-6}$ cm^2/sec, particle travels a root mean squared distance of 10 Å in $2{\cdot}3 \times 10^{-9}$ sec, 1000 Å in $2{\cdot}3 \times 10^{-5}$ sec and 1 μ in $2{\cdot}3 \times 10^{-3}$ sec.
(*ii*) For $a = 1\ \mu$, $D = 2{\cdot}15 \times 10^{-9}$ cm^2/sec, particle travels a root mean squared distance of 10 Å in $2{\cdot}3 \times 10^{-6}$ sec, 1000 Å in $2{\cdot}3 \times 10^{-2}$ sec and 1 μ in 2·3 sec.

The effect of variations in viscosity and temperature are readily estimated; it requires large changes in η and T to have really significant effects on the distances covered by diffusion.

From these ideas we can estimate the minimum time required for two particles to be in collision in the absence of any other force affecting their mutual approach.

Assuming monodisperse spherical particles originally spaced at the corners of a cubic lattice, Mysels[31] shows that the average time of encounter is of the order of

$$t = 5{\cdot}2\eta a^3/kT\varphi^{4/3} \tag{1.18}$$

where k is the Boltzmann constant and φ the volume fraction of the particles in the system. For water at 20°C this becomes

$$t = 1{\cdot}3 \times 10^{12} a^3/\varphi^{4/3} \tag{1.19}$$

and hence for a dispersion containing a volume fraction of particles of 0·1% with radius 1000 Å the time is about 13 sec. Hence the time involved in particle collisions may be short but increases rapidly with particle radius and slowly with dilution.

The theory of rapid flocculation was first proposed by Smoluchowski[32] who treated the problem as one of diffusion (Brownian motion) of the spherical particles of an initially monodisperse dispersion, with every collision, in the absence of a repulsive force, leading to a permanent contact. The total number N of particles (singlets, doublets, triplets, etc.) that are present at time t (sec) is given by

$$N = N^0/(1+t/t_{1/2}) \tag{1.20}$$

where N^0 is the number initially and $t_{1/2}$ the time required to reduce the total number by one-half. The number concentration of aggregates containing k monomers is

$$N_k = \frac{N^0(t/t_{1/2})^{k-1}}{(1+t/t_{1/2})^{k+1}} \tag{1.21}$$

In general we may express the rate of flocculation, *i.e.* the rate of decrease of the total number of particles, as

$$-dN/dt = k'N^2 \tag{1.22}$$

where k' is the rate constant, and this becomes, when integrated with $N = N^0$ at $t = 0$

$$1/N = 1/N^0 + k't \tag{1.23}$$

For the most rapid rate of disappearance of particles of all types Smoluchowski's theory gives

$$k_0' = 8\pi Da \tag{1.24}$$

and a theoretical value for k_0' may be evaluated using the Einstein expression $D = kT/6\pi\eta a$. Hence the reciprocal of the total number of particles should increase linearly with time during flocculation and the slope of the $1/N$ against t plot for rapid flocculation might be compared with that predicted by the Smoluchowski theory. Such linear plots have been recorded[33–37] and the k_0' values are of the expected order of magnitude. Experimental k_0' values are sometimes larger and sometimes smaller than theory predicts. Polydispersity and non-spherical form both lead to greater rates although the effects are not significant unless gross departures from the ideal are involved. However, Smoluchowski's theory does not take into account the long range force of attraction (van der Waals) between the particles (only attraction on contact is considered). Taking account of this force[38] leads to

$$-\frac{dN}{dt} = \frac{4\pi DN^2}{\int_{2a}^{\infty} \frac{\exp(V_A/kT)}{R^2}\, dR} \tag{1.25}$$

where R is the distance between particle centres and V_A the potential energy of attraction; this expression becomes that of Smoluchowski when $V_A = 0$. The factor by which the Smoluchowski rate is increased depends on those parameters which determine the magnitude of the attractive force, *i.e.* particle radius and nature, and the type of medium. The correction factor would normally be $\sim$1–2 but might take values as high as 4 under optimum conditions.[38] Some discrepancies reported in the literature between experimental and theoretical rapid rates would be removed by introducing this factor.

When the experimental rapid rate is appreciably smaller than that predicted, as frequently found with dispersions in aqueous media, then another mechanism must be invoked. One looks for a possible source of a repulsive force which is not removed under the condition of the experiment; the electrical repulsive force associated with charged particles is annulled by the high concentrations of electrolyte used in rapid-rate experiments with aqueous dispersions. Structuring of the medium around the particles has been proposed,[39] such structured liquid having to be displaced before particles make contact, and this involves energy; the magnitude of this effect has yet to be accurately assessed.

One important observation which may be made from Smoluchowski's theory is that for other than the most dilute dispersions, flocculation would be expected to be fast on any practical time scale, *e.g.* for aqueous dispersions at 25°C with initially N^0 particles per cm^3, $t_{1/2} \approx 2 \times 10^{11}/N^0$ sec, and therefore instant separation of the disperse phase of a practical system (containing ~1–10% solids) would occur provided no other factors oppose the collision process.

The foregoing discussion assumes the dispersion to be stationary and all collisions result from Brownian motion. For a polydisperse system under the influence of the gravitational field or, of greater importance, when in forced convection (shaking), the disparity in velocity between large and small particles become important. The effect is autocatalytic since on flocculation the clusters grow and the disparity increases. Under such conditions we would not expect any theory based strictly on diffusion to apply since shear flow is involved. This effect is called *ortho-kinetic* flocculation; relatively little effort has been made in its investigation, but some theoretical treatments have been reported.[40–43]

Forces of interaction between particles

At least three major types of interaction are involved in the approach of colloidal particles, namely

(*i*) the London–van der Waals force of attraction,
(*ii*) the Coulombic force (repulsive or attractive) associated with charged particles, and
(*iii*) the repulsive force arising from solvation, adsorbed layers, etc.

These forces originate from entirely different sources and therefore may be evaluated separately.

The interplay of (*i*) and (*ii*) form the basis of the classical theory of flocculation of lyophobic dispersions, first proposed by Deryaguin and Landau[44] in Russia and independently by Verwey and Overbeek[45] in the Netherlands, and now known as the DLVO theory. The force (*iii*) is less well defined. Some attempts have been made to evaluate the effect of the adsorbed layer on the attractive force between particles, and also of the repulsive force when two adsorbed layers on approaching particles interact with each other. More complications arise when the adsorbed species are polymeric ions. The question of whether the medium in the vicinity of the particle plays any significant role in flocculation behaviour, whether through electrical effects (*e.g.* change in dielectric constant) or viscosity effects due to structuring, has yet to be resolved.

In general a colloidal dispersion will be stable for a reasonable period of time if (*ii*) or (*iii*) exceeds (*i*) by such a factor that an energy barrier exists which is of magnitude several orders higher than the thermal energy (kT) of the particles. When very much larger than kT the flocculation rate will be virtually zero and the system would be considered indefinitely stable, although the time scale is somewhat arbitrary. When (*i*) exceeds (*ii*) or (*iii*) at all distances between particles, the dispersion will flocculate as fast as the particles can diffuse together.

The nature and interplay of these three forces will now be considered in more detail.

London–van der Waals attractive forces

London–van der Waals forces are based on electric interactions; they are due to the interaction of dipoles within the particles and these may be the permanent dipoles of polar particles or the dipoles that may be induced in non-polar particles which are nevertheless polarisable. The interaction between the dipoles is electromagnetic in character, and it can readily be shown that this force is always attractive.

There are two methods of calculating the attractive force between solid particles. The first, the microscopic approach, assumes that the interaction between individual atoms or molecules are additive so that integrating over all pairs of atoms or molecules leads to the total energy for the macroscopic bodies. This method was used by Hamaker[46] and leads to the following expression for the attractive potential energy between two spheres of equal radius in a vacuum

$$V_A = -\frac{A}{6}\left[\frac{2}{s^2-4}+\frac{2}{s^2}+\ln\left(\frac{s^2-4}{s^2}\right)\right] \tag{1.26}$$

where the distance parameter $s = R/a$, and A is termed the Hamaker constant given by $\pi^2 q^2 B$. q is the number of atoms (molecules) per cm^3 and B the London constant in the equation for the attraction between two atoms

$$V = -B/r^6 \tag{1.27}$$

where r is the distance between the two atoms. Hence provided a value for A is available the relationship of V_A to distance may be calculated. The attraction between two isolated atoms is effective only over a very short range (of the order of atomic dimensions) but on addition, as in this treatment for colloidal particles, the range is much longer and is of the order of the dimensions of the particles, *i.e.* hundreds to thousands of Ångstroms.

The evaluation of A is a major problem. The London constant may be estimated (only) theoretically by calculating certain molecular constants from optical data. For this calculation the Slater–Kirkwood expression as modified by Moelwyn–Hughes[47] is probably the most satisfactory

$$B = (3/4)Z^{1/2}h\upsilon_v{\alpha_0}^2 \tag{1.28}$$

where Z is the number of electrons in the outer shell of one molecule, h is Planck's constant, υ_v the characteristic frequency of the molecules and α_0 the static polarisability. From optical data υ_v may be calculated using the dispersion formula

$$n^2 - 1 = C\upsilon_v^2/(\upsilon_v^2 - \upsilon_L^2) \tag{1.29}$$

where n is the refractive index at frequency υ_L and C is a constant. α_0 is calculated from

$$\alpha_0 = \frac{3M}{4\pi N_0 \rho}\cdot\frac{n^2 - 1}{n^2 + 2}\cdot\frac{\upsilon_v^2 - \upsilon_L^2}{\upsilon_v^2} \tag{1.30}$$

where ρ is the density of material of molecular weight M. Values of A of $10^{-12} - 10^{-13}$ erg are obtained but their precision is in doubt. Indirectly values of A may be obtained from flocculation experiments but agreement between the two methods must not necessarily be taken as indicative of the correctness of the A value. Theoretical and experimental A values have been compared by Lyklema.[48]

The second approach (the macroscopic approach) for the calculation of the attractive force as proposed by Lifshitz,[49,50] begins with the optical properties of the interacting macroscopic bodies, and derives the attraction from the imaginary parts of their complex dielectric constants particularly in the far ultraviolet. Unfortunately the appropriate data are available for only a very limited number of materials so that at present we are left to make estimates using the London theory. The microscopic and macroscopic approaches have been critically compared by Krupp,[51] and the reader is referred to his article for further details on the Lifshitz theory and examples of the van der Waals constants obtained from it.

When a medium separates the interacting particles the attractive force is weaker than that calculated by eqn. (1.26) because of the interaction between the molecules of the medium and between the particles and the medium. Hamaker showed that for the two phase system the total A is given by

$$A = A_{pp} + A_{mm} - 2A_{pm} \tag{1.31}$$

while the distance function in the brackets of eqn. (1.26) remains the same. A_{pp} and A_{mm} are the constants for the particles and medium (*in vacuo*) respectively and A_{pm} that associated with the particle–medium interaction. In the absence of more precise information it is generally assumed that $B_{pm} = (B_{pp}B_{mm})^{\frac{1}{2}}$ which is predicted by London for two single electronic oscillators having the same characteristic frequency (υ_v) but different polarisabilities; this is normally not strictly true and tends to overestimate the effect. Hence we may write as a good approximation

$$A = (A_{pp}^{\frac{1}{2}} - A_{mm}^{\frac{1}{2}})^2 \tag{1.32}$$

Often the values of A_{pp} and A_{mm} are of similar magnitude and since these values are certain to be in error, it is not possible to estimate A for a system to much better than an order of magnitude.

Another method of estimating the Hamaker constant A has been proposed by Fowkes[52] based on interfacial tension data. Fowkes demonstrates that the interaction energies due to dispersion forces at an interface may be readily predicted by the geometric mean of the dispersion force contribution to the surface free energy (γ^d) of the two materials. Directly or indirectly (*e.g.* through contact angle data) values of γ^d for a variety of solids and liquids were obtained using the defining relation

$$\gamma_{12} = \gamma_1 + \gamma_2 - 2\sqrt{\gamma_1^d\gamma_2^d} \tag{1.33}$$

which states that the interfacial tension γ_{12} is equal to the sum of the separate tensions of the components minus a term due to the London interaction between surface molecules of 1 and bulk molecules of 2 and *vice versa*. γ_1^d is related to the Hamaker constant A_1 for component 1 by

$$A_1 = 6\pi r_{11}{}^2\gamma_1^d \tag{1.34}$$

where r_{11} is the intermolecular distance. For a two-phase system

$$A = 6\pi r_{11}{}^2\,(\sqrt{\gamma_1^d} - \sqrt{\gamma_2^d})^2 \tag{1.35}$$

and Fowkes has calculated values of A for a variety of solids in contact with water assuming $6\pi r_{11}{}^2 = 1{\cdot}44 \times 10^{-14}$ for water and systems with volume elements such as oxide ions, metal atoms, $-CH_2$ or $-CH$ groups which have about the same size. Fowkes shows that the agreement between these values and some derived from other experiments is good. Furthermore the basic idea permits accurate calculation of several other important surface chemical parameters which also compare favourably with experiment,[52] lending weight to the validity of the principle.

Two refinements to the Hamaker treatment should be considered. The first involves transmission of the London force through the medium; theoretically we would expect the force to be reduced by a factor equal to the dielectric constant at electronic frequencies, this being to allow for the influence of the dielectric permeability of the medium on the propagation of the force. The factor used is the square of the refractive index of the medium for wavelengths in the visible region of the spectrum at which no light absorption occurs, *e.g.* $n^2 = 1{\cdot}8$ for water. The second refinement is concerned with the finite time taken for the London force to reach a distant atom, *i.e.* at distances comparable with, or greater than, the characteristic wavelength of the electrons ($\sim 10^{-5}$ cm). Again a reduction in the force is apparent. For large spheres with distances of separation much smaller than the radius, Schenkel and Kitchener[53] have derived useful approximate formulae for estimating the retarded force. But for other situations the calculation becomes complex[54] and subject to so many uncertainties that it is probably not worthwhile.

Despite all the difficulties attractive forces are calculable. For spherical particles of radius ≤ 1000 Å the Hamaker equation (eqn. 1.26) with A corrected for transmission effects, is probably satisfactory, and Fowkes' values of A seem reliable. For larger particles retardation effects should be considered. Formulae are also available[55] for flat plates which might be more appropriate in certain cases.

The coulombic repulsive force

Since in a medium containing ions a charged particle with its electric double layer is electrically neutral (Chapter 3), no net coulombic force exists between charged particles at large distances from each other. As the particles approach the diffuse parts of the double layers interpenetrate giving rise to a repulsive force which increases in magnitude as the distance between the particles decreases. The distance at which the repulsive force becomes significant increases with the thickness of the double layer ($1/\kappa$) and the force increases with the surface potential (ψ_0).

For particles that are sufficiently far apart for there to be no interaction the characteristics of each individual particle–double layer system are determined by the surface charge (or surface potential) and the ionic strength and properties of the medium. When two particles interact some changes in these parameters must occur, but in the estimation of the interaction energy either constancy in surface potential (implies rapid establishment of adsorption equilibrium) or in surface charge (implies slow desorption) is assumed. Constancy in surface potential is normally considered to be appropriate to systems such as the classical silver halide sols where the potential is determined by the silver and halide ion concentration in bulk solution, but recent experiments[56] cast some doubt on this approach and suggest the constancy in charge assumption is more appropriate. The latter is also probably more appropriate to systems stabilised by adsorbed surface active agents. Probably neither are strictly true in any one situation but the final results for both are not significantly different. The treatment of interacting double layers is published in detail in Verwey and Overbeek's book,[45] and only a brief account is given here. The theory is not simple; it turns out that the interaction of two flat parallel double layers is the easiest to handle and is the only one to be considered here, but relevant working formulae for spheres will be given.

For a flat double layer we may write

$$\frac{d^2\psi}{dx^2} = \frac{8\pi nze}{\varepsilon}\sinh\left(\frac{ze\psi}{kT}\right) \quad (1.36)$$

where ψ is the potential at distance x from the surface, n the concentration of electrolyte containing ions of valency $z_+ = z_- = z$, e is the electronic charge and ε the dielectric constant. Substituting for

$$y = ze\psi/kT, \qquad w = ze\psi_0/kT, \qquad \kappa^2 = 8\pi ne^2z^2/\varepsilon kT$$

we obtain

$$d^2y/dx^2 = \kappa^2 \sinh y \quad (1.37)$$

the solution of which for $\kappa x > 1$, *i.e.* large distance from the surface, is

$$y = 4\gamma \exp(-\kappa x) \quad (1.38)$$

where $\gamma = \exp[(w/2) - 1]/\exp[(w/2) + 1]$. It may readily be shown that this solution is a very good approximation for distances down to $\kappa x = 1$ for all values of ψ_0.

To derive an exact expression for the potential distribution between two flat plates separated by a distance $2d$, it is necessary to solve eqn. (1.37) to satisfy the boundary conditions $\psi = \psi_0$ when $x = 0$ and $x = 2d$, and $d\psi/dx = 0$ when $x = d$. The solution involves an elliptic integral which can only be solved numerically. However, when $\kappa d > 1$ (*i.e.* the interaction is not too great) it is sufficiently accurate to assume that the potential may be built up additively from the potentials of the two single double layers. Hence, from eqn. (1.38) the potential ψ_d half-way between the plates is given by

$$u = 8\gamma \exp(-\kappa d) \quad (1.39)$$

where $u = ze\psi_d/kT$ and $\kappa d > 1$.

The simplest way to calculate the force of repulsion between two interacting double layers is to follow Langmuir in considering the force as arising from the osmotic pressure of the excess ions in the space where the double layers overlap. The ionic concentrations (n_+ and n_-) at $x = d$ are given by the Boltzmann equation

$$n_+ = n\exp(-ze\psi_d/kT) \quad \text{and} \quad n_- = n\exp(ze\psi_d/kT) \quad (1.40)$$

and the osmotic pressure at this plane is

$$kT(n_+ + n_-) = nkT[\exp(u) + \exp(-u)] = 2nkT\cosh u \quad (1.41)$$

Outside the field of the double layers where $\psi = u = 0$, the osmotic pressure is $2nkT$. Hence the difference between these two pressures are a measure of the force p acting on the unit area of each plate,

$$p = 2nkT(\cosh u - 1) \quad (1.42)$$

The repulsive potential energy is given by the work done against this force when the plates are brought together from a large distance. This work per unit area of plate is given by

$$V_R = -2\int_{\infty}^{d} p\,dx = -2\int_{\infty}^{d} 2nkT(\cosh u - 1)\,dx \tag{1.43}$$

but in the general case this integral cannot be readily evaluated because the relation between u and x is complex. However, if we again assume small interaction, *i.e.* u is small and $(\cosh u - 1) = u^2/2$ then

$$V_R = -2\int_{\infty}^{d} nkT\,64\gamma^2 \exp(-2\kappa x)\,dx \tag{1.44}$$

or

$$V_R = 64nkT\gamma^2[\exp(-2\kappa d)]/\kappa \tag{1.45}$$

The exact numerical solutions for the interaction of parallel plates have been computed and published by Devereux and de Bruyn.[57] The corresponding tables for spherical particles are not generally available and it is normal practice to use approximate formulae which are valid for limited conditions. The two equations most commonly used were derived for small ψ_0 and prove most useful in many practical situations.

For systems in which $\kappa a \gg 1$ (large particles in aqueous systems with moderate electrolyte concentrations) and weak interaction (shortest distance Δ between surfaces of spheres is large compared with $1/\kappa$).

$$V_R = \tfrac{1}{2}\varepsilon a\psi_0^2 \ln[1 + \exp(-\kappa\Delta)] \tag{1.46}$$

The equation gives values which are close to those from the exact treatment (using graphical integration) provided $w(=ze\psi_0/kT) \ngtr 2$ and $\kappa a \nless 10$.

For systems in which $\kappa a \ll 1$ (small particles in non-aqueous media or in aqueous systems with very low electrolyte concentration)

$$V_R = \frac{\varepsilon a^2\psi_0^2}{\Delta + 2a}\beta \exp(-\kappa\Delta) \tag{1.47}$$

and is quite accurate up to $\kappa a = 1$ and $\psi_0 \sim$ 50–60 mV. β is a factor to allow for loss of spherical symmetry in the double layers as they overlap and is defined in Verwey and Overbeek's book.[45] For the treatment of the intermediate region of κa and for higher potentials the reader is referred to the original literature.

The total interaction energy

Since the double layer repulsion and the van der Waals attraction operate independently the total potential energy V_{tot} for the system is given by the sum of the two terms; the form of the resulting potential energy against distance relationship will be dependent upon the relative magnitudes of the two forces. V_R decreases exponentially with distance while V_A shows an approximate inverse relationship with the square of the distance. Attraction predominates at short distances (other than immediately adjacent to the surface at atomic distances when repulsion dominates—little is known

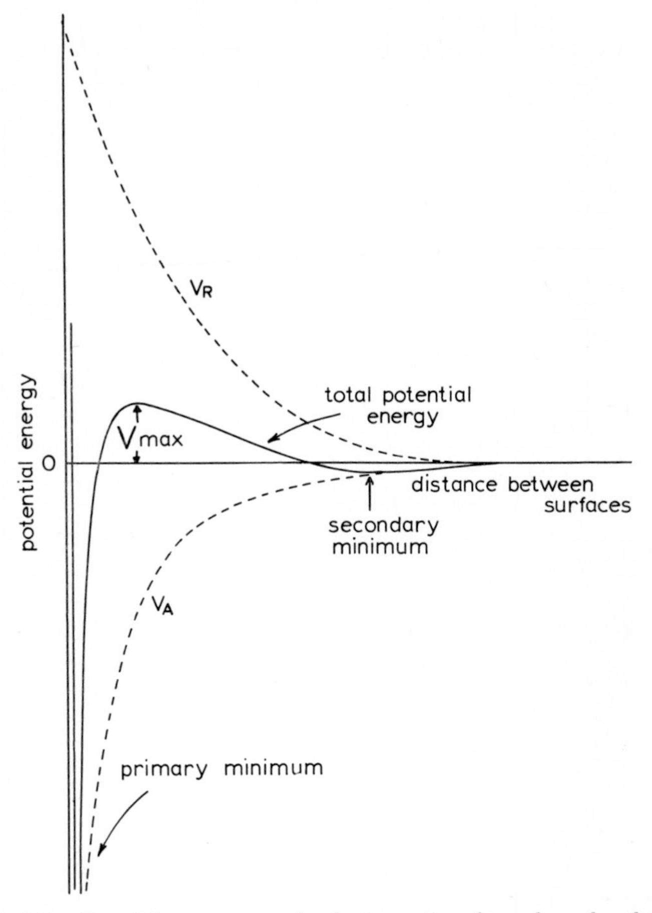

Fig. 1.2. Potential energy curves for the interaction of two charged surfaces.

about this effect and in colloid problems it is usually ignored). Otherwise the form of the V_{tot} curve depends to a large extent on the V_R term. Figure 1.2 illustrates the type of plot we might expect for particles of radius 0·1–1 μ in an aqueous system containing about 0·01 M of 1:1 electrolyte for which the range of attractive and repulsive forces are similar.

Three important characteristics are shown in Fig. 1.2 and these are directly related to flocculation behaviour. The potential energy barrier must be surmounted before the particles make lasting contact in the primary minimum. Provided the barrier is considerably larger than the thermal energy of the particles relatively few will make contact and the system should be stable. But if the secondary minimum is of depth $\gg kT$ then the particles would flocculate with a liquid film between them in the cluster. Since both the attractive and repulsive forces are approximately proportional to the particle radius, the secondary minimum should become

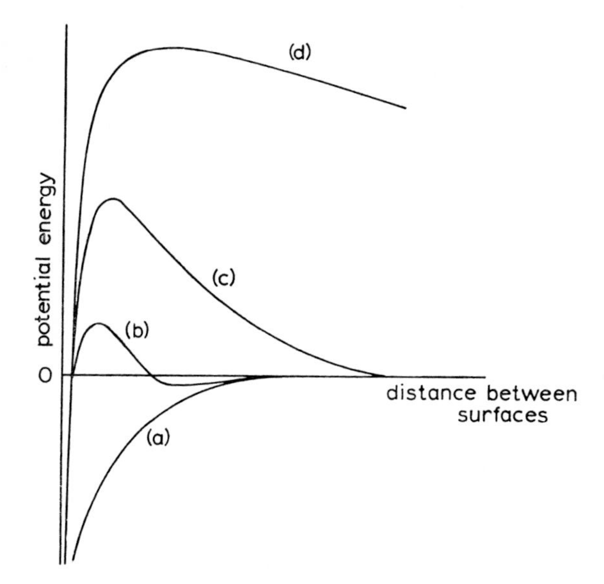

Fig. 1.3. *Influence of electrolyte concentration on the total potential energy of interaction of two spherical particles of radius* 1000 Å *in aqueous media.* (*a*) $1/\kappa = 10^{-7}$ *cm,* (*b*) $1/\kappa = 10^{-6}$ *cm,* (*c*) $1/\kappa = 10^{-5}$ *cm,* (*d*) $1/\kappa = 10^{-4}$ *cm.*

increasingly significant with increasing particle size, and particularly so with parallel plates. The effect will also increase with increasing electrolyte concentration which reduces the distance over which the repulsive forces operate, but this also reduces the energy barrier which again would promote flocculation. Systems which have flocculated into the secondary minimum tend to be *reversible*, *i.e.* they can be readily redispersed (peptised) with shaking; those in the primary minimum need considerably more energy to redisperse. In some cases particles might on standing pass from the secondary to the primary minimum, a phenomenon yet to be investigated.

The effect of reducing the electrolyte concentration (increasing $1/\kappa$ at constant ψ_0) on the total potential energy is shown in Fig. 1.3 and illustrates the difference between dispersions in aqueous and non-aqueous solutions of a surface active agent of the same stoichiometric concentration, *e.g.* Aerosol OT (sodium di-2-ethylhexyl sulphosuccinate) which dissolves in both water and hydrocarbons, but ionises to markedly different extents in the two solvents. When $1/\kappa$ is large the secondary minimum disappears since the range of V_R is then considerably greater than that of V_A, and under these conditions the accuracy of the V_A values becomes much less important.

Rate of slow flocculation

The energy relationships tell us whether or not we might expect a system to be stable, but gives no information on the rate of flocculation. We now consider the kinetics of the flocculation of a system for which there is an energy barrier. The rate is a function of the probability of particles having sufficient energy to overcome this barrier. The problem is again one of diffusion and was first solved by Fuchs[58] in 1934 for aerosols and later applied to aqueous dispersions by Deryaguin.[59]

Fuchs' theory leads to a factor W by which the rapid rate (Smoluchowski) is reduced by the presence of a repulsive force; W is called the *stability ratio* and is related to the height of the potential energy barrier by

$$W = 2\int_{2}^{\infty} \frac{\exp(V_{tot}/kT)}{s^2}\, ds. \tag{1.48}$$

However, Smoluchowski's value for the rapid rate does not assume the existence of long range attractive forces. Introduction of the appropriate term leads to the following equation.[38]

$$W = \frac{\int_{2}^{\infty} \frac{\exp(V_{tot}/kT)}{s^2}\, ds}{\int_{2}^{\infty} \frac{\exp(V_A/kT)}{s^2}\, ds} \tag{1.49}$$

so that $W = 1$ when $V_{tot} = V_A$ at all distances. Given the appropriate potential energy diagrams from, say, the DLVO theory, the stability ratio may be calculated by graphical or numerical integration and then compared with experimental values of $W = k_0'/k'$, the ratio of the experimental rate constants for rapid and slow flocculation. Such a comparison is a severe test of the applicability of theory to practical cases, and the observed deviations, although often not appreciable, reflect the assumptions and approximations which are necessary in the calculation of the potential energy terms.

Application of the DLVO theory

The application of the DLVO theory to flocculation problems has become common practice despite all the limitations. There is no doubt that in principle the theory is correct but refinements are obviously necessary in many cases. We shall consider here some examples of theory related to practice for the purpose of illustrating the applicability and shortcomings of the theory.

In making theoretical predictions of stability it is necessary to assign values to three important parameters:

(*i*) the Hamaker constant,
(*ii*) the Stern potential (*see* Chapter 3) which determines the characteristics of the diffuse part of the double layer, the region particularly relevant to colloid stability, and
(*iii*) the particle dimensions.

Evaluation of the Hamaker constant has already been discussed. Since the Stern potential cannot be determined by direct measurement it is usual to use the zeta potential obtained from an electrokinetic experiment (Chapter 3), and this has been successful in a number of cases.[53,60–62] Measurement of particle dimensions is itself a major problem, hence the desirability of using monodisperse systems when the validity of the theory is being assessed. Nevertheless, for a dispersion containing only one type of particle, the deviations from monodispersity and spherical geometry may effectively be small enough so that an appreciable error is not involved when the theory for monodisperse spheres is applied. For powders, electron microscopy is the most attractive technique for assessing particle dimensions. The problems of aqueous and non-aqueous systems will now be treated separately. For aqueous systems the potential energy relationships are sensitive to all three parameters, but in hydrocarbons an absolute value of A is not required.

Aqueous systems

It is well known that the addition of sufficient indifferent electrolyte to a stable aqueous hydrophobic sol usually leads to flocculation. The familiar Schulze–Hardy rule states that the effect is determined by the counterion and increases markedly with the valency. A great deal of early work supported this hypothesis and it is also predicted by the DLVO theory.

Although in principle the transition from slow to rapid flocculation is continuous, *i.e.* there is no sharp flocculation point, the relation between W and electrolyte concentration usually shows a fairly sharp transition at a *critical* concentration value and in practice such a sharp transition is observed; it is a question of rate plus energy barrier and not of the barrier alone. The critical electrolyte concentration n' at which rapid flocculation takes place is taken as that for which the maximum in the total potential energy curve is zero, *i.e.* when $V_{tot} = 0$ and dV_{tot}/dd are satisfied by the same value of d. Hence, for two flat plates

$$64n'kT\gamma^2 [\exp(-2\kappa d)]/\kappa - A/48\pi d^2 = 0 \qquad (1.50)$$

and

$$dV_{tot}/dd = -2\kappa V_R + 2V_A/d = 0. \qquad (1.51)$$

The term $A/48\pi d^2$ is the appropriate form of the attractive potential energy.[55]

So $\kappa d = 1$ and

$$64n'kT\gamma^2 \exp(-2)/\kappa = A\kappa^2/48\pi \qquad (1.52)$$

or

$$n' = 107\varepsilon^3 k^5 T^5 \gamma^4 / A^2 e^6 z^6 \qquad (1.53)$$

which approaches $n' = \text{constant}/z^6$ when ψ_0 is high ($\gamma \rightarrow 1$), *i.e.* the critical electrolyte concentration is inversely proportional to the sixth power of the valency of the counterion. Mean flocculation concentration of mono-, di-, and trivalent cations for a negative silver iodide sol are 142, 2·43 and 0·068 mM respectively, which are in the ratio 1:0·017:0·0005. The theoretical z^{-6} ratio is 1:0·016:0·0013. Deviations for ions of high valency may be attributed to neglect of adsorption and ion-size effects, besides chemical changes (hydrolysis) that may have occurred with the electrolyte. A secondary, but not negligible, effect is that associated with varying the counterion at a given valency[63] (the lyotropic series), the concentration decreasing as the unhydrated ionic radius increases. Such specific effects are not explicable using the Gouy theory, which is non-specific, so Stern theories or modifications to the Poisson–Boltzmann relation must be introduced.

Some interesting observations have been made comparing experimental W values with those predicted by the theory. Rate constants are measured for slow and rapid flocculation of aqueous dispersions of polymer latices, which may be prepared in a virtually monodisperse spherical form and make ideal systems for such comparison. Obviously it is necessary to study the initial stages of the process where only doublets are formed, since the appearance of larger aggregates would alter the potential energy relationships. The experiment involves adding indifferent electrolyte (containing no potential determining ions) to the stable dispersion and following changes in turbidity, particle number, etc. W is then the relative rate of flocculation compared with that observed with high concentrations of electrolyte corresponding to rapid flocculation (when $W = 1$). Plots of log W against log n for the experimental data are compared with the linear relationship predicted by the DLVO theory. Published data[61,64–66] on a variety of systems meet the requirement and usually extrapolate at $W = 1$ to values of the critical electrolyte concentration n' that are in agreement with the theory. However, on more critical examination, involving variations in particle size or surface potential, it has become apparent that the theory does not fully account for the facts. For example, theory predicts an increase in slope with particle size but Ottewill and Shaw[65] found that for polystyrene latices covering a range of particle diameters from 600 to 4230 Å the slopes are almost independent of particle size. Experiments carried out in the author's laboratory[67] with silver iodide sols also fail to confirm the expected increase of slope with surface potential, but the observed effect can be explained by introducing the discreteness of charge effect.[68] (Chapter 3).

In practical systems particles consisting of aggregates of smaller primary particles are often involved, and it is of particular interest to know which value to assign to the effective radius, that of the primary particle or that of the total aggregate. This would be a crucial factor in the case where the double layer thickness is small, *i.e.* at reasonably high electrolyte concentrations. A recent study[62] of the flocculation behaviour of two carbon blacks in aqueous solutions of sodium dodecyl sulphate containing 0·1 M NaCl (for which $1/\kappa = 10$ Å) admirably demonstrates this point. Graphon and Sterling MTG are two graphitised blacks with surface areas of 78·9 and 7·7 m^2/gm and mean primary particle radius of approximately 125 Å and 1250 Å, respectively. In dispersion Sterling MTG exists primarily as individual particles but Graphon is mainly aggregated with a peak in the size distribution at about 1000 Å. The adsorption of the surface active agent is identical on both carbons when put on a unit area basis. Theory predicts that under the conditions prevailing the Sterling MTG particles should flocculate into a secondary minimum even at high zeta potentials, and indeed the systems were all unstable. Similar behaviour would be expected for Graphon if the aggregate size determines the stability, but in fact the experimental log W against zeta potential relationship closely followed that predicted by theory for $a = 125$ Å (the primary particle radius) and $A = 5 \times 10^{-13}$ erg (the value estimated by Fowkes[52]).

Non-aqueous systems

Extension of the DLVO theory to non-aqueous systems has attracted relatively little attention. For media of intermediate dielectric constant, *e.g.* alcohols, ketones, etc., we might expect the behaviour to be similar to that already established with aqueous systems, but the actual ionic concentrations being one or two orders of magnitude lower. Studies of a fundamental nature on such systems are scarce,[69] but some attention has been paid to dispersion in hydrocarbon solutions of surface active agents for which the ionic concentrations are extremely small, *e.g.* $\sim 10^{-10}$ M corresponding to $1/\kappa \sim 10\ \mu$.[37] Hence κa falls almost to zero, the exponential term in eqn. (1.47) tends to unity and V_R is given, to a very good approximation, by

$$V_R = \frac{\varepsilon a^2 \psi_0^2 \beta}{\Delta + 2a}. \tag{1.54}$$

Since $1/\kappa$ is large the capacity of the double layer is small and only a small surface charge density is necessary to obtain an appreciable surface potential. Furthermore, the slow decay in potential from the surface means that the zeta potential, readily obtained from electrophoresis experiments,[70,71] may be equated with considerable accuracy to the surface potential. Another simplification compared with aqueous systems is that an absolute value of A is not required since the V_R term dominates the V_{tot} against distance relationship. Only the particle radius presents problems (as usual).

It is therefore possible to predict, with some confidence, the stability of a dispersion from readily obtainable parameters, and this has been done for hydrocarbon media for a range of particle size and surface potential.[72] Computations based on eqns. (1.26), (1.49) and (1.54) using β values

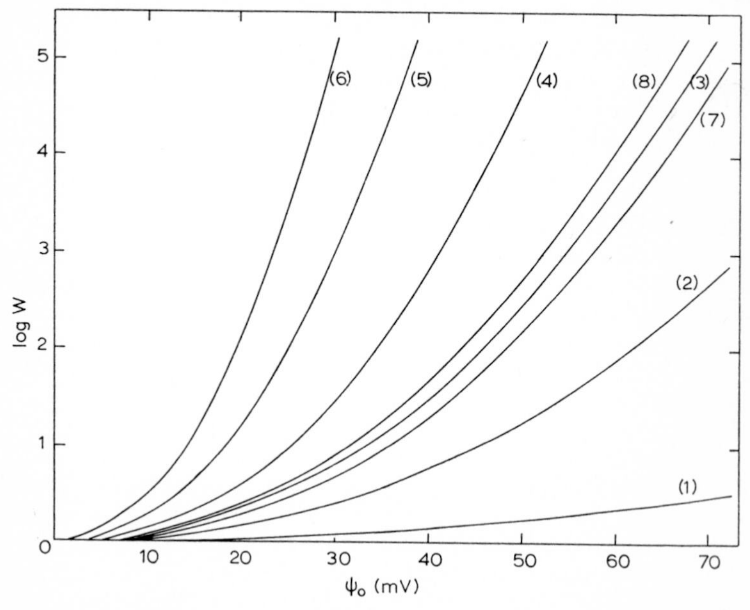

Fig. 1.4. Theoretical curves of stability ratio against surface potential for systems in which $\kappa a \ll 1$. A = $2{\cdot}5 \times 10^{-12}$ *erg*, a = 200 Å (1), 1000 Å (2), 1800 Å (3), 3200 Å (4), 6000 Å (5), 10 000 Å (6). a = 1800 Å, A = 1×10^{-11} *erg* (7), A = 5×10^{-13} *erg* (8).

calculated from Verwey and Overbeek's equation[73] assuming $\kappa a = 0$ and $\varepsilon = 2{\cdot}24$ are illustrated in Figs. 1.4–6 for a range of a, ψ_0 and A, and Fig. 1.7 shows how the half-life of $t_{1/2}$ of a dispersion is related the stability ratio and potential energy maximum V_{max}. The half-life was calculated from the the expression

$$t_{1/2} = \frac{6\eta \int_2^\infty \frac{\exp V_{tot}/kT}{s^2}\, ds}{4kTN^0} \qquad (1.55)$$

using $N^0 = 10^{10}$ particles/ml for Fig. 1.7. Figure 1.5 demonstrates the importance of particle size, Fig. 1.4 the relative insensitivity to the A value, and Fig. 1.6 shows how the stability depends on V_{max}. From Fig. 1.7 the stability of any particular dispersion may be estimated by making the appropriate allowance for particle concentration (eqn. (1.55)). Table 1.1 shows how for 10^{10} particles/ml the various factors, commonly used to express colloid stability, are related.

A correlation between zeta potential and stability for dispersion in non-polar media and the criteria for accurate measurement of electrophoretic mobility was first established by van der Minne and

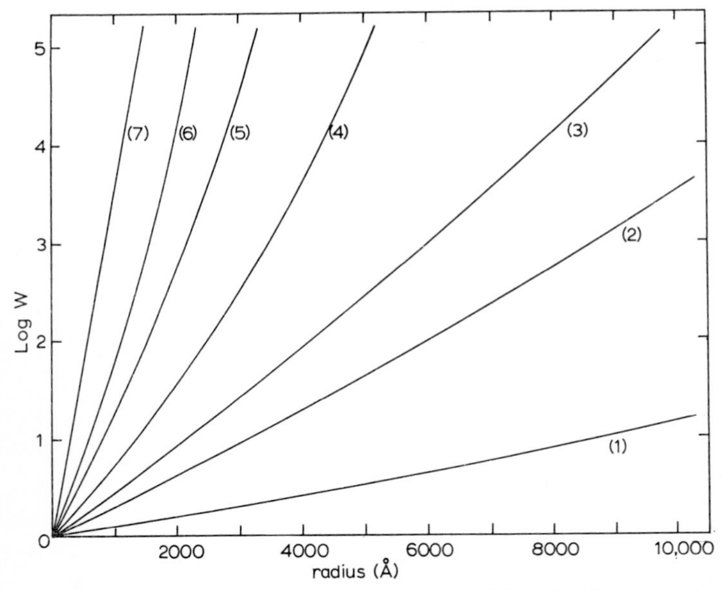

Fig. 1.5. *Theoretical curves of stability ratio against particle radius for systems in which* $\kappa a \ll 1$. A $= 2{\cdot}5 \times 10^{-12}$ *erg*, $\psi_0 = 15$ *mV* (1), 25 *mV* (2), 30 *mV* (3), 40 *mV* (4), 50 *mV* (5), 60 *mV* (6), 80 *mV* (7).

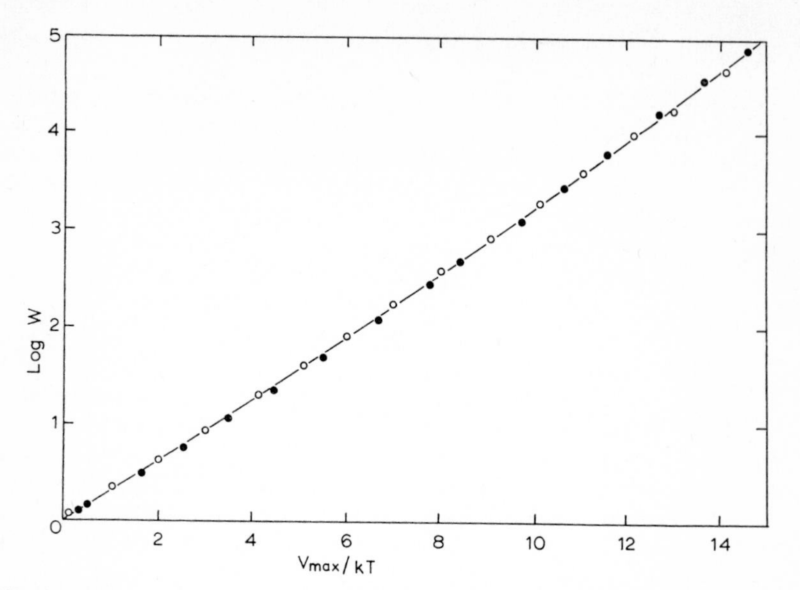

Fig. 1.6. *Theoretical values of stability ratio in relation to the height of the potential energy barrier, for various particle radii in the range* 200–10 000 Å *and for* $\kappa a \ll 1$. $A = 1 \times 10^{-11}$ *erg* (○), 5×10^{-13} *erg* (●).

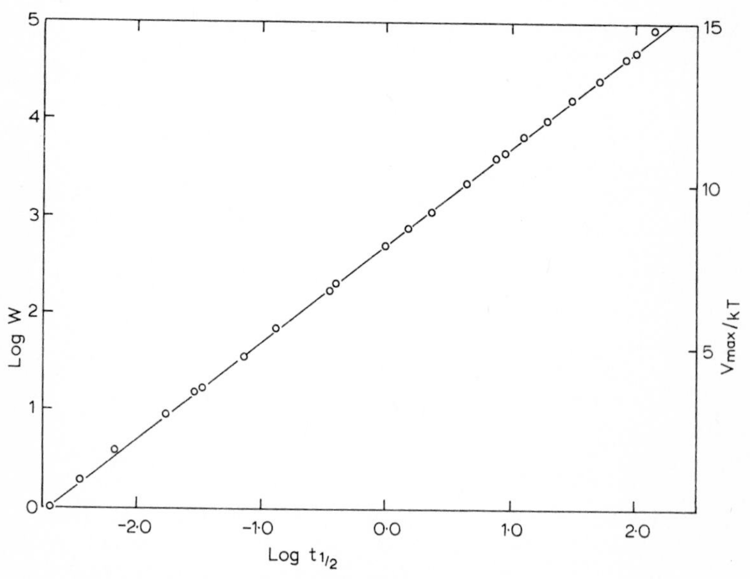

Fig. 1.7. *Theoretical values of the stability ratio in relation to the half-life (in hours) of a dispersion and the potential energy maximum, for various radii in the range* 200–10 000 Å *and* $\kappa a \ll 1$. $A = 2{\cdot}5 \times 10^{-12}$ *erg and particle concentration* 10^{10} *per ml.*

Hermanie.[70,74] The validity of the DLVO theory has been demonstrated both by qualitative[75,76] and by rigorous experimental tests[37,77] on systems in which there is no significant influence of the interaction of adsorbed layers of surface active material, *i.e.* only the charge mechanism is operative. It has also been shown that the presence of trace amounts of water may have a profound effect on the stability of hydrophilic powders in non-aqueous media,[69,78,79] presumably through changes in surface energy and/or charge but the details have not yet been worked out.

TABLE 1.1

V_{max}/kT	0	5	10	15	25
log W	0	1·6	3·3	5·0	8·3*
$t_{1/2}$ (*hours*)	0·002	0·1	7	400	10^6*

* extrapolated

The particle concentration effect

For practical systems, when the particle concentration is high, the validity of the DLVO theory is open to question. Particle–particle interaction might not begin at 'infinite' distance of separation, *i.e.* one particle is, on average, sufficiently close to another to reduce the effective energy barrier which must be overcome for the particles to come into contact (not considering at this time the secondary minimum effect). Albers and Overbeek[80] have made a preliminary attempt to solve this problem, their purpose being to explain their flocculation data for the charged droplets of water-in-oil emulsions. The two-particle interaction calculations were abandoned and replaced by a model based on twelve nearest neighbours, situated on a spherical shell around a central particle, with all other particles randomly distributed outside this shell. It was shown that the calculation of the interaction energy of the one particle with all other particles leads to an energy barrier which is significantly lower than that for the simple interaction of two particles, and the effect increases with particle concentration. The magnitude of the effect will be directly related to the thickness of the double layer, since double-layer overlap is the cause of the repulsion. Hence it will be much more significant in non-aqueous media. The calculations show, for example, that when $1/\kappa = 10\ \mu$ a particle concentration of 10% would correspond to an energy barrier about 50% of that for an infinitely dilute dispersion and at 50% concentration the barrier would be reduced by an order of magnitude. Unfortunately quantitative experimental data are not available for a worthwhile comparison to be made with this theory but practical experience tends to support the general conclusion.

Heteroflocculation

So far we have considered only the interaction between particles which are identical in nature, size and surface potential, *i.e.* only one value for each of a, A and ψ_0 are applicable. As already mentioned when only one type of particle is present it is often adequate to assume that the deviations from monodispersity and spherical geometry may be effectively small enough that little error is involved in applying the DLVO theory. When more than one type of particle is present or if there is a large disparity in size, the situation becomes more complex. Interaction between dissimilar particles may be termed *heteroflocculation*.

A theory of heteroflocculation was proposed originally by Deryaguin,[81] and Devereux and de Bruyn[57] included the interaction of dissimilar double layers in their calculations on flat plates. A more recent theoretical treatment[82] is mathematically simpler because it is based on the approximate DLVO theory for spheres with $\kappa a > 10$ and values of ψ_0 less than 50 mV. The potential energy of repulsion for two particles of radius a_1 and a_2 and potentials ψ_{01} and ψ_{02} is given by

$$V_R = \frac{\varepsilon a_1 a_2(\psi_{01}^2 + \psi_{02}^2)}{4(a_1 + a_2)} \left[\frac{2\psi_{01}\psi_{02}}{(\psi_{01}^2 + \psi_{02}^2)} \ln\left(\frac{1 + \exp(-\kappa\Delta)}{1 - \exp(-\kappa\Delta)}\right) + \ln(1 - \exp(-2\kappa\Delta)) \right] \quad (1.56)$$

which becomes eqn. (1.46) when $a_1 = a_2$ and $\psi_{01} = \psi_{02}$. Combining this with the appropriate expression of Hamaker[46] for the attraction between two spheres gives the total potential for the interaction from which energy barriers, stability ratios, etc., may be evaluated. The effect of changes in radius and potential on the potential energy diagrams are illustrated in Fig. 1.8 and Fig. 1.9. Two facts emerge from the theoretical calculations.

Firstly, it is the particles with the smaller radius and/or potential which determine the height of the potential energy barrier. Secondly, although the different particles may have the same sign of surface charge and potential (but different magnitude) they may nevertheless attract each other when the double layers overlap; this is a consequence of the relative changes which occur in the charge and potential as the particles approach.[83]

This analysis is of great importance to practical systems but as yet no fundamental study of heteroflocculation has been reported which would permit a worthwhile comparison with the calculations. The relevance to flooding and flotation in paint systems is obvious.[84] (*See* Chapter 7.)

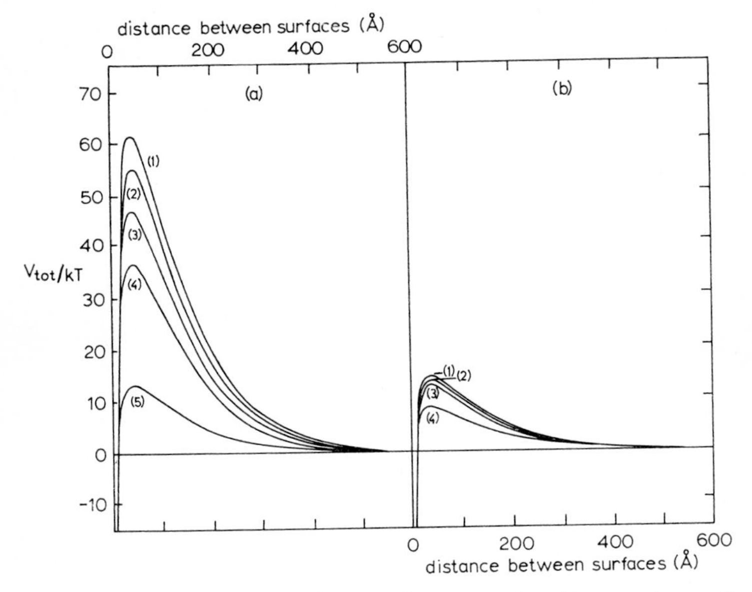

Fig. 1.8. *Theoretical curves of total potential energy against distance of separation of two spherical particles of radius* a_1 *and* a_2 *and equal surface potential* (35·86 *mV*). $A = 5 \times 10^{-13}$ *erg*, $1/\kappa = 10^{-6}$ cm^{-1}, $\varepsilon = 78{\cdot}5$.

(*a*) $a_1 = 1250$ Å, $a_2 = 1250$ Å (1), 1000 Å (2), 750 Å (3), 500 Å (4), 125 Å (5).

(*b*) $a_1 = 125$ Å, $a_2 = 1250$ Å (1), 1000 Å (2), 750 Å (3), 125 Å (4).

Stability arising from adsorbed layers

The presence of an adsorbed layer of surface active agent on the surface of solid particles dispersed in a liquid medium may affect the stability to flocculation in a number of different ways, each differing in principle but often overlapping in practice, hence making difficult the interpretation of observed phenomena.

This discussion will be arbitrarily divided into two sections, one dealing with completely non-ionic (uncharged) systems and the other with ionic systems. Some of the steric factors involved in the former will also be important in the latter, *e.g.* with an ionised polymer adsorbed on a charged surface in an aqueous medium.

Uncharged systems

Essentially there are two factors which must be considered to explain the stability of uncharged systems. The first concerns the effect of the adsorbed layer on the attractive force between particles. With the appropriate choice of surface-active agent a significant reduction in the magnitude of the

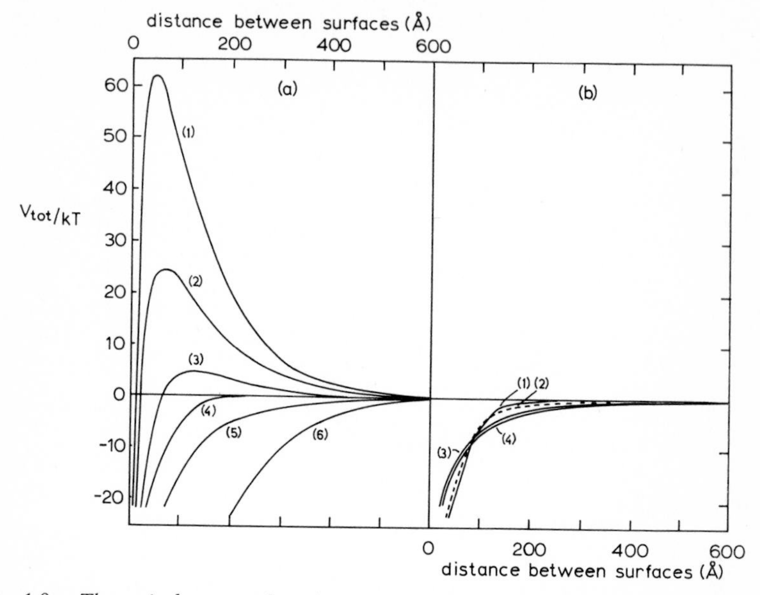

Fig . 1.9. Theoretical curves of total potential energy against distance of separation of two spherical particles of radius $a_1 = a_2 = 1250$ Å and varying surface potential. $A = 5 \times 10^{-13}$ *erg*, $1/\kappa = 10^{-6}$ *cm*$^{-1}$, $\varepsilon = 78{\cdot}5$.

(*a*) $\psi_{01} = +35{\cdot}86$ *mV*, $\psi_{02} = +35{\cdot}86$ *mV* (1), $+20{\cdot}48$ *mV* (2), $+10{\cdot}24$ *mV* (3), $+5{\cdot}12$ *mV* (4), 0 *mV* (5), $-35{\cdot}86$ *mV* (6).

(*b*) $\psi_{01} = +5{\cdot}12$ *mV*, $\psi_{02} = +35{\cdot}86$ *mV* (1), $+20{\cdot}48$ *mV* (2), $+ 10{\cdot}24$ *mV* (3), $+ 5{\cdot}12$ *mV* (4).

attractive force is possible. When particles collide the distance between the surfaces is increased by approximately twice the thickness of the adsorbed layer so that provided the adsorbed layer is compact and no desorption occurs, and that there is negligible attraction between the material of the adsorbed layers, the stability will depend on the magnitude of V_A corresponding to this distance. The second factor arises when the attractive force at this distance is still sufficiently large that interaction of the adsorbed layers may occur. The result is a change in the Gibbs free energy of interaction ΔG_R, which is related to the enthalpy (ΔH_R) and entropy (ΔS_R) changes by the equation

$$\Delta G_R = \Delta H_R - T\Delta S_R$$

The sign of ΔG_R is determined by the relative magnitudes of the enthalpy and entropy terms—if negative, flocculation is (energetically) promoted; if zero, flocculation proceeds as if no adsorbed layers were involved; if positive, the particles are protected from flocculation. For ΔG_R to be positive we require that

(*i*) both the entropy ($T\Delta S_R$) and enthalpy terms are negative, with the former being the larger—termed *entropic* stabilisation indicating a dominating effect due to reduction in the number of configurations of the molecules in the adsorbed layer,

or

(*ii*) both terms are positive with the enthalpy change predominating—termed *enthalpic* stabilisation,

or

(*iii*) ΔH_R is positive and ΔS_R negative, and therefore both terms contribute to stability—termed *enthalpic–entropic* stabilisation.

Napper[85] has discussed how entropic and enthalpic effects may be distinguished in practice. In (*i*) the magnitude of the entropy term decreases as the temperature is reduced, hence entropically stabilised dispersions would flocculate on cooling. Enthalpically stabilised systems normally flocculate on heating, since in this case the $T\Delta S_R$ term is less than ΔH_R but the difference between them decreases as the temperature is raised. But both entropy and enthalpy terms can be temperature dependent, so the temperature change criterion is not definitive, and there is no unambiguous way of experimentally defining the controlling factor. For further discussion the reader is referred to Napper's article.[85]

The two factors will now be considered in more detail.

Effect of adsorbed layer on V_A.—Some calculations made by Vold[86] demonstrate the relation between the thickness and nature of the adsorbed layer and the force of attraction between two spherical particles. Three cases were considered, one being particles covered with a solvate layer of bound medium of thickness δ, one with a homogeneous adsorbed layer of surfactant molecules, and the third with an adsorbed layer of oriented amphipathic molecules. The general expression derived by Vold is

$$V_A = -\frac{1}{12}\left[(A_{mm}^{\frac{1}{2}} - A_{ss}^{\frac{1}{2}})^2 H_s + (A_{ss}^{\frac{1}{2}} - A_{pp}^{\frac{1}{2}})^2 H_p + \right.$$
$$\left. + 2(A_{mm}^{\frac{1}{2}} - A_{ss}^{\frac{1}{2}})(A_{ss}^{\frac{1}{2}} - A_{pp}^{\frac{1}{2}}) H_{ps}\right] \qquad (1.57)$$

where the subscript s refers to the solvate layer, and A_{mm}, A_{ss}, and A_{pp} are the Hamaker constants for the medium, solvate layer and particle respectively. H_s, H_p and H_{ps} are the appropriate forms of the distance function in the more general form of the Hamaker equation for two spheres of radii a_1 and a_2

$$V_A = -\frac{A}{12} H(x, y) \qquad (1.58)$$

where $H(x, y)$ is given by

$$H(x, y) = \frac{y}{x^2 + xy + x} + \frac{y}{x^2 + xy + x + y} + \\ + 2 \ln \frac{x^2 + xy + x}{x^2 + xy + x + y} \quad (1.59)$$

and $x = \Delta/2a$, and $y = a_2/a_1$. Δ is the shortest distance between the surfaces, which is assumed when the particles are in contact to be the minimum distance between surface atom centres beyond which electronic repulsion becomes important. Thus H_s refers to two spheres of radius $a + \delta$ with the surfaces of the solvate layers separated by Δ, H_p to two unsolvated spheres of radius a separated by $\Delta + 2\delta$, and H_{ps} to one sphere of radius a and one of radius $a + \delta$ separated by $\Delta + \delta$. The parameters in eqn. (1.59) are as follows:

for H_s, $y = 1$ and $x = \Delta/2(a+\delta)$
for H_p, $y = 1$ and $x = (\Delta+2\delta)/2a$
for H_{ps}, $y = (a+\delta)/a$ and $x = (\Delta+\delta)/2a$

and with these functions and eqn. (1.57) values of V_A at contact (Vold assumes $\Delta = 3$ Å) and as a function of interparticle distance may be calculated.

The reader is referred to the original paper for tabulated examples of the Hamaker functions, and to more recent papers[84,87] which show graphically how the V_A against distance relationship is determined by the value of δ and the relative magnitudes of the Hamaker constants. In general the following conclusions may be drawn, assuming the optimum combination of Hamaker constants.

(*i*) Only very small particles ($a \sim 100$ Å) can be stabilised by solvate layers of bound medium of thicknesses ~5–10 Å that are normally associated with such layers. The existence of solvate layers of considerable thickness (500–1000 Å) has been established for a range of polar liquids adjacent to glass and quartz surfaces by Deryaguin,[88] and in some cases we might expect the stability to flocculation to be considerably enhanced. Experimental information on this latter aspect is very limited,[39,89–91] but the implications are significant.

(*ii*) A homogeneous adsorbed layer which differs in composition for the medium is, for equal values of δ, more effective in reducing V_A than bound medium alone, but if the adsorbed layer contains oriented amphipathic molecules the effect may be even larger.

(*iii*) The effect is dependent on particle radius; large values of $\delta(\sim 100$ Å) would be required to stabilise particles of radius $\sim 1\ \mu$.

Interaction of adsorbed layers.—The magnitude of the forces that come into play during the interaction of two adsorbed layers on the surfaces of approaching colloidal particles depend on a number of factors, and the problem of estimating the effect is not a simple one. Surface coverage (density of packing of adsorbate, configuration and chemical nature of adsorbed molecules), and the thickness of the adsorbed layer are the major factors, but in many systems neither are well defined from experiment particularly where polymers are involved. However, some attempts have been made, albeit few, to assess the magnitude of the repulsive force when the layers interact and these will now be briefly reviewed.

Ottewill[92] has suggested that the total change in Gibbs free energy when two particles approach from infinity (no interaction) to a distance Δ between the surfaces is composed of three components: $\Delta G_{\text{surface}}$ associated with the reduction in the number of configurations (decrease in entropy) of the adsorbed molecules, $\Delta G_{\text{osmotic}}$ which arises from the excess osmotic pressure caused by overlapping of the layers (which may be regarded as having the same origin as a swelling pressure), and $\Delta G_{\text{chain elasticity}}$ associated with the compression of the adsorbed layers when particles collide. A reasonable estimate of $\Delta G_{\text{surface}}$ may be made from published theoretical treatments based on acceptable, but simple models, but the other two factors involve parameters for which data are not readily available. The following discussion will review briefly the basic ideas that have been put forward.

Estimation of $\Delta G_{\text{surface}}$—The existence of a repulsive force arising from loss of configurational entropy when two adsorbed layers interact was first recognised by Mackor,[93] who treated the simple case of two flat plates to which are attached rigid rods (representing the adsorbed molecules) anchored at one end by a freely hinged joint. All orientations of the rod are assumed to have equal probability and therefore the model is only appropriate to low surface coverages. When the plates are brought close together the gyration of the rod on the surface is hindered and the total number of configurations reduced, and this, Mackor demonstrated, would lead to a repulsive force. The idea was extended by Mackor and van der Waals[94] to higher surface coverages by use of a quasi-lattice model, and by assuming desorption occurs on interaction leading to an increase in free energy which is equivalent to a repulsion. They showed that the magnitude of the repulsive term was sufficient, when added to the attractive energy, to give rise to a potential energy of such a magnitude that stabilisation would occur. However, the applicability of this approach is limited to short chains and even then the parameters are difficult to evaluate from experimental data. Furthermore, for adsorbed layers of thickness δ no entropic repulsion would exist until $\Delta < 2\delta$, and for short chains ($\delta \sim 30$ Å) the value of V_A at this distance would in most cases be sufficiently large that there would be a secondary minimum in the potential energy diagram of sufficient depth that flocculation would be guaranteed (theoretically).

A more tractable method for longer chains, therefore applicable to polymeric adsorbates, has been worked out by Clayfield and Lumb.[95] The calculation applies to an irreversibly adsorbed molecule with an adsorbable group at one end only, contained in a solvent having no net interaction with the unadsorbed part of the molecule; the authors suggest that a hydrocarbon solution of a block copolymer comprising a hydrocarbon chain and a polar chain, is an appropriate basis for the model. The reduction in entropy of the adsorbed polymer molecule on compression is calculated, not the interpenetration of molecules on opposing surfaces. Sphere–sphere and sphere–flat plate interactions are considered and the relative orders of magnitude of the effect compared. The theory predicts that the greater the freedom of movement of the polymer chain, the more effective it is as a stabilising agent. Furthermore, there is an optimum size of molecule which will prevent adhesion of particles of a given size, and increasing the length of the polymer chain with the object of increasing δ may actually increase the possibility of flocculation. This is a result of the relative magnitudes of the potential energy maxima and secondary minima, which vary with particle size and chain length. Experimental support for this conclusion is at present very limited.[96] For further detail the reader is referred to the authors' original papers.[95,97]

Another theoretical treatment of entropic repulsion has been proposed by Meier.[98] It is assumed that the adsorbed molecules are sufficiently long and flexible that the chains obey random-flight statistics, and the configurational statistics, and hence free energies, evaluated. In addition the possible change in free energy of mixing of polymer and solvent as the density of chain segments changes when the adsorber layers interact, is evaluated. Meier concludes that at low coverages the loss of chain configurations is the most important of the two effects, but at high coverages the mixing term increases in significance particularly for polymers in 'good' solvents. From the theory it is possible to predict the effect of molecular weight, surface coverage, etc.

To assess the magnitude of the repulsive potential energy arising from these effects it is necessary to have information on the adsorbed layer, such as surface coverage, fraction of polymer segments attached directly to the surface, effective thickness, etc. Such data have been recently made available by the application of infrared absorption and ellipsometry. The use of infrared in the study of polymers adsorbed on solid surfaces was first reported by Fontana and Thomas,[99] and makes use of the fact that on adsorption there is a shift in the frequency of the characteristic band which permits the fraction of segments adsorbed to be estimated. With a knowledge of the fraction of groups bound directly to the surface and adsorption data it is possible to obtain information on the structure of the adsorbed layer.[100] Many structures are possible for a looped or coiled macromolecule attached in part to a surface, ranging from those which yield a relatively flat and compressed layer to those which give adsorbed layers highly extended away from the interface. The number and arrangement of attached portions and the size and distribution of the unattached loops define the conformation of the polymer molecule. Ellipsometry provides a method for estimating the extension of the loops away from the surface. The change in the state of polarisation of light upon reflection from a film covered surface is measured and used to calculate the thickness and refractive index of the film, the latter giving information on the concentration of polymer in the adsorbed layer.[100,101]

Estimation of $\Delta G_{\text{osmotic}}$ *and* $\Delta G_{\text{chain elasticity}}$—The remaining free energy terms ($\Delta G_{\text{osmotic}}$ and $\Delta G_{\text{chain elasticity}}$) are less well established but nevertheless would appear to make some contribution to the overall process. Fischer[102] considered the magnitude of the osmotic term; he related the excess osmotic pressure Π_e generated in an overlapping volume δV to the surface concentration c_s and the second virial coefficient B of the adsorbate, by

$$\Pi_e = BN_0kTc_s^2. \tag{1.60}$$

The free energy charge associated with the overlapping adsorbed layers is

$$\begin{aligned}\Delta G_{\text{osmotic}} &= 2\int_0^{\delta V} \Pi_e \,\delta V \\ &= 2\Pi_e \,\delta V \\ &= 2\Pi_e \frac{2\pi}{3}\left(\delta - \frac{\Delta}{2}\right)^2 \left(3a + 2\delta + \frac{\Delta}{2}\right) \\ &= kTBN_0{c_s}^2 \frac{4\pi}{3}\left(\delta - \frac{\Delta}{2}\right)^2 \left(3a + 2\delta + \frac{\Delta}{2}\right)\end{aligned} \tag{1.61}$$

indicating that if B is positive $\Delta G_{\text{osmotic}}$ is a repulsive energy term, and knowing (estimated) a value of B the contribution of the osmotic effect to the overall process may be assessed.

Jäckel[103] derived an expression for the free energy associated with chain elasticity in terms of the elastic modules of the adsorbed layer, the elastic energy obtained by compression of the adsorbed layers constituting a repulsive energy. But this and the osmotic term are subject to the gross uncertainties in the values of the constants involved and therefore any estimates of their contribution to the overall process can at best be only considered as very approximate.

Charged systems

The action of adsorbed high molecular weight materials is somewhat more complex in this case since both ionic and non-ionic mechanisms may be operative. The situation that arises in any particular system will depend on the magnitude of the surface potential and ionic strength, the ionic character of the adsorbed material, the amount adsorbed and the thickness of the adsorbed layer. The variety of phenomena observed may be roughly divided into two categories, one associated with stabilisation (protective action) and the other with flocculation (flocculating action) brought about by the material itself.

Protective action.—Gums, starch, and cellulose derivatives, proteins and polyacrylates are effective agents for protecting charged colloidal particles in aqueous media against flocculation by electrolytes. Several factors are involved including those already discussed (DLVO theory, entropic repulsion, reduced attraction by adsorbed layer).[92] Mathai and Ottewill[87] studied the addition of electrolytes to negative aqueous silver iodide sols stabilised by various homogeneous non-ionic surfactants of the poly-oxyethylene type. From plots of log W against log n critical electrolyte flocculation concentrations n' were obtained (*see* above) which for any particular polymer increased with concentration, the increase being considerably close to the critical micelle concentration in which range there was a rapid increase in adsorption. Above the critical micelle concentration the critical electrolyte concentration remains almost constant and the sols are very stable. Increasing the length of the ethylene oxide chain for the same alkyl chain length results in a decrease in the extent of adsorption, hence for the same molar concentration $C_{16}E_9$ is less effective in stabilising the sol against flocculation by cations than $C_{16}E_6$ (C_{16} indicates a hydrocarbon chain containing 16 carbon atoms and nine ethylene oxide units are indicated by E_9). Hence, the protection is related directly to the adsorption and the hydrocarbon chain length.

The protective action of polyoxyethylene glycols on gold sols is also related to the molecluar weight. Heller and Pugh[104] followed the colour change (red to blue) on the addition of potassium chloride to sols protected by the polymers and suggested that the flexible polymer chains sterically prevented the particles approaching close enough for the attractive forces to lead to flocculation.

The fundamental parameters of the electric double layer will also be changed by adsorption of high molecular weight materials. Experiments on the modification of the Stern potential at the mercury–electrolyte solution interface by the adsorption of polyoxyethylene compounds indicates[105] that there is a change in the potentials at short distances from the surface which results in a decrease in the magnitude of the electrostatic repulsion term. Furthermore, for the adsorption of amphipathic surfactants with the charged end-group on the solution side of the adsorbed layer, *e.g.*, the sulphate ion of sodium dodecyl sulphate adsorbed on a carbon black from aqueous solution,[62] it is necessary to modify the distance parameters used in the V_R calculation since the polar group giving rise to the surface potential does not physically coincide with the surface. This is particularly important when at moderately high ionic strengths, the double layer thickness is of the same order of magnitude as the adsorbed layer thickness.[62]

Flocculating action.—The same additives that at high concentration (and strongly adsorbed) give protective action might give rise to flocculation at lower surface coverages. For example, suspensions of clay in solutions of gelatin of concentration around 1–10 ppm flocculate but 0·1% solutions stabilise such materials. In certain cases when the colloidal particles and agent are oppositely charged the dispersion may become unstable as a result of charge neutralisation associated with the adsorption of the additive whereas at higher concentration (and adsorption) protection results, *e.g.* silica with polyethylenimine of low molecular weight. Another mechanism of action of polymeric materials is that in which parts of the adsorbed molecules are attached to two or more particles bringing about flocculation by a bridging mechanism; this may only occur when the surface coverage is low leaving parts of the surface available for further adsorption, hence bridging, when the particles come into close proximity. At higher additive concentrations protection action is obtained.

Some considerable effort is being put into the understanding of the flocculating action of high molecular weight polymers, a process which is of great importance in water clarification. The important question is often the nature of the interaction of the polymer with the surfaces of the particles. The appropriate points of attachment both on the polymer molecule and the surface must be determined, and to this end such studies as the effect of *p*H and ionic strength on adsorption, flocculation, electrokinetic effects and viscosity of polymer in solution, and studies on the character of the surface by infrared spectroscopy, etc., have been directed. Each situation has its own peculiarities and controlling factors; in most cases the details of the behaviour are not well understood. The reader is referred to papers by Healy and La Mer,[106] and by Kitchener *et al.*[107] for further discussion on the polymeric flocculants.

Concentrated dispersions

Reference has already been made to the application of the DLVO theory to systems containing high particle concentrations as found in normal practical dispersions containing pigments. At present a simple answer to this problem is not available. However, it has been suggested[108] that neither in aqueous or non-aqueous media will the electrostatic charge mechanism assure stabilisation of concentrated dispersions, and to prevent flocculation some sort of protective action by polymeric species is required. This protective action could arise from the effects associated with adsorption (already discussed) or through the polymer in the bulk forming a gel structure through which diffusion of the particles is restricted. In principle there is a range of possible mechanisms from that directly associated with the properties of the adsorbed layer to that which is entirely dependent on bulk solution properties, and hence the whole problem becomes complex and lacks quantitative precision.

Rehbinder[109] has devoted considerable effort to the study of concentrated disperse systems, and believes that effective stabilising action by the adsorbed layer is only possible if the layer is structured, *i.e.* it has a much higher viscosity than the medium and thus serves as a mechanical barrier. It is, of course, necessary for the cohesive energy between the external surfaces of the layers to be low.

Colloid stability in aqueous and non-aqueous media was the subject of a Faraday Society Discussion in 1966. The reader is referred to the published volume[110] containing the papers and discussion remarks, and in particular to the introductory and concluding remarks by J. Th. G. Overbeek and B. V. Deryaguin respectively.

REFERENCES

1. W. Gerstner, *J. Oil and Colour Chem. Assoc.*, **49** (1966) 954.
2. V. K. La Mer, *J. Colloid Sci.*, **19** (1964) 291.
3. H. J. Osterhof and F. E. Bartell, *J. Phys. Chem.*, **34** (1930) 1399.
4. A. Dupré, *Ann. Chimie*, (4) **6** (1865) 274.
5. J. A. Kitchener, *Proc. 3rd Int. Congress on Surface Activity*, **2** (1960) 426.
6. W. J. Dunning, *J. Oil and Colour Chem. Assoc.*, **48** (1965) 509.
7. D. M. Gans, *J. Paint Technology*, **38** (1966) 322.
8. D. M. Gans, *J. Paint Technology*, **39** (1967) 501.
9. A. W. Adamson, *Physical Chemistry of Surfaces*, Interscience, New York, Second Edition, 1967, Chapter 7.
10. P. M. Heertjes and N. W. F. Kossen, *Powder Technology*, **1** (1967) 33.
11. P. M. Heertjes and W. C. Witvoet, *Powder Technology*, **3** (1970) 339.
12. E. D. Washburn, *Phys. Rev.*, **17** (1921) 374.
13. V. T. Crowl and W. D. S. Wooldridge, *Wetting*, SCI Monograph No. 25, 1967, 200.
14. R. J. Good, *Chem. and Ind.* (1971) 600.
15. R. J. Good, *Aspects of Adhesion*, **8** (1971).
16. W. A. Zisman, *Contact Angle, Wettability and Adhesion*, Adv. in Chem. Series No. 43, ACS Washington, 1964.
17. D. M. Gans, *J. Paint Technology*, **41** (1969) 515.
18. F. G. Greenwood, G. D. Parfitt, N. H. Picton and D. G. Wharton, *Adsorption from Solution*, Adv. in Chem. Series No. 79, ACS, Washington, 1968.

19. G. D. Parfitt and D. G. Wharton, *J. Colloid and Interface Sci.* **38** (1971) 431.
20. T. M. Doscher, *J. Colloid Sci.*, **5** (1950) 100.
21. K. Meguro, *J. Chem. Soc. Japan* (*Ind. Chem. Sect.*), **58** (1955) 905.
22. K. Tamaki, *J. Japan Oil Chem. Soc.*, **9** (1960) 426.
23. F. F. Aplan and D. W. Fuerstenau, *Froth Flotation*, Ed. D. W. Fuerstenau, New York, American Institute of Mining and Metallurgical Engineering, (1962).
24. G. A. H. Elton, *Proc. 2nd Int. Congress on Surface Activity*, **3** (1957) 161.
25. P. A. Rehbinder, *Colloid J. USSR*, **20** (1958) 493.
26. P. A. Rehbinder and V. I. Likhtman, *Proc. 2nd Int. Congress on Surface Activity*, Butterworths, London, 1957, No. 3, 563.
27. E. D. Shchukin and P. A. Rehbinder, *Colloid J. USSR*, **20** (1958) 601.
28. A. T. DiBenedetto, *The Structure and Properties of Materials*, McGraw-Hill, New York, 1967.
29. G. M. Bartenev, I. V. Iudena and P. A. Rehbinder, *Colloid J. USSR*, **20** (1958) 611.
30. G. S. Khodakov and P. A. Rehbinder, *Colloid J. USSR*, **22** (1960) 375.
31. K. J. Mysels, *Introduction to Colloid Chemistry*, Interscience, New York, 1959, Chapter 5.
32. M. Von Smoluchowski, *Zeits. Physik. Chem.*, **92** (1917) 129.
33. A. E. van Arkel and H. R. Kruyt, *Rec. Trav. Chim.*, **39** (1920) 656.
34. P. Tuorilla, *Kolloidchen. Beih.*, **22** (1926) 191.
35. W. I. Higuchi, R. Okada, G. A. Stelter and A. P. Lemberger, *J. Amer. Pharm. Ass.*, **52** (1963) 49.
36. R. H. Ottewill and D. J. Wilkins, *Trans. Faraday Soc.*, **58** (1962) 608.
37. K. E. Lewis and G. D. Parfitt, *Trans. Faraday Soc.*, **62** (1966) 1652.
38. D. N. L. McGown and G. D. Parfitt, *J. Phys. Chem.*, **71** (1967) 449.
39. G. A. Johnson, S. M. A. Lecchini, E. G. Smith, J. Clifford and B. A. Pethica, *Disc. Faraday Soc.*, **42** (1966) 120.
40. D. L. Swift and S. K. Friedlander, *J. Colloid Sci.*, **19** (1964) 621.
41. G. M. Hidy, *J. Colloid Sci.*, **20** (1965) 123.
42. G. M. Hidy and D. K. Lilly, *J. Colloid Sci.*, **20** (1965) 863.
43. D. S. Jovanovic, *Kolloid Zeits.*, **203** (1965) 42.
44. B. V. Deryaguin and L. D. Landau, *Acta Physiocochim. URSS*, **14** (1941) 633.
45. E. J. W. Verwey and J. Th. G. Overbeek, *Theory of the Stability of Lyophobic Colloids*, Elsevier, Amsterdam, 1948.
46. H. C. Hamaker, *Physica*, **4** (1937) 1058.
47. E. A. Moelwyn-Hughes, *Physical Chemistry*, 2nd edition, Pergamon, London, 1961.
48. J. Lyklema, *Pont. Acad. Sci. Scr. Varia*, **31** (1967) 181.
49. E. M. Lifshitz, *Soviet Phys. JETP*, **2** (1956) 73.
50. E. E. Dzyaloshinkskii, E. M. Lifshitz and L. P. Pitaevskii, *Soviet Phys. JETP*, 37 (1960) 161.
51. H. Krupp, *Adv. in Colloid and Interface Science*, **1** (1967) 111.
52. F. M. Fowkes, *Ind. Eng. Chem.*, **56** (1964) 40.
53. J. H. Schenkel and J. A. Kitchener, *Trans. Faraday Soc.*, **56** (1960) 161.
54. H. B. G. Casimir and D. Polder, *Phys. Rev.*, **73** (1948) 360.
55. Ref. 45, p. 101.
56. G. Frens, D. J. Engels and J. Th. G. Overbeek, *Trans. Faraday Soc.*, **63** (1967) 418.
57. O. F. Devereux and P. L. de Bruyn, *Interaction of Plane Parallel Double Layers*, M.I.T. Press, Cambridge, USA, 1963.

58. N. Fuchs, *Zeits. Physik.*, **89** (1934) 736.
59. B. V. Deryaguin, *Trans. Faraday Soc.*, **36** (1940) 203.
60. S. N. Srivastava and D. A. Haydon, *Trans. Faraday Soc.*, **60** (1964) 971.
61. A. Watillon and A. M. Joseph-Petit, *Disc. Faraday Soc.*, **42** (1966) 143.
62. G. D. Parfitt and N. H. Picton, *Trans. Faraday Soc.*, **64** (1968) 1955.
63. J. Th. Overbeek, *Colloid Science,* Ed. H. R. Kruyt, Elsevier, Amsterdam, 1948, Vol. 1, Chapter 8.
64. H. Reerink and J. Th. G. Overbeek, *Disc. Faraday Soc.*, **18** (1954) 74.
65. R. H. Ottewill and J. N. Shaw, *Disc. Faraday Soc.*, **42** (1966) 154.
66. R. H. Ottewill and D. J. Wilkins, *Trans. Faraday Soc.*, **58** (1962) 608.
67. G. D. Parfitt and A. L. Smith, unpublished data.
68. S. Levine, J. Mingins and G. M. Bell, *J. Electroanalyt. Chem.*, **13** (1967) 280.
69. L. A. Romo, *Disc. Faraday Soc.*, **42** (1966) 232.
70. J. L. van der Minne and P. H. J. Hermanie, *J. Colloid Sci.*, **7** (1952) 600.
71. G. D. Parfitt, *J. Oil and Colour Chem. Assoc.*, **51** (1968) 137.
72. D. N. L. McGown and G. D. Parfitt, *Kolloid Zeits.*, **219** (1967) 48.
73. Ref. 45, p. 155.
74. J. L. van der Minne and P. H. J. Hermanie, *J. Colloid Sci.*, **8** (1953) 38.
75. H. Koelmans and J. Th. G. Overbeek, *Disc. Faraday Soc.*, **18** (1954) 52.
76. D. N. L. McGown, G. D. Parfitt and E. Willis, *J. Colloid Sci.*, **20** (1965) 650.
77. D. N. L. McGown and G. D. Parfitt, *Disc. Faraday Soc.*, **42** (1966) 225.
78. F. J. Micale, Y. K. Lui and A. C. Zettlemoyer, *Disc. Faraday Soc.*, **42** (1966) 238.
79. D. N. L. McGown and G. D. Parfitt, *Kolloid Zeits.*, **220** (1967) 56.
80. W. Albers and J. Th. G. Overbeek, *J. Colloid Sci.*, **14** (1959) 510.
81. B. V. Deryaguin, *Disc. Faraday Soc.*, **18** (1954) 85.
82. R. Hogg, T. W. Healy and D. W. Fuerstenau, *Trans. Faraday Soc.*, **62** (1966) 1938.
83. A. Bierman, *J. Colloid Sci.*, **10** (1955) 231.
84. V. T. Crowl, *J. Oil and Colour Chemists Assoc.*, **50** (1967) 1023.
85. D. H. Napper, *Ind. Eng. Chem. Prod. Res. Develop.*, **9** (1970) 467.
86. M. J. Vold, *J. Colloid Sci.*, **16** (1961) 1.
87. K. G. Mathai and R. H. Ottewill, *Trans. Faraday Soc.*, **62** (1966) 759.
88. B. V. Deryaguin, *Disc. Faraday Soc.*, **42** (1966) 109.
89. G. D. Parfitt and E. Willis, *J. Colloid Sci.*, **22** (1966) 100.
90. See discussion remarks, *Disc. Faraday Soc.*, **42** (1966) 134–42.
91. A. Fawcett, G. D. Parfitt and A. L. Smith, *Nature*, **204** (1964) 775.
92. R. H. Ottewill, *Non-ionic Surfactants*, Ed. M. Schick, M. Dekker, New York, 1967, Chapter 19.
93. E. L. Mackor, *J. Colloid Sci.*, **6** (1951) 492.
94. E. L. Mackor and J. H. van der Waals, *J. Colloid Sci.*, **7** (1952) 535.
95. E. J. Clayfield and E. C. Lumb, *J. Colloid and Interface Sci.*, **22** (1966) 269.
96. E. J. Clayfield and E. C. Lumb, *Disc. Faraday Soc.*, **42** (1966) 285.
97. E. J. Clayfield and E. C. Lumb, *Macromolecules*, **1** (1968) 233.
98. D. J. Meier, *J. Phys. Chem.*, **71** (1967) 1961.
99. B. J. Fontana and J. R. Thomas, *J. Phys. Chem.*, **65** (1961) 480.
100. R. R. Stromberg, E. Passaglia and D. J. Tutas, *J. Res. Natl. Bur. Std.*, **67A** (1963) 43.
101. R. R. Stromberg, E. Passaglia and D. J. Tutas, *Ellipsometry for the measurement of surfaces and thin films*, N.B.S. Misc. Publication, 256 (1964) 281.
102. E. W. Fischer, *Kolloid Zeits.*, **160** (1958) 120.
103. K. Jäckel, *Kolloid Zeits.*, **197** (1964) 143.
104. W. Heller and T. L. Pugh, *J. Polymer Sci.*, **47** (1960) 203.

105. A. Watanabe, F. Tsuji and S. Veda, *Kolloid Zeits.*, **193** (1963) 39.
106. T. W. Healy and V. K. La Mer, *Rev. Pure and Appl. Chem.* (*Australia*), **13** (1963) 112.
107. R. W. Slater, J. P. Clark and J. A. Kitchener, *Proc. Brit. Ceramic Soc.* (1968).
108. P. A. Rehbinder and A. B. Taubman, *Colloid J. USSR*, **23** (1961) 301.
109. P. A. Rehbinder, *Colloid J. USSR.* **20** (1958) 493.
110. *Discussions of the Faraday Society*, Vol. 42, 1966.

CHAPTER 2

PROPERTIES OF THE SOLID–LIQUID INTERFACE OF RELEVANCE TO DISPERSION

M. J. JAYCOCK

FORCES BETWEEN ATOMS, IONS AND MOLECULES[1-3]

The state of any solid–liquid interface is essentially dependent upon the forces which exist between the atoms, ions and molecules that are present in the neighbourhood of the junction between solid and liquid phases. The theoretical description of such forces represents one of the few ideas common throughout the various branches of physical science, and it would therefore seem appropriate to consider briefly the basic concepts involved.

Attractive and repulsive forces between molecules

Let us consider, for the sake of simplicity, a pair of molecules of a monatomic gas. The forces which exist between the pair are usually termed van der Waals' or dispersion forces. If there is a preferred separation of the molecules it will be at the distance corresponding to a minimum in the potential energy φ of the pair. At short range repulsive forces predominate, increasing with decreasing distance of separation a causing an increase in the potential energy. These are associated with electron shell interaction and are sometimes called Born repulsive forces. At longer range attractive forces predominate, also increasing with decreasing distance of separation, causing a decrease in the potential energy of the pair. These result from the effects of instantaneous dipoles upon polarisable molecules. The position of the potential energy minimum will be determined by the relative magnitudes of the attractive and repulsive energies, and will correspond to the position of zero force. A diagrammatic representation of the total potential energy as a function of the distance of separation is given in Fig. 2.1.

The total potential energy may be represented as[4,5]

$$\varphi_{\text{total}} = \varphi_{\text{repulsive}} + \varphi_{\text{attractive}} \tag{2.1}$$

or

$$\varphi = \frac{A}{a^n} - \frac{B}{a^m} \tag{2.2}$$

where A and B are positive constants, and n and m are integers where n is greater than m.

For a pair of molecules, eqn. (2.2) is often rewritten as either

$$\varphi = 4D_e \left[\left(\frac{\sigma}{a}\right)^{12} - \left(\frac{\sigma}{a}\right)^{6} \right], \tag{2.3}$$

or

$$\varphi = \frac{27}{4} D_e \left[\left(\frac{\sigma}{a}\right)^{9} - \left(\frac{\sigma}{a}\right)^{6} \right] \tag{2.4}$$

where D_e is the minimum value of φ, and σ is the value of a when $\varphi = 0$. These expressions are often called the 12:6 and 9:6 potentials, respectively, but eqn. (2.3) is even more commonly called the Lennard-Jones potential.

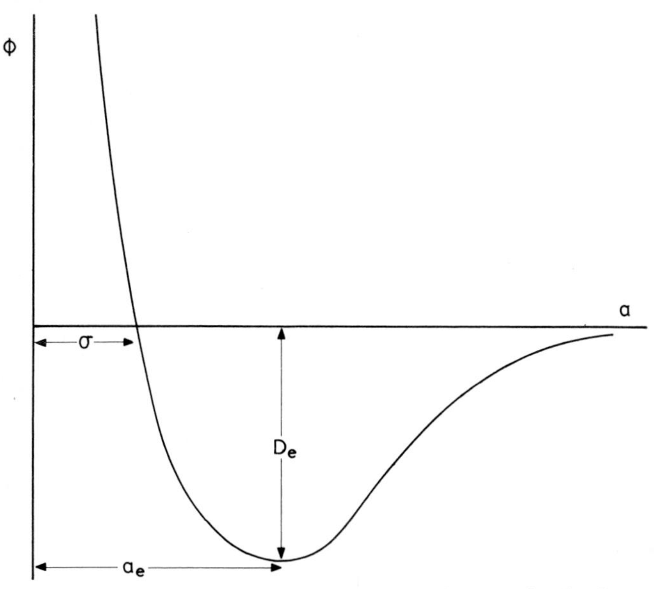

Fig. 2.1. The total potential energy curve of a pair of molecules.

The general equations (2.1) and (2.2) may be applied to systems other than a pair of molecules, providing the correct expression for the type of force involved is used. The expressions presented here are not necessarily the most accurate equations to describe these forces, and for a detailed consideration of the problem the reader is referred to ref. 1.

Coulombic forces between ions

Coulomb experimentally established that the force f between two charges, e_1 and e_2, a distance a apart in a medium of dielectric constant ε, is given by*

$$f = \frac{e_1 e_2}{\varepsilon a^2} \tag{2.5}$$

and the potential energy by

$$\varphi(a) = -\int \frac{e_1 e_2}{\varepsilon a^2}\, da = \frac{e_1 e_2}{\varepsilon a}. \tag{2.6}$$

The comparative strength of the electrostatic forces between ions in aqueous solution can be estimated from the fact that at 25°C the potential energy of attraction between two univalent ions of opposite sign is approximately equal to kT when their distance is 7 Å, and approximately ten times this value if the separation is reduced to 2·25 Å. It is this type of force which is the major factor at comparatively long range.

Interaction between a permanent dipole and an ion

The interaction between a permanent dipole and an ion may be treated quite simply by adding the Coulombic terms due to the interaction of the ion with each pole-charge in turn, making allowance for the orientation of the dipole. When the length of the dipole is small compared with the distance of separation, the force between the ion and the dipole, of moment μ, is given by

$$f = \frac{2e\mu \cos \theta}{a^3}, \tag{2.7}$$

and the potential energy by

$$\varphi(a) = \frac{e\mu \cos \theta}{a^2}, \tag{2.8}$$

where θ is the angle between the axis of the dipole and the line joining the ion to the centre of the dipole. Thus the interaction may be either repulsive or attractive according to the sign of the ionic charge and the orientation of the dipole.

Interaction between two permanent dipoles

The energy of interaction between two permanent dipoles may be obtained by applying Coulomb's Law to the four fractional charges concerned. The resulting expression is more complicated because of the

* If SI units (rationalised MKS) are used, equation (2.5) becomes

$$f = \frac{e_1 e_2}{4\pi\varepsilon a^2} \tag{2.5a}$$

The factor 4π is included so that its occurrence or otherwise in derived expressions would accord with geometric expectations and not vice versa, *e.g.* the equation for a parallel plate condenser becomes $C = \varepsilon/d$ and that for an isolated spherical condenser $C = 4\pi\varepsilon a$.

necessity of describing the orientation of one dipole with respect to the other. If the distance apart is large compared with the length of either dipole, then

$$\varphi = \frac{\mu_1\mu_2}{a^3}\,[2\cos\omega_1\cos\omega_2 - \sin\omega_1\sin\omega_2\cos(\chi_1-\chi_2)], \qquad (2.9)$$

where ω_1 and ω_2 are the angles of inclination of the dipolar axes to the lines of centres, reflected in some plane, and χ_1 and χ_2 are the angles subtended between the dipolar axes and the perpendiculars to the plane at the dipolar centres. Equation (2.9) predicts that the system will have maximum potential energy when the axes of the dipoles are in the same plane with like poles facing each other, of $2\mu_1\mu_2/a^3$. The position of minimum energy will also be when the dipoles are in the plane but with unlike poles facing each other, when the potential energy will be $-2\mu_1\mu_2/a^3$.

Polarisability

A symmetrical molecule such as methane or argon possesses no permanent dipole moment. However, in the presence of an external electric field electrons may be displaced from their usual positions and the molecule acquires an induced dipole. It is usually postulated that at moderate field strengths the induced moment μ_i is directly proportional to the field strength F thus

$$\mu_i = \alpha F \qquad (2.10)$$

This equation defines the polarisability α. A whole series of relationships may be derived to describe the interaction of an induced dipole with an ion, or a dipole, or an induced dipole using Coulomb's Law, and for further information one of the advanced texts should be consulted.

DESCRIPTION OF AN INTERFACE

The boundary between two homogeneous bulk phases should not be thought of as a plane of zero thickness, but more realistically as a separate surface phase of definite thickness. The definition of a surface phase as described by Gibbs[6,7] can be readily understood with the aid of Fig. 2.2 which represents a section perpendicular to the surface. At all points in and above plane AA' the physical and chemical properties are those associated with the bulk phase α, whereas at all points in and below plane BB' the physical and chemical properties are those associated with the bulk phase β. In the surface phase σ the properties gradually change from those of phase α at plane AA', to those of β at plane BB'; in any plane in the surface phase parallel to AA' or BB' the properties will everywhere be the same.

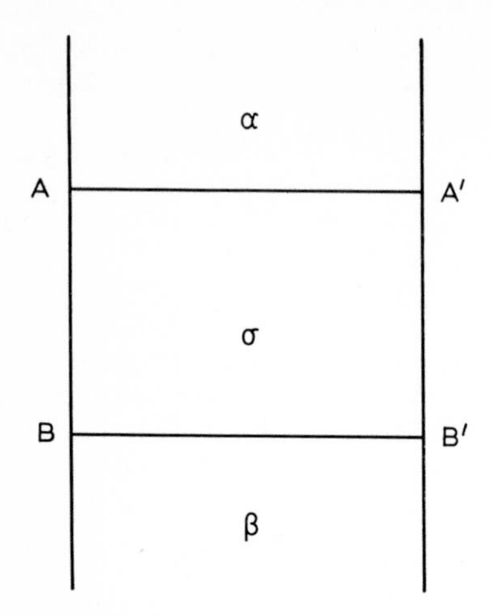

Fig. 2.2. Definition of a surface phase.

Surface tension

Drops of liquid behave as if they are surrounded by an elastic skin which prevents them from spreading. Young[8] attempted an explanation of this *surface tension* in terms of the forces existing between the molecules of the liquid. The cohesive forces must be greater than those due to thermal motion otherwise the drop would vaporise. The molecules at the surface will be less attracted to their neighbours than molecules in the bulk, since a molecule at the surface has fewer neighbours. Therefore since the energy of a surface molecule will be greater than that of a similar molecule in the bulk phase, the surface of such a pure phase will tend to contract spontaneously, reducing the free energy of the system to a minimum.

The surface tension γ_0 is the force per unit length contracting such a surface.[9] If P, V, A, S, and T represent the pressure, volume, surface area, entropy and absolute temperature of the system of n moles of chemical potential μ, then

$$dF = -SdT - PdV + \gamma_0 dA + \mu dn \quad (2.11)$$

where F is the Helmholtz free energy or work function of the system. At constant values of T, V and n this reduces to

$$\gamma_0 = \left(\frac{\partial F}{\partial A}\right)_{T,\, V,\, n.} \quad (2.12)$$

If we define ${}^sF$ as the Helmholtz free energy per unit area, then $dF = d(A{}^sF) = {}^sFdA + Ad{}^sF$, and therefore on substitution

$$\gamma_0 = {}^sF + A\left(\frac{\partial\, {}^sF}{\partial A}\right)_{T,\, V\, .n} \quad (2.13)$$

However, in a one-component system ${}^sF$ is only dependent upon the configuration of the surface molecules, therefore it must be constant, hence

$$\gamma_0 = {}^sF \tag{2.14}$$

when T and V are constant. It can also be shown that

$$\gamma_0 = {}^sG \tag{2.15}$$

the Gibbs free energy per unit area, when T and P are constant. Since surface changes are accompanied by only small changes in pressure and volume, ${}^sF$ and ${}^sG$ are virtually identical.

Total surface energy

The total surface energy per unit area ${}^sU$ may be calculated from

$${}^sF = {}^sU - T{}^sS \tag{2.16}$$

where the superscript s has the same significance as before. The entropy term ${}^sS$, may be found by using the relation

$${}^sS = -\left(\frac{\partial^s F}{\partial T}\right)_{n,V} = -\left(\frac{\partial \gamma_0}{\partial T}\right)_{n,V} \tag{2.17}$$

so that

$${}^sU = \gamma_0 - T\left(\frac{\partial \gamma_0}{\partial T}\right)_{n,V} \tag{2.18}$$

The quantity ${}^sU$ is practically temperature independent.

Surface entropy

As has been mentioned previously, a molecule in the surface has fewer neighbours than a molecule in a bulk phase. So that compared with a bulk liquid there is a new possibility of randomness. Davies[9] has shown that the two possibilities of a molecule being near the surface or in the bulk phase, give rise to an entropy increase of approximately $R \ln 2 = 1{\cdot}4$ e.u., this being a standard entropy change associated with forming a surface. The temperature coefficient of surface tension must consequently be negative.

Interfacial tension

Although we have talked about surface tension we mean by this the interfacial tension between a gaseous phase and the liquid phase. Whether the gaseous phase is the pure vapour of the liquid, or air saturated with vapour makes no measurable difference to the value of the interfacial tension, and consequently the use of the name surface tension since, for practical purposes, we can consider the measured value of γ_0 as being a property solely of the liquid. This surface tension or interfacial tension is also fairly readily measured by a number of standard methods.[9,10]

The situation in respect of liquid–liquid interfaces is somewhat similar since, for example, a drop of water in oil behaves similarly to a drop of water in air; the surface area being reduced to a minimum under the influence of the interfacial tension γ_i. Furthermore, interfacial tensions may be fairly readily measured by methods[9,10] essentially similar to those for the measurement of surface tension. However, in this case both liquids often influence the interfacial tension. For example, for *n*-butanol against water at 25°C, γ_i is only 1·6 dyne/cm, a low value being characteristic of oils with polar or hydrophilic groups. Since the surface tension of *n*-butanol at 25°C is 24 dyne/cm the butanol molecules must have their –OH groups oriented towards the interface, where the repulsion between these molecules prevents γ_i attaining a high value.

Further examples illustrating this point can be found in Table 2.1, namely *n*-octanol and diethyl ether, although the values of γ_i are higher. As Bikerman[12] has noted the mutual insolubility of the oil and water runs parallel with the interfacial tensions.

TABLE 2.1

SURFACE TENSIONS OF PURE LIQUIDS AGAINST AIR,[11] AND INTERFACIAL TENSIONS BETWEEN WATER AND PURE LIQUIDS[11]

Liquid	*Temp. (°C)*	γ_0 *(dyne/cm)*	γ_i *(dyne/cm)*
Water	20	72·8	
	25	72·0	
Bromobenzene	25	35·75	38·1
Benzene	20	28·88	35·0
	25	28·22	34·71
n-Octanol	20	27·53	8·5
Carbon tetrachloride	20	26·9	45·1
Diethyl ether	20	17·01	10·1

In this section we have confined attention to liquid surfaces and have not considered the solid surface or interfacial tensions involving a solid, which is the subject of the next section.

THE SOLID SURFACE[13]

The nature of a solid prevents the application of the methods suitable for a gas–liquid or liquid–liquid interface to the problem of measuring the interfacial tension* when one of the phases is a solid. However, under special conditions the methods may occasionally be applied. For example, a copper wire at temperatures near its melting point tends to shorten, being sufficiently plastic to flow under the influence of surface tension. By applying stress to reduce the strain to zero, a surface tension of 1370 dyne/cm was calculated,[14] which is much higher than the values quoted for γ_0 and γ_i in Table 2.1.

Surface mobility

Sintering is possible with certain solids because under the correct conditions they show a certain amount of bulk and surface mobility. In the formation, say, of a sintered glass disc, powdered glass is heated under some pressure to a temperature just below its melting point, when the particles of glass tend to fuse where they touch each other and form a porous disc. Surface tension forces certainly play a part in what is a rather complicated process. Because of irregularities in the surface of the particles on a submicroscopic or microscopic scale, the area of contact is likely to be very small, and quite low external pressures applied to the mass as a whole will develop local pressures great enough to cause plastic deformation.

Furthermore, bulk and surface diffusions are likely to be significant factors at temperatures high enough to cause sintering.[15] Scratches on a silver surface disappear at temperatures near the melting point,[16] or if a silver ball is placed on the surface of the metal at this temperature then the area of contact enlarges and a weld results.[17] Sintering becomes important at temperatures above about three-quarters of the melting point expressed in degrees Kelvin; it is characterised by an increase in apparent density,[18,19] and a decrease in absorptive power and chemical reactivity.[20]

It is possible to get some idea of surface mobility by applying gas kinetic theory to the equilibrium between evaporation and condensation. Adamson[21] has calculated that for the equilibrium between water and water vapour at room temperature the average life-time of a molecule in the surface is 10^{-6} sec. Similar calculations for a solid surface, say of a metal, show that if the saturation vapour pressure is 10^{-40} atm the average lifetime of a surface atom would be 10^{37} sec. On the other hand[22] for copper at threequarters of its melting temperature (725°C) the estimated vapour pressure is 10^{-8} mm Hg which leads to an average surface lifetime of about 1 sec. Similar circumstances apply to bulk diffusion, and for copper at 725°C[23] the bulk self-diffusion coefficient is about 10^{-11} cm^2/sec, which would correspond to a mean displacement of 450 Å in one

* The use of the term *surface tension* and also the term *surface energy* for an interface involving a solid has given rise to considerable confusion. The equation relating γ to ${}^sF$ is $\gamma = {}^sF - \Sigma\mu_i\Gamma_i$. If a dividing surface is chosen such that $\Sigma\mu_i\Gamma_i = 0$, then and only then would it be legitimate to term γ the *specific surface free energy* but this situation is not usual (see Ref. 15, p. 8, and Ref. 90).

second, whereas at room temperature the mean displacement per second is negligible, since there is an apparent activation energy of 54 kcal/mole.

When considering surface diffusion the picture is rather different. This has been studied directly by means of the field-emission microscope.[24] The tip of a thin metal wire cathode is located at the centre of a hollow glass sphere, the inner surface of which is coated with a suitable fluorescent material. When a high potential is applied, electrons are emitted from the tip and cause a fluorescent pattern of light and dark areas to appear representing a magnified image of the emission pattern of the tip. Magnifications of up to 10^7 may be obtained. It has been shown by this method that surface migration becomes important at about 40% of the melting point in degrees Kelvin. Noticeable surface diffusion has been found on copper[25] at 700°C, when the activation energy is low.

For most high melting point solids at room temperature the surface must be considered essentially static with the atoms or molecules merely vibrating around equilibrium or quasi-equilibrium positions. However, as the melting point is approached, firstly there is an increase in lateral mobility, and then an increase in bulk diffusion, and finally vaporisation.

Conditioning of solid surfaces

The comparative immobility of the surface of a solid under normal conditions implies that the physical state of a surface will be largely dependent on the immediate history of that surface. Thus one would expect to find differences between a freshly cleaved surface, a polished surface, a ground surface, and a heat treated surface.

Polishing surfaces drastically affects their nature. The layer on the surface formed during this process is usually known as the Beilby layer, and appears to be amorphous under the microscope,[26] having the appearance of a liquid viscous film which had flowed into all irregularities in the surface.[27] Electron diffraction studies of this surface layer show diffuse rings typical of the amorphous state or of the unordered liquid state. Raether[28] has concluded that the Beilby layer is in fact microcrystalline, but that the crystallites are so small that they give rise to diffuse rings with electron diffraction. However, there is also evidence of oxide being dragged into the surface region[29] together with impurities from the polishing media.[30] The fact that the Beilby layer is not stable has been illustrated by Cochrane[31] who deposited a thin layer of gold on nickel. The gold layer was polished and then removed and studied by electron diffraction, which showed a gradual transition from microcrystalline to crystalline. Electropolished surfaces are more nearly normal than polished surfaces, usually showing crystalline electron diffraction patterns.[32]

The above discussion illustrates some of the reasons for surface variation, not forgetting that contamination might be more important, in an individual case, than any of them.

Surface tension and surface free energy

The surface free energy of a solid is not necessarily equatable to the surface tension. Whereas a newly formed liquid surface will rapidly take up a uniform equilibrium state, the same is not true of a solid. The newly formed solid is likely to possess a considerable range of values of surface free energy, varying from region to region of the surface. Moreover, for a liquid the surface tension is the same in all directions, but this may not be true for a solid. If we resolve the surface tension into two directions at right angles, we can represent these partial surface tensions by γ_1 and γ_2.

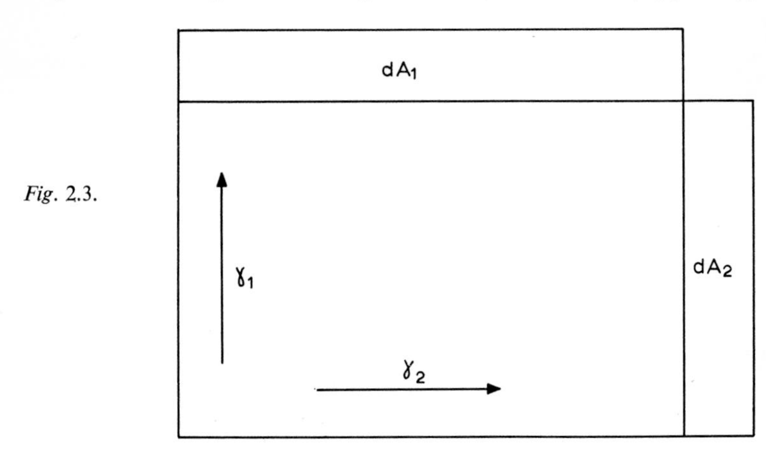

Fig. 2.3.

For an anisotropic solid, if the area is increased in two directions by dA_1 and dA_2, as shown in Fig. 2.3 then the total increase in free energy is given by the reversible work[33] against the stresses γ_1 and γ_2

$$d(A^sF) = \gamma_1 dA_1 + \gamma_2 dA_2 \tag{2.19}$$

where ${}^sF$ denotes the free energy per unit area. If $\gamma_1 = \gamma_2$ then

$$\gamma = \frac{d(A\,{}^sF)}{dA} = {}^sF + A\left(\frac{d\,{}^sF}{dA}\right) \tag{2.20}$$

which is the same as eqn. (2.13) and implies the same condition. In this case, however, $(d^sF/dA) \neq 0$ as a general rule: it could only be so for a homotattic (perfectly uniform) surface.[34,35] Thus if the surface achieved some uniform equilibrium state ${}^sF = \gamma$; in all other cases ${}^sF$ and γ will be different from their equilibrium values and different from each other.[36]

The Laplace equation describes the pressure gradient ΔP across a curved surface, in terms of the two radii of curvature, r_1 and r_2, thus

$$\Delta P = \gamma\left[\frac{1}{r_1} + \frac{1}{r_2}\right]. \tag{2.21}$$

For a nearly spherical isotropic crystal, the pressure difference causes compression. If β is the compressibility then the volume change ΔV is approximately

$$\frac{\Delta V}{V} = 3\frac{\Delta \mathrm{r}}{\mathrm{r}} = -\Delta P.\beta$$

or

$$\Delta \mathrm{r} = -2\beta\gamma/3. \tag{2.22}$$

This effect has been investigated[37] by means of X-ray powder photographs. For magnesium oxide values of Δr near the expected value of 0·6 Å, *i.e.* a 0·1 % change in lattice distance for a 600 Å crystal, were found using a calculated value of 6573 dyne/cm for the surface tension. The results, however, were thought to be affected by adsorbed gases and surface contamination. The problem has been reinvestigated by Guilliatt and Brett[38] who have confirmed that adsorbed films produce marked effects. Traces of moisture produced lattice dilation, whereas in the absence of water vapour or adsorbed gases a lattice contraction was observed, the magnitude of which increased with decreasing crystallite size, and was substantially in agreement with the theoretical results of Anderson and Scholtz.[39]

Calculated values of surface tension and surface energies

The method of calculating surface energies varies with the nature of the forces that exist between the atoms or ions or molecules which constitute the surface. These forces were considered separately earlier in this chapter. The simplest case is that of a covalently bonded crystal, such as the diamond, since in this case it is possible to ignore long range forces and merely consider valency forces. Harkins[40] calculated the surface energy at 0°K as one-half of the energy of the number of bonds broken in forming the surface.

$$^{s}U = \tfrac{1}{2}U_{\mathrm{cohesion}}. \tag{2.23}$$

If the cleaved surface of diamond is the (111) plane, and if the interplanar atomic separation is 2·32 Å, then $1{\cdot}83 \times 10^{15}$ bonds are broken per cm^2 of surface. The covalent bond energy is 5400 erg/cm^2. The corresponding calculation for the (100) plane is 9140 erg/cm^2. It should be noted that since we are calculating for 0°K these values will be equal to the surface free energy at that temperature. Harkins calculated by means of eqn. (2.17) and the Eötvös equation[41]

$$\gamma(M/\rho)^{2/3} = K(T_c - T) \tag{2.24}$$

where M is the molecular weight, ρ the density, T_c the critical temperature, and K is a constant, that the entropy contribution at 25°C to the surface energy was negligible; so that the above values are also approximately those of the surface free energy at 25°C. This method of calculation can only be approximate since no allowance is made for surface distortion, and the surface considered is an ideal one.

The simplest crystals involving long range forces for which calculations are possible are those of the rare gases, which are normally face-centred cubic lattices. The principle of the calculation is essentially the same as for the diamond crystal, namely, the calculation of the net energy of interaction across the cleavage plane, that is the surface energy at 0°K is taken as being the excess potential energy of the molecules near the surface. Shuttleworth's[42] calculations (Table 2.2) were based on a form of

TABLE 2.2

SURFACE ENERGIES AT 0°K FOR RARE GAS CRYSTALS[42]

Rare gas	^sU (*erg/cm*^{2}) (111) *plane*	(100) *plane*
Ne	17·2	17·9
A	41·1	42·7
Kr	53·3	55·3
Xe	60·7	63·0

eqn. (2.3) but included allowance for surface distortion by allowing the atoms of the surface plane to move to the position of minimum energy. Although this lattice distortion term was less than 1% of the total it represented a considerable change in the first plane position.

The calculation of surface energies for ionic crystals is more complex since there are both long range forces and coulombic forces to consider, and a more complicated potential energy function must be employed. Early calculations employing once more the calculation of the potential energy function for a cleavage plane of alkali halide crystals were made by Born *et al.*[43] This type of calculation was developed by Lennard-Jones, Taylor and Miss Dent[44,45,46] who allowed for surface distortion by allowing movement of atoms in the surface plane, which they found represented a contraction of the distance of the outer plane from that beneath it of about 5%, to the position of minimum energy. They found $^sU = 77$ erg/cm^2 for sodium chloride at 0°K.

Subsequently developments have been made in several directions. The potential energy function has been refined,[47]

$$\varphi(a) = z_1 z_2 \frac{e^2}{a} - \frac{A'}{a^6} - \frac{B'}{a^8} + C' \exp(-a/\rho), \qquad (2.25)$$

where z_1 and z_2 are the charges on the ions and A', B' and C' are constants; also a more complete treatment of surface distortion has been proposed, associated with the unsymmetrical electric field at the surface. Verwey[48] stated that besides surface polarisation, the outer plane distortion could not be treated by movement of the plane as a whole, but that there were differences between positive and negative ions. The positive ions move in and the negative ions out from the bulk lattice plane separation. This leads to a distortion energy of about 80 erg/cm^2 tending to decrease the surface energy, and giving a value[49] of ^sU of about 50 erg/cm^2. The situation has been well reviewed[49] and it was concluded that the calculated values of the surface energy of the (100) plane of sodium chloride vary from 33 to 129 erg/cm^2 depending on which assumptions are made. This would seem to underline the current unsatisfactory position. However, in order to give some idea of the order of magnitude and of the difference between compounds, some values for the alkali halides[42] and the alkaline earth oxides and sulphides[44] are listed in Table 2.3. The experimental

TABLE 2.3

SURFACE ENERGIES FOR IONIC CRYSTALS (erg/cm^2)

Cation		*Anion*							
		F^-	Cl^-	Br^-	I^-	O^{2-}		S^{2-}	
		(110)	(110)	(110)	(110)	(110)	(100)	(110)	(100)
Na^+	calc.	171	155	145	132				
	expt.	335	190	178	138				
K^+	calc.	160	134	124	111				
	expt.	242	173	159	139				
Ca^{2+}	calc.					1030	2850	359	1440
Mg^{2+}	calc.					1360	3940	358	1730

values are for the liquids at their melting points. The results for the divalent ions should be treated with just as much caution as the alkali halide results. Benson and McIntosh[50] have shown that by using different potential energy functions to calculate the surface energy of magnesium oxide, values vary from -298 to 1362 erg/cm^2.

Calculations of the distortion in the surface region of a 100 face of a sodium chloride crystal have been extended by Benson *et al.*[51] The model employed allowed ions in the first five layers at a free 100 face of a hemicrystal to relax in a direction normal to the face and to be polarised by the electric field in the surface region. The equilibrium positions were determined by minimising the energy of the system, and corresponded to a value of $-107{\cdot}4$ erg/cm^2 for the total correction to the surface energy due to surface distortion. The calculated equilibrium positions show several interesting characteristics. The negative ions in each of the first five layers are displaced outwards towards the surface from their expected bulk crystalline positions, the displacement increasing as the surface is approached. The displacement of the positive ions on the other hand alternates. Those in the first, third and fifth layers are displaced inwards, whereas those in the second and fourth layers move outwards. Thus these calculations suggest that there is considerable difference between the surface atomic conformation of a real crystal and that existing in the bulk of the material. Clearly this type of approach should be extended to adsorption potential calculation, but because of the considerable extra complexity involved in also introducing lateral relaxation, this has not so far been attempted.

The calculations so far considered have assumed that the crystals of ionic solids are completely ionic, that is that the bonds possess no covalent character. In an attempt to allow for this with lithium fluoride[52] a value of ${}^{s}U = 557$ erg/cm^2 for the (110) plane was obtained, a value markedly higher than approximately 200 erg/cm^2 obtained by a purely ionic model calculation. The calculated values of surface energy usually are very low compared with experimental estimates.[13] The experimental values quoted in Table 2.3 are those for the liquid halides at their freezing points, and one would expect the values for the solids to be higher than for the liquids, yet the tabulated calculated values are lower.

In view of the uncertainty of the values of surface energy of solids it is scarcely surprising that calculated values of the surface tension differ widely, since in general the method of calculation depends on eqn. (2.20). Values for the surface tension of alkali halide crystals vary from positive values[44,45] of several thousand dyne/cm to negative values of several hundred.[42] As mentioned earlier the change in X-ray spacings for fine magnesium oxide and sodium chloride at least indicates that the surface tension values should be positive.

Energies associated with edges and corners

Those atoms or molecules or ions which are nearest an edge obviously have fewer nearest neighbours than those on a surface or those in the bulk

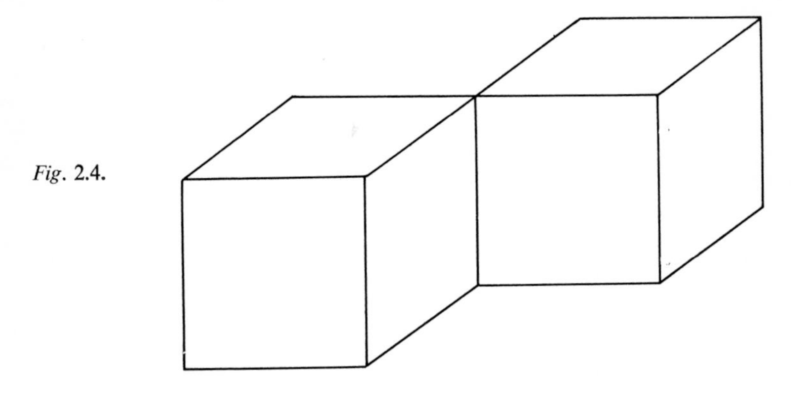

Fig. 2.4.

of the solid. By similar methods to those used to calculate surface energies, it is possible to calculate edge energies. Edge energies for two cubes having a common edge (Fig. 2.4) have been calculated[44] and a value of 10^{-5} erg/cm obtained. More recently[53] a value of $2{\cdot}9 \times 10^{-6}$ erg/cm for the edge energy between (110) planes of sodium chloride has been obtained. Little confidence can be placed in the absolute magnitude of these quantities since no attempt was made to allow for the distortion which one would expect to be present at an edge.

Energies associated with single edges and corners can be calculated, but the problem of allowing for distortion is even more acute, and at present one would have no confidence in the answers.

Specific surface area

The specific surface area is defined as the total surface area of 1 g of material, and is a frequently determined parameter of powdered substances. In effect it is a way of expressing the state of subdivision, and in many respects it is a method of expressing particle size without the need to determine the size distribution.

The BET method

The most popular experimental method of measuring specific surface area is that of gas adsorption.[54] In essence, if the amount of gas required to cover the surface with a complete monolayer can be evaluated, the specific surface area can be calculated, if the area occupied by a single molecule is known. Brunauer[55] classified experimental isotherms into five types according to their shape, as shown in Fig. 2.5, where the volume adsorbed V (*i.e.* the equivalent volume at STP of the amount of gas adsorbed) is plotted as a function of the relative pressure, P/P_0, where P is the equilibrium pressure and P_0 the saturation vapour pressure of the adsorbate at the isotherm temperature. The majority of adsorption isotherms for gases such as nitrogen on non-porous solids have the form of Type II, where the equivalent volume at STP of the gas adsorbed, is plotted as a function of the relative pressure. In 1935 Brunauer and Emmett[56] considered that the central nearly linear portion of a Type II isotherm corresponded to the formation of a second layer, and that the monolayer equivalent volume could be evaluated by extrapolating this linear portion back to the V axis, this value being subsequently termed point A. In a later paper Emmett and Brunauer[57] considered a number of other points on the isotherm which might be used to evaluate the monolayer volume and concluded that the point B, the beginning of the linear section of the isotherm was more consistent, if a range of gases were considered, in the value of the specific surface area. In Table 2.4 some of their values for the surface area of catalysts are given assuming each point in turn is equivalent to the monolayer volume, and calculating the area

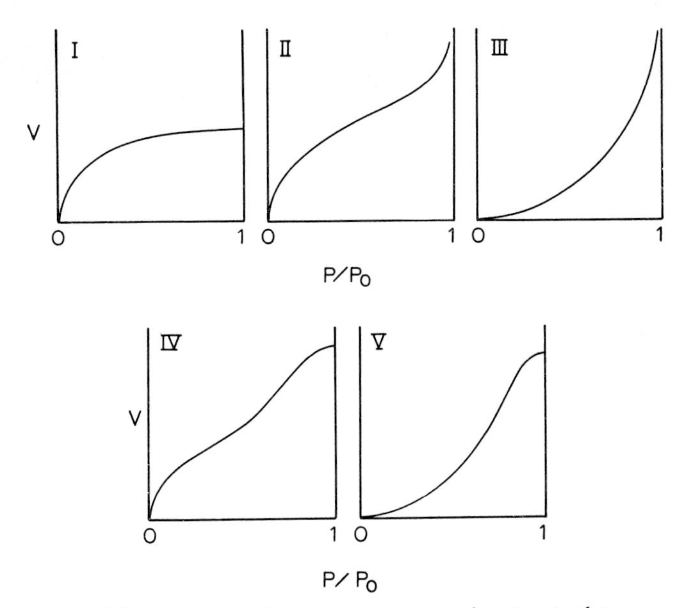

Fig. 2.5. Brunauer's five types of gaseous adsorption isotherm.

occupied by a single molecule from both the solid (S) and liquid (L) densities. The point B method offered the most concordant series of values, and on this evidence together with further evidence from the heats of adsorption, the point B method became a frequently used method. Practical identification of this point is not always easy, and the subsequent development of the Brunauer–Emmett–Teller[58] (BET) equation represents an analytical way of locating point B,[59] despite the considerable theoretical shortcomings of the equation. The equation expresses the

TABLE 2.4

Gas adsorbed	Temp. of isotherm (°C)	b.p. of gas (°C)	Point A			Point B		
			V_m (ccs at STP)	Surface area (m^2) (S)	(L)	V_m (ccs at STP)	Surface area (m^2) (S)	(L)
N_2	−183	−195·8	4·9	18·3	22·5	5·5	20·5	25·3
N_2	−195·9	−195·8	4·8	17·9	21·0	5·5	20·5	24·1
CO	−183	−192·0	5·0	18·5	21·5	5·7	21·1	25·9
Ar	−183	−185·7	5·2	18·0	20·3	5·8	20·1	22·6
O_2	−183	−183·0	4·7	15·4	17·9	5·5	18·0	21·0
CO_2	− 78·5	− 78·5	4·7	17·9	23·4	5·95	22·7	27·4
n-C_4H_{10}	0·0	0·3	2·0	17·3	17·3	2·2	19·0	19·0

equivalent volume adsorbed at relative pressure P/P_0, using a constant C, and the monolayer capacity V_m

$$V = \frac{V_m CP}{(P_0 - P)(1+(C-1)P/P_0)} \tag{2.26}$$

which may be rearranged to give the practically useful form

$$\frac{P}{V(P_0 - P)} = \frac{1}{V_m C} + \frac{C-1}{V_m C}\frac{P}{P_0}. \tag{2.27}$$

Hence, if $P/V(P_0 - P)$ is plotted against P/P_0, the value of V_m can be calculated from

$$V_m = 1/(slope + intercept). \tag{2.28}$$

The BET equation is capable of describing more or less adequately most Type II and Type III isotherms, the type being a function of the value of the constant C, but in general only Type II isotherms corresponding to high C values yield reliable values of V_m.

The mathematical character of the BET equation can perhaps be more readily seen if it is rearranged in the form

$$\frac{V}{V_m} = \frac{1}{1-x} - \frac{1}{1+(C-1)x} \tag{2.29}$$

where $x = P/P_0$. Thus the adsorption isotherm of V/V_m against x consists of the difference between the upper branches of two rectangular hyperbolae representing the first and second terms on the right hand side of eqn. (2.29). The first has asymptotes at $V/V_m = 0$ and $x = 1$, while the second has asymptotes at $V/V_m = 0$ and $x = -1/(C-1)$, and also cuts the V/V_m axis at $V/V_m = 1$. The difference between these two curves gives the BET isotherm, the shape of which is strongly dependent on C. For C values greater than 2 an isotherm of Type II results, whereas if C is less than 2 the resulting isotherm is of Type III. The point of inflection on the Type II curve is not necessarily at a value of $V/V_m = 1$. It can be shown that this is the case when $C \simeq 9$ and when $C \to \infty$. However, the discrepancy is only marked when $C < 9$.

The calculation of the specific surface area from V_m requires a value for the area occupied per molecule in the complete monolayer σ_m. The assumption is made that the value of σ_m for a particular adsorbate at a specific adsorption temperature is independent of the adsorbent. Obviously this represents an oversimplification of any practical situation, however the value of 16·2 Å^2 per molecule for nitrogen at $-195{\cdot}8$°C, based on the liquid density, has become a 'standard' value for this type of measurement, and values of σ_m for other adsorbates are usually obtained by reference to this 'standard'. The variability of the values of σ_m obtained in this way is due to the disturbance of the ideal liquid structure of the adsorbed film, which has been assumed, by the surface of the adsorbent. Table 2.5 illustrates this point for the case of krypton adsorption at $-195°C$, and it should be noted that variations of up to ± 2 Å^2 were found in several groups of solids.[60,65] This degree of uncertainty in σ_m should be borne in mind in assessing the absolute validity of any calculated value of specific surface area, despite the reproducibility which can be achieved in the BET value of V_m with one solid sample and one adsorbate at one adsorption temperature.

TABLE 2.5

VALUES OF σ_m OBTAINED BY COMPARISON WITH NITROGEN ADSORPTION AT $-195°C$ BY THE BET METHOD ASSUMING σ_m = 16·2 Å^2 PER MOLECULE

σ_m *(Kr)* (Å^2 *per molecule*)	*No. of solids*	*Reference*
19·5	5	60
19·5	3	61
19·5	2	62
20·8	6	63
21·8	1	64
22·6	2	65
15·2 from liquid density		60
14·0 from solid density		60

The BET equation (2.26) applies to those cases where the adsorbed film can build up to an infinite thickness, but in the case of porous solids, the pore size may restrict the possible thickness of the adsorbed film. In this case the usual form of the BET equation (2.27) does not fit the experimental data at high pressures, and a satisfactory value for V_m cannot always be obtained by this method. If adsorption at saturation is restricted to n-layers, the BET methods lead to the isotherm[58,66]

$$V = \frac{V_m Cx}{(1-x)} \frac{(1-(n+1)x^n + nx^{n+1})}{(1+(C-1)x - Cx^{n+1})} \tag{2.30}$$

where $x = P/P_0$ and the remaining terms have their previous significance. This equation is not easily applied since there are three unknown constants. A graphical method of solution has been described by Joyner, Weinberger and Montgomery.[67] Equation (2.30) can be written in the form

$$V = \frac{V_m C\Phi_{(n,x)}}{1 + C\Theta_{(n,x)}} \tag{2.31}$$

where

$$\Phi_{(n,x)} = \frac{x(1-x)^n - nx^n(1-x)}{(1-x)^2}$$

and

$$\Theta_{(n,x)} = \frac{x(1-x^n)}{1-x}.$$

Equation (2.31) can be put in the linear form

$$\frac{\Phi_{(n,x)}}{V} = \frac{1}{V_m C} + \frac{\Theta_{(n,x)}}{V_m} \tag{2.32}$$

Tables have been compiled[67] containing values of $\Phi_{(n,x)}$ and $\Theta_{(n,x)}$, and using these tables the procedure is to select, by a least-squares method, the value of n which gives the most linear plot of $\Phi_{(n,x)}/V$ against $\Theta_{(n,x)}$ and then to calculate V_m and C from the slope and intercept. However, this procedure is time consuming and tedious. With the aid of a modern computer statistical methods may be applied directly to eqn. (2.30) to evaluate V_m and C for the 'best' value of n. The n-layer equation is often found to fit over a wider range of values of x than the 0·05 to 0·35 usually found with the ∞-form of the BET equation. This may be ascribed to the flexibility conferred by the three constants in the equation, and does not reflect any increased validity in the theoretical background of the equation. The method is no longer frequently employed since, as will be discussed later, there are more convenient and useful methods of interpreting the adsorption data on porous solids. The remaining use of the method is in dealing with those cases where adsorption data give a curved normal BET plot and a standard isotherm is not available to permit the use of the 't-plot' method. For example the adsorption of krypton on stannic oxide at -195°C is such a case, and the author has been able to compare the area of various samples using the Joyner *et al.* procedure.

Area from electron microscopy

The electron microscope offers an obvious way of determining the size and shape of particles and hence their surface area. If powdered materials existed normally in precise geometric shapes, and all the particles in a sample were of identical size, then the problem of calculating the surface area would be simple. In practice the irregularity of the particles and the wide distribution of particle size frequently encountered, make the problem of calculating surface area a very difficult one. Usually a large number of particles have to be investigated in order to gain some idea of the size distribution, and considerable approximations made in the shape in order to calculate an area. The method is comparatively time consuming, but being a direct method is still frequently used. Because of the difficulties of allowing for surface roughness, areas calculated by this method are frequently lower than those obtained by other methods.

Surface area by measurement of adsorption from solution[69]

Although adsorption from solution is only of secondary importance in determining surface areas, it has the attractive features of comparative simplicity of technique and quickness. However, if the method is to be reliable it must be possible to determine the monolayer coverage unambiguously, and to assign an accurate value to the area occupied by the adsorbed species, which normally means that the orientation of the adsorbed species must be known: thus in many respects the problems are similar to those encountered with the BET method. Generally sparingly soluble solutes are used, but adsorption from solution may give a complete monolayer on one solid substrate, but not on another at any practicable concentration.

The most formidable objection to solution methods lies in the problem of assigning a satisfactory value to the area occupied by a single adsorbed molecule, which is commonly done by using a 'standard' solid whose surface area has been determined by the BET method. Thus the determined value has all the uncertainties of the BET method together with those of adsorption from solution.

The adsorption of long-chain fatty acids from organic solution has been frequently used since it has been thought that the molecular areas are accurately known from work on insoluble monolayers on aqueous substrates, in other words, it is assumed that a close-packed vertically oriented monolayer is formed. The method has been used satisfactorily to measure the surface area of titania,[71] alumina[72] and some[73,74] metals.

Dyestuffs are attractive because of the ease of their colorimetric determination. Methylene blue has been quite widely used, but many different values have been ascribed to the area occupied by an adsorbed molecule, 54 $Å^2$,[75] 78 or 138 $Å^2$,[76] 102 and 108 $Å^2$,[77] and 197 $Å^2$.[78] The range of these values obviously make the method unsuitable as a general method.

Giles[79] has considered the requirements for a satisfactory general solute, and has claimed considerable success with *p*-nitrophenol[80] from water, or from hydrocarbons. This method should be approached with caution. It should be remembered that the size of an adsorbed molecule determines the magnitude of the surface imperfections which will be followed; for example a pore that will be penetrated by a nitrogen molecule in a BET determination, may be too small to allow a methylene blue molecule to enter.

The surface of real solids

Impurities and heterogeneity

Ideally the surface of a solid is considered to be smooth, but in actual fact[81] it is probable that although the total surface energy would be a minimum, this situation would also correspond to minimum entropy. Thus the normal equilibrium surface with the minimum attainable energy is rough with irregularities perhaps several hundred Ångströms in height representing a compromise between entropy and energy considerations. Even if we can exclude the possibility of contamination, which can obviously give rise to surface variation, the surface of a real solid is irregular because of surface defects, growth spirals, edges, and corners.

Earlier when considering the effects of edges and corners we considered the effect on surface energy of a decrease in the number of nearest neighbours of a particular atom or molecule. The surface equivalents of Shottky and Frenkel defects will produce similar effects, and a growth spiral effectively represents a spiral edge on face. All these factors and a number of others[82] contribute to surface heterogeneity. If the surface of an uncontaminated real crystal is likely to be heterogeneous, then it is scarcely surprising that impurities may also be a cause of surface heterogeneity. In fact surface heterogeneity means that calculated values of surface energy can only approximate to experimental values. Thus different preparations of the same substance are likely to give rise to different experimental values of surface energy, as are different methods of evaluation.

Before leaving the subject of surface energy, the effect of variations in surface tension and particle size on solubility might be mentioned. The Kelvin equation may be written as

$$RT \ln \left(\frac{a}{a_0}\right) = \frac{2\gamma \bar{V}}{r} \tag{2.33}$$

for a spherical, isotropic particle where $\bar{V}$ is the molar volume, and a the activity of the solid. The value of a determines the solubility of the solid in relation to a_0, the value for a plane surface. The increased solubility of small particles is considerable. The results obtained by Dundon and Mack[83]

TABLE 2.6

EFFECT OF SURFACE TENSION ON THE SOLUBILITY OF SALTS

Salt	*Increase in solubility %*	*Particle size*	*Surface tension*	*Relative hardness Moh's scale*
PbI_2	2	0·4	130	Very soft
$CaSO_4 2H_2O$	4–12	0·2–0·5	370	ca. 2
Ag_2CrO_4	10	0·3	575	ca. 2
PbF_2	9	0·3	900	ca. 2
$SrSO_4$	26	0·25	1400	3·0–3·5
$BaSO_4$	80	0·1	1250	2·5–3·5
CaF_2	18	0·3	2500	4

are shown in Table 2.6, and it is worth noting the parallel between the surface tension, calculated from the increase in solubility, and hardness.

Porosity

An extreme case of surface irregularity is the pore. Porous solids show certain differences from non-porous ones in terms of their surface properties. A porous solid has a high specific surface area without necessarily having a small particle size. Thus practical adsorbents are often porous materials such as charcoals, and offer the advantage of large specific surface area without the difficulties caused by small basic particle size.

It is convenient to divide the pores in an adsorbent into three types in terms of size: macropores, transitional pores and micropores.[84] In principle the upper limit of the radius of curvature of a macropore is large, but the lower limit may be taken conventionally as 1000 to 2000 Å. This lower limit is chosen so that the volume filling of macropores by the capillary condensation of vapour takes place only at relative pressure values very close to unity. Thus conventional gas adsorption methods cannot detect the difference between the surface of a macropore and a plane surface. Experiments on the forcing of mercury into macropores permit an estimate of the volume distribution of pores and their specific surface area to be obtained for the range of effective radii from 10^6 to 10^3 Å.

The dimensions of transitional pores are still considerably larger than molecular dimensions, but they are small enough for capillary condensation to occur at relative pressure values easily distinguishable from unity in a gas adsorption experiment. Thus the adsorption isotherm usually shows hysteresis, that is the desorption branch is above the adsorption branch at higher values of relative pressure (Fig. 2.6). The reason for this can be seen by considering the liquid meniscus which will be formed when the pore is full. The Laplace equation (2.21) suggests that a pressure drop will exist across the meniscus at the neck of a capillary, and that if the pressure inside the capillary is to drop below P_0 then the external pressure will have to be less than $(P_0 - \Delta P)$. Thus the height of the hysteresis loop will give an indication of the pore volume, and the width of the pore.[85] The effective radii of transitional pores at the maxima of the distribution curves usually fall within the range 40 to 200 Å.

The mercury porosimeter will also give information about transitional pores as well as the macropores previously mentioned. If a liquid is added to a porous solid, menisci will be formed in the pores. For contact angles less than 90° the liquid enters the pores spontaneously, but for contact angles greater than 90° a positive pressure will be necessary to force the liquid into the pores. If the pores are assumed to have circular cross-sections then the Laplace equation may be written as

$$\Delta P = 2\gamma \cos \theta / r \qquad (2.34)$$

where r is the radius of the pore and θ the contact angle. Thus by knowing the volume of mercury forced into the pores and the pressure it is possible to calculate a distribution of pore volume as a function of pore radius.[86]

The smallest pores are usually known as micropores and have dimensions commensurate with molecular dimensions. It is known that the adsorption energy in micropores is considerably higher than the corresponding values in transitional pores or on a plane surface. The absence

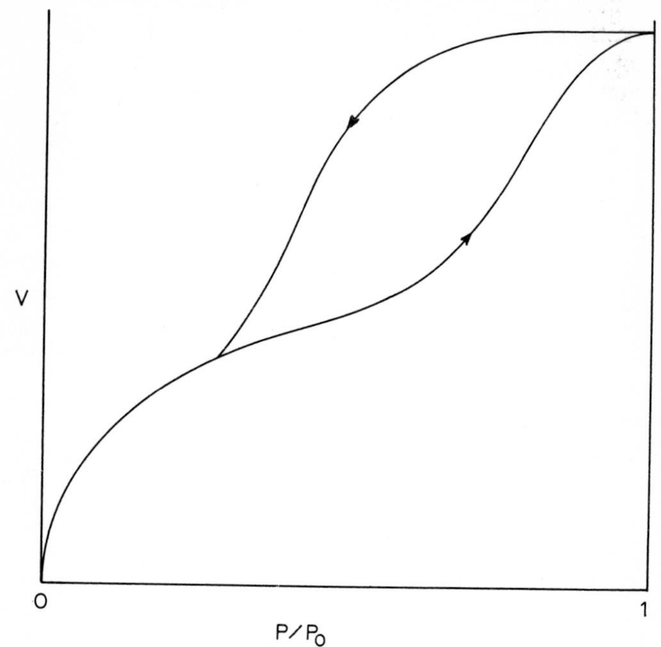

Fig. 2.6. Brunauer type IV isotherm showing hysteresis.

of layer-by-layer filling and capillary condensation means that the concept of surface area has little meaning and pore volume cannot be obtained by hysteresis measurements. Dubinin[84] has developed a method for obtaining the pore volume from adsorption data, and de Boer's 't-plot' has permitted somewhat similar information to be obtained.[87]

Recently considerable attention has been devoted to determining experimental standard isotherms for non-porous adsorbents. At present these have been established for a number of oxides using nitrogen as adsorbate at 77·5°K, and methods have been developed to use these as a basis for evaluating the surface area and porous characteristics of other samples of the same oxides. The standard curve is established by determining the adsorption isotherm on a number of non-porous samples and expressing them all in terms of the plot V/V_m against P/P_0, where V_m is usually determined by the BET method.

This isotherm may also be expressed in terms of the statistical thickness, t, of the adsorbed layer. De Boer[87] calculated that the adsorbed film would be 3·54 Å thick for a complete monolayer of nitrogen at 77·5°K, thus the V/V_m scale may be directly converted into one in terms of t (Fig. 2.7a). Use is made of the standard in the following manner. For each

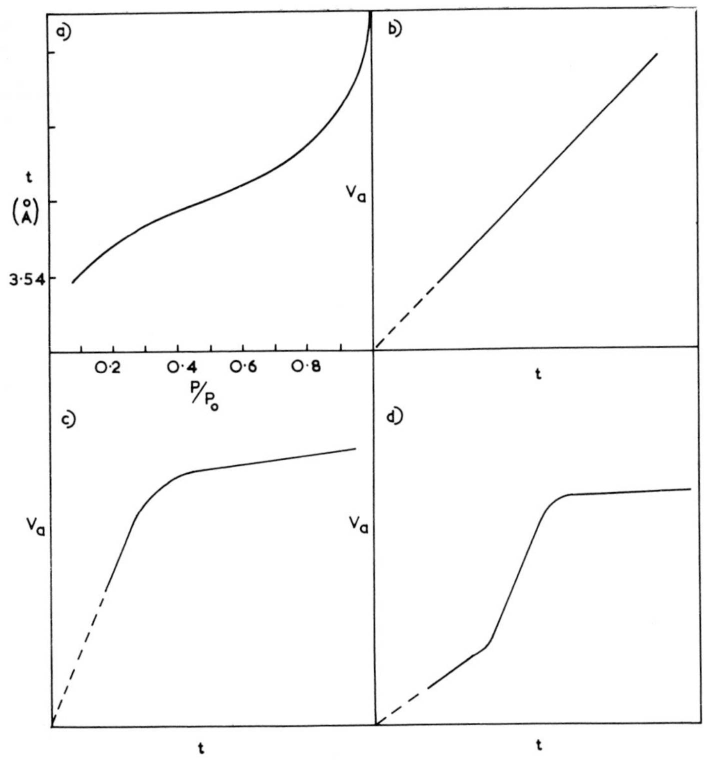

Fig. 2.7. The de Boer 't'-plot method. (a) The standard isotherm, (b) t-plot for a non-porous sample, (c) for a microporous sample and (d) for a sample with transitional pores.

experimental point on the experimental isotherm, V_a vs P/P_0, a value of t is read off from the standard isotherm for each experimental value of P/P_0, and the experimental data replotted in terms of V_a versus t. If the sample is non-porous then a straight line results (Fig. 2.7b), and the slope of the line is directly proportional to the surface area. Microporous samples are characterised by plots of the type shown in Fig. 2.7c, and those with transitional pores give rise to curves of the type shown in Fig. 2.7d.

Certain objections to the t-plot procedure have been put forward by Sing.[88] Most significant among these are that the method is not independent of the BET method, and cannot be readily applied to cases where the non-porous standard sample gives a Type III isotherm or a Type II isotherm with a very small knee at low pressure. He proposes that t be replaced by $(V/V_x)_s$, which is termed α_s, where V_x is the amount adsorbed at a selected value $(P/P_0)_x$. The location of $(P/P_0)_x$ is not completely arbitrary, usually being chosen at a value of 0·4 since micropore filling and monolayer coverage usually occur at lower pressures, and capillary condensation, at least when in association with hysteresis, takes place at higher pressures.

WETTING AND CONTACT ANGLES

If a drop of liquid is placed on a perfectly flat and uniform solid surface, it need not necessarily spread completely over the surface, but its edge may make a contact angle θ with the solid. Figure 2.8 illustrates three possible cases in order of increasing attraction between molecules of the liquid and solid. If the equilibrium tensions are resolved horizontally, then

$$\gamma_{S/V} = \gamma_{S/L} + \gamma_{L/V} \cos \theta \tag{2.35}$$

where the subscripts S, L, V refer to the solid, liquid and vapour phases respectively. If we apply Dupré's methods[89] to define the work of adhesion between solid and liquid $W_{S/L}$ we have

$$W_{S/L} = \gamma_{S/V} + \gamma_{L/V} - \gamma_{S/L} \tag{2.36}$$

and hence by combining the two equations we arrive at Young's equation:[90]

$$W_{S/L} = \gamma_{L/V} (1 + \cos \theta). \tag{2.37}$$

This derivation has assumed that the vertical force due to $\gamma_{L/V}$, that is $\gamma_{L/V} \sin \theta$, produces no deformation of the solid surface. In certain cases the presence of this force has been demonstrated, using a drop of mercury on a sheet of mica 1 μ thick, the sheet of mica deforming all round the edge of the drop.[91]

Magnitude of contact angles

Young's equation indicates that the magnitude of the contact angle will depend on the relative values of the adhesion between solid and liquid, and the mutual cohesion of the liquid which is related to $\gamma_{L/V}$. An angle of 180° would indicate zero adhesion, but this angle has never been observed in practice even with hydrophobic materials and water. The contact angle[9] between water and polythene is 94°, and between water and paraffin wax up to 110°. The values just quoted refer to smooth surfaces soon after formation. The contact angle is also dependent on the roughness of the surface and on any contamination of the solid surface or of the liquid surface, or possible rearrangement of the solid surface caused by the presence of the liquid phase.

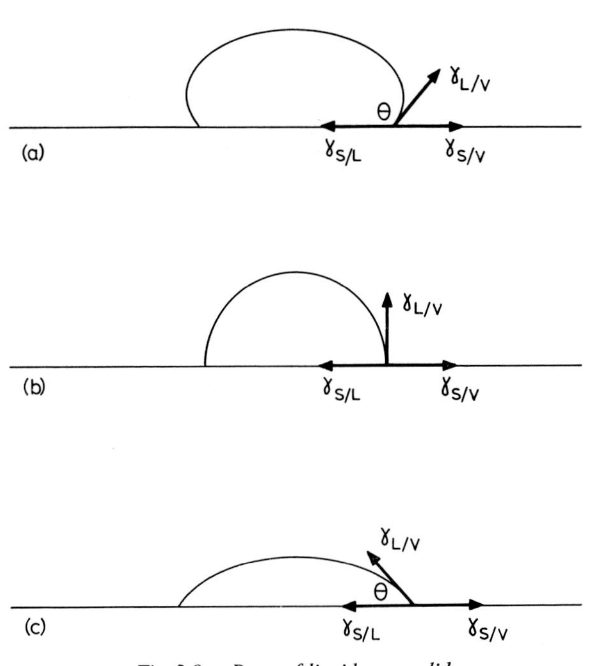

Fig. 2.8. Drop of liquid on a solid.

Surface roughness has the effect of making the contact angle further from 90°. If the smooth solid gives a contact angle greater than 90°, the angle is increased by roughening the surface; if the contact angle is less than 90°, then roughness will decrease it. For example, rough paraffin wax[92] has a contact angle of 132°. Wenzel[93] has proposed that surface roughness may be measured by

$$R = \cos\theta'/\cos\theta \tag{2.38}$$

where R is the roughness ratio, and θ' is the average contact angle on a rough surface as opposed to θ on a smooth surface. This relation is only applicable to sub-microscopic roughness, since for course roughness the edge becomes ragged and R is no longer a constant.

Contact angles are best measured where possible by the inclined plate method in which a plate of solid is immersed in the liquid and tilted so that the meniscus on one side of the plate is horizontal, as illustrated in Fig. 2.9. The angle of inclination of the plate is taken as being equal to the contact angle.

It is often difficult to obtain reproducible results, and in particular the contact angle when the solid surface is advancing into the liquid is frequently greater than that when the solid is being withdrawn, so that there is hysteresis of the contact angle. It has been found that very clean and smooth surfaces show no hysteresis of the contact angle[92,94,95] and therefore it seems probable that this phenomenon is associated with surface contamination and roughness.

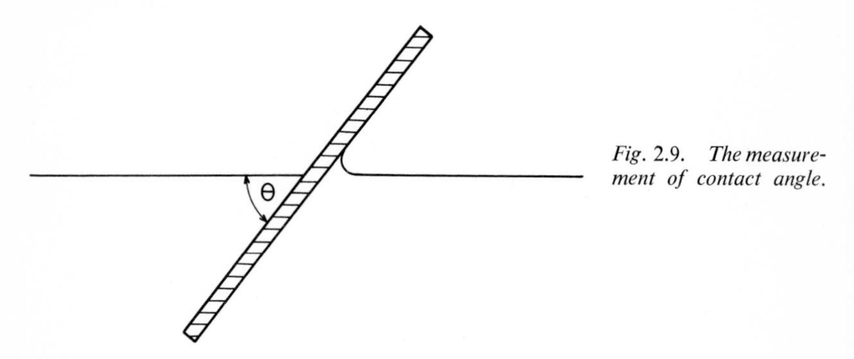

Fig. 2.9. The measurement of contact angle.

It is also possible to get some idea of the contact angle for a powdered sample by measuring the pressure required to force a liquid which will wet the powder through a sample ($\theta = 0°$), and then the pressure required to force a liquid with unknown contact angle through the powder. Thus by applying eqn. (2.34) to both cases and eliminating r, the contact angle may be calculated. The approximate nature of the method is obvious. This and other methods have been described,[96] each subject to uncertainties.

Spreading coefficients

The balance of surface tensions at equilibrium has been described by eqn. (2.35). When a liquid is spreading and wetting a solid then the advancing contact angle of the very thin film must tend to zero and there is a resulting force causing spreading. This is usually described in terms of a spreading coefficient S

$$S = \gamma_{S/V} - \gamma_{S/L} - \gamma_{L/V}. \tag{2.39}$$

If an equilibrium contact angle can be maintained then

$$S = \gamma_{L/V}(\cos\theta - 1) \tag{2.40}$$

which is valid when $\theta > 0°$, and is often used to calculate spreading coefficients from values of θ and $\gamma_{L/V}$.

Oils spread readily on chromium and S is positive. For 7-butyltridecane S is 25 dyne/cm although by increasing the chain branching S may be reduced to about 17 dyne/cm.[9] The importance of the surface history can be shown since if the chromium surface is cleaned by ion bombardment and at once wetted with water, oil will not spread and therefore S must be negative.[97] Lubricating oils spread readily on steels particularly if they are somewhat oxidised to polar compounds.[98]

Many metal surfaces adsorb a monolayer of the spreading liquid and hence the value of θ measured, and the value of S so derived, will be for the liquid on a monolayer coated solid.[97,99,100] To obtain a value for the clean surface a correction must be applied for the work of adhesion.

The adhesion of liquids to solids

The work of adhesion of a liquid is given by equation (2.36). If this is combined with equation (2.39), then

$$W_{S/L} = S + 2\gamma_{L/V}. \qquad (2.41)$$

The value for the adhesion of water to paraffin wax is about 46 erg/cm^2. If measured values are used then if the solid is likely to be covered with an adsorbed monolayer, and if values for a clean surface are required, then correction must be made.[99,100]

Flotation[101,102]

The essential step in froth flotation is the attachment of a particle, already completely wetted by the solution around it, to an air bubble of sufficient size to carry it from the bulk to the surface of the solution. In practice the particles are small in comparison to the air bubbles, and many of them may be attached to each bubble.

Let us consider the gravity-free model illustrated in Fig. 2.10. The particles of solid are assumed to be spherical of radius r and all three phases of the same density. If the particle is entirely in phase A, then

$$F_A = 4\pi r^2 \gamma_{S/A} \qquad (2.42)$$

where F_A is the Helmholtz free energy of the surface, and $\gamma_{S/A}$ the specific surface free energy. Similarly if the particle is completely in phase B, then

$$F_B = 4\pi r^2 \gamma_{S/B}. \qquad (2.43)$$

Finally the particle may be at the interface, in which case both $\gamma_{S/A}$ and $\gamma_{S/B}$ contribute to the surface free energy, and part of the interface between phases A and B has been eliminated. If there is to be an equilibrium

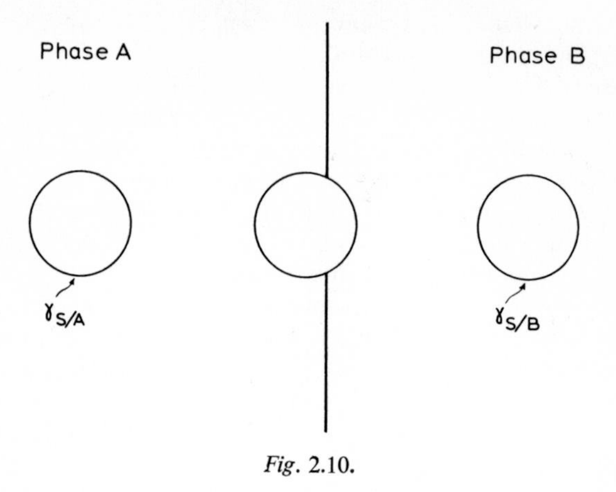

Fig. 2.10.

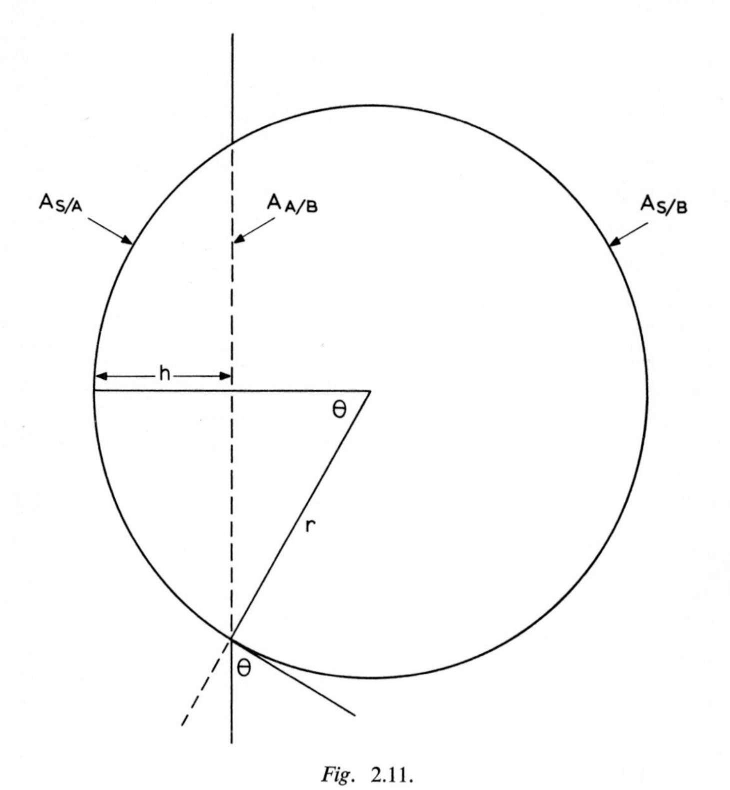

Fig. 2.11.

position for the particle at the interface then the total surface free energy must be at a minimum, therefore

$$dF = \gamma_{S/A} dA_{S/A} + \gamma_{S/B} dA_{S/B} + \gamma_{A/B} dA_{A/B} = 0. \tag{2.44}$$

Considering Fig. 2.11 shows that

$$\gamma_{S/A}(2\pi r dh) + \gamma_{S/B}(-2\pi r dh) - \gamma_{A/B}\pi(2r-2h)\, dh = 0 \tag{2.45}$$

which simplifies to

$$\gamma_{S/A} - \gamma_{S/B} = (1-h/r)\, \gamma_{A/B} \tag{2.46}$$

but

$$\cos\theta = 1 - h/r$$

therefore

$$\gamma_{S/A} - \gamma_{S/B} = \gamma_{A/B} \cos\theta \tag{2.47}$$

which is the Dupré equation (2.35) for contact angle equilibrium. Consequently a particle will seek a stable position in the interface such that the angle θ becomes the equilibrium contact angle. In the ordinary circumstances when gravity is present, there will be a slight displacement from this position until the surface tension restoring forces are equal to the gravitational pull. Thus particles forming a finite contact angle at the air–liquid contact angle can be floated, whereas those wetted by the solution can not.

Separation of ores by flotation involves adding reagents to control the contact angle such that it has a large value on one ore and the other is wetted.

Wetting and non-wetting

A solid is completely wet with a liquid if the contact angle is zero. If an equilibrium is established it will be described by eqn. (2.35), which may be rearranged as

$$\cos\theta = \frac{\gamma_{S/V} - \gamma_{S/L}}{\gamma_{L/V}}. \tag{2.48}$$

If the right hand side is equal to unity, or exceeds unity under non-equilibrium conditions, the liquid will wet the solid. Thus $\gamma_{S/V}$ should be large and $\gamma_{S/L}$ and $\gamma_{L/V}$ small. Often $\gamma_{S/L}$ is large when surfaces are rough or particulate. Wetting agents not only lower $\gamma_{L/V}$ but also allow the water to penetrate the surface lowering $\gamma_{S/L}$ also. Many wetting agents, such as sodium di-*n*-octylsulphosuccinate, have irregularly shaped molecules.[103]

Non-wetting is a term usually applied to the case where the contact angle is greater than, or equal to 90°. On rough or hairy surfaces many liquids have high contact angles and are non-wetting. If we have a composite surface where f_1 and f_2 are the fractions which are liquid–solid and liquid–air respectively then[104]

$$\cos\theta = f_1 \cos\theta_1 - f_2. \tag{2.49}$$

Ducks have a structure of fine hairs on their feathers which makes them difficult to wet because f_2 is large as well as θ_1. In a surface active agent solution θ_1 has a low value and hence the term $f_1 \cos\theta_1$ changes from negative to positive, and if f_2 is not too large the duck is wetted.

ADHESION[103,105,106]

Adhesive forces

It is possible to show[103] that for perfect surfaces adhesive strengths should be very high. If one substitutes in Young's equation (2.37) the value $\theta = 0°$, that is the adhesive just wets the solid, and $\gamma = 30$ dyne/cm, the force required to separate the surfaces by a molecular dimension is $(30 \times 2)/10^{-7}$ dyne/cm which is equal to approximately 600 kg/cm^2 or 4 ton/in^2. However, the values obtained in practice are usually of the order of 30 kg/cm^2, which is attributed to plastic and viscous flow which may take place before fracture, and the presence of flaws and dislocations in the surface of the solid.[105,106]

If two pieces of hard metal are carefully cleaned and then pressed together, the adhesive force is very small if the air is clean and dry. This is because under these conditions the surfaces undergo elastic deformation at the points where they come into contact, resulting in a force which tends to separate the surfaces again. However, in a humid atmosphere, the adhesion is affected by the extent of the adsorption of water vapour at the surface, and increases towards a limit at the saturation vapour pressure, which is the adhesion in the presence of liquid water.

This type of system is usually studied experimentally by using a sphere in contact with a plane surface. If it is assumed that the sphere and the plane surface are smooth, and that the liquid collects around the point of contact, then we have the situation as shown in Fig. 2.12. The pressure inside the liquid will be less than atmospheric, and by applying the Laplace equation (2.21), the pressure difference will be approximately γ/r_m. The total force over the whole liquid pool is thus $\pi r^2 \gamma/r_m$, and since as a first approximation $r^2 = 2R \,.\, 2r_m$, we have

$$\text{adhesive force} = 4\pi R\gamma. \tag{2.50}$$

This result shows that the adhesion should be directly proportional to R, and independent of the thickness of the liquid film.

Flat, parallel surfaces may also adhere strongly by means of a liquid film. If we consider two flat circular discs, each 1 cm^2 in area and that a film of oil which wets the discs separates them by 1000 Å, then the radius of curvature of the meniscus at the edge of the discs will be 500 Å. The Laplace equation indicates a pressure difference of $25/500 \times 10^{-8} = 5 \times 10^6$ dyne/cm^2 if the surface tension of the oil is 25 dyne/cm. The adhesive force would be equal to 5·1 kg which may cause distortion of quite thick discs, for example microscope cover glasses.

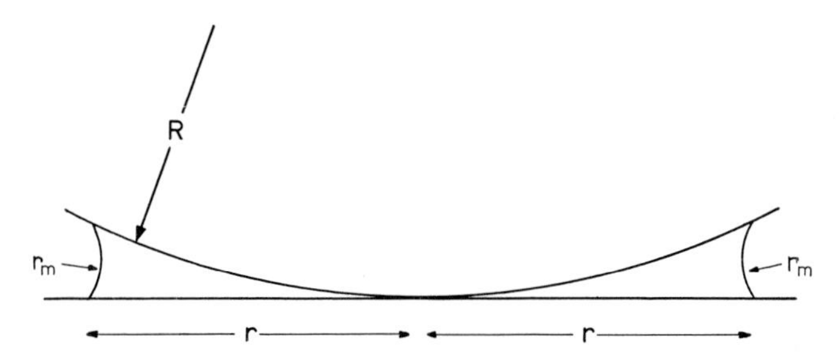

Fig. 2.12. Adhesion of a hard sphere, radius R, *to a hard solid plane surface, with water present at the point of contact as a pool of radius* r. *The radius of curvature of the meniscus is* r_m, *the separation of the solid surfaces at the meniscus being approximately* $2r_m$.

The adhesive forces between pieces of indium are very considerable, even in the absence of a liquid film. This is attributed to the lack of elastic stresses in the solid. The formation of junctions between such surfaces may be reduced by liquid lubricants, and reduced often to negligible proportions by covering the surfaces with a monolayer of a substance such as lauric acid.[105]

The stickiness of particles

Powders frequently adhere strongly when they are damp. With comparatively small amounts of water present interparticle adhesion may become comparable with the weight of the particles, since the weight varies as the cube of the radius, and the adhesive force as the radius. For two spherical particles, each of radius R, eqn. (2.50) must be modified[103] so that

$$\text{adhesive force} = 2\pi R\gamma. \tag{2.51}$$

The 'stickiness' of particles completely immersed in a medium is a different phenomenon since no 'lenses' of a third phase around the points of contact are involved. There are two main reasons for this phenomenon. Firstly, high molecular weight, polymeric materials may be adsorbed onto the surface of the particles, thereby increasing the collision radius and decreasing the electrical repulsion between the particles. Such materials are added to fine suspensions to increase their filterability, and to electrically stabilised emulsions to cause creaming without coalescence. The macromolecules may form bridges between the surface preventing direct contact, and also independent movement. The whole subject of stability is more fully discussed in Chapter 1. The second mechanism involves bridging by polyvalent ions between charged groups on the surface of two particles. Not only do these ions form bridges but they also reduce the Stern potentials of the individual particles, and hence the mutual repulsion between them. The particle surfaces may approach very closely, but the bridging is sensitive to the presence of sequestering and complexing agents which react with the polyvalent ions concerned. If the surfaces are strongly hydrophobic, the reduction of the electrical charges by specific absorption of polyvalent ions will result in very strong adhesion. The theory behind the electrical double layer phenomena just mentioned will be considered in Chapter 3.

Sliding friction and lubrication

Sliding friction is the force opposing the movement of one surface over another. This is of importance in dispersion since it affects the amount of work required to disperse a solid in a liquid, and is in turn affected by the lubricating properties of the medium. This process involves the shearing of any welded junctions formed at points of contact, and the energy of elastic deformation is not recovered. The coefficient of friction μ is defined as the ratio of the frictional force to the normal force holding the surfaces together. In many circumstances μ is a constant, which is known as Amonton's Law. The tangential force required to induce sliding is known as the static friction, and the lesser force required to maintain the sliding motion is known as the kinetic friction.

The presence of an oxide film on the surface of metals such as nickel, tungsten or copper reduces the coefficient of friction from the high value of 5 to a value of the order of 0·5. A thick layer of oil between the surfaces, fluid lubrication, will reduce the value still further to 0·002; this very low value being due to the fact that the only resistance to motion is due to the viscosity of the oil itself. Usually in practice it is not possible to maintain a thick lubricating layer between the surfaces, and the film will be reduced until the surfaces are separated by adsorbed films of molecular thickness, which is known as boundary lubrication. In this case the coefficient of friction is about 0·1, the exact value depending on the nature of the surface and the chemical composition of the lubricant; the bulk viscosity of the lubricant being of lesser significance. The addition of a little fatty acid to a non-polar oil can reduce the coefficient of sliding friction of two cadmium surfaces by about two orders of magnitude, effective lubrication occurring with boundary films of only one or two molecular thicknesses of fatty acid.

However, single layers are rapidly worn away, and must be rapidly replaced, therefore clearly the rate of adsorption is important.

Rolling friction

The coefficient of rolling friction is generally less than the coefficient of sliding friction. Interfacial slip scarcely contributes to the resistance. Consequently lubricants have little effect on the coefficient of rolling friction, but may greatly reduce the wear occurring during rolling.

ADSORPTION FROM SOLUTION

Many early experiments on adsorption from solution were concerned with adsorbents of technological importance. Isotherms were measured for dilute solutions and only recently have more concentrated solutions been investigated.[107] For solutions of solids such as stearic acid in benzene a certain modicum of success has been obtained in interpreting experimental isotherms in terms of the Freundlich equation.

$$w/m = Ac^{1/n} \tag{2.52}$$

where w is the weight of adsorbate adsorbed, m the weight of adsorbent, c the concentration of adsorbate in solution at equilibrium, and A and n are constants. The Langmuir equation has been similarly employed

$$w/m = \frac{Bc}{1 + Dc} \tag{2.53}$$

where B and D are constants. These equations are often applied to isotherms of the shape illustrated in Fig. 2.13(a), but they cannot describe the shapes shown in Fig. 2.13(b) and (c). These two shapes are characteristic

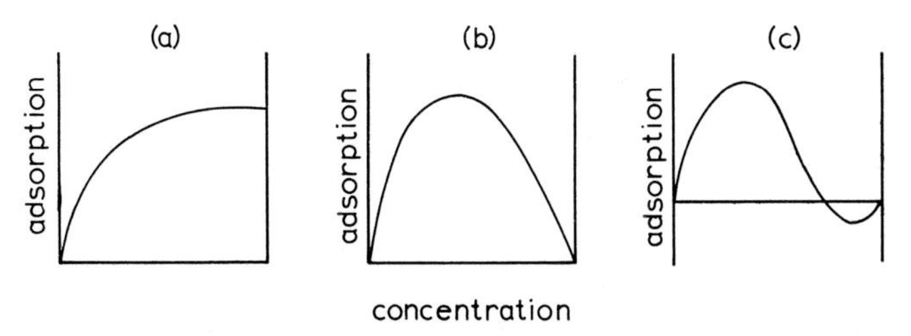

Fig. 2.13. Shapes of adsorption isotherms from solution.

of adsorption from solutions of completely miscible liquids. Adsorption on charcoal of acetic acid in water gives the former shape, while ethyl alcohol in benzene the latter.

Experimentally isotherms are determined by measuring the change in concentration of a solution when an adsorbent is added, due to adsorption.

It is possible to express the adsorption in two different ways. The first is given directly by the experimental method just described, and is usually termed preferential or selective adsorption, which corresponds to the term surface excess, frequently employed for the liquid–vapour interface. The second is absolute adsorption, and is defined as the surface concentration of the adsorbed component.

The composite isotherm

Since adsorption from solution of necessity involves more than one component, the composite isotherm represents the result of combining the individual isotherms of the components.[107]

When a weight m of adsorbent is brought into contact with n_0 moles of a two component solution, the mole fraction of component 1 decreases by Δx. This change in concentration is the result of the adsorption on the surface of ${}^s n_1$ moles of component 1 and ${}^s n_2$ moles of component 2 per unit weight of adsorbent. At equilibrium there are n_1 and n_2 moles of the two components in the liquid phase giving fraction x of component 1 whose initial mole fraction was x_0. Then

$$n_0 = n_1 + n_2 + {}^s n_1 m + {}^s n_2 m \tag{2.54}$$

and

$$x_0 = \frac{n_1 + {}^s n_1 m}{n_0}, \qquad x = \frac{n_1}{n_1 + n_2} \qquad \text{and} \qquad (1-x) = \frac{n_2}{n_1 + n_2}$$

therefore

$$\Delta x = x_0 - x$$

$$= \frac{n_2 {}^s n_1 m - n_1 {}^s n_2 m}{(n_1 + n_2)\, n_0}.$$

Thus

$$\frac{n_0 \Delta x}{m} = {}^s n_1 (1-x) - {}^s n_2 x. \tag{2.55}$$

The quantity $n_0 \Delta x / m$ is the quantity usually plotted against x to give the composite isotherm. Many examples of composite isotherms are given by Kipling.[107]

Adsorption from completely miscible liquids

The study of adsorption from this type of solution is characterised by the multiplicity of models which have been used to interpret isotherms. In fact a considerable amount of data has been published without detailed analysis which may reflect the current uncertainty which is felt about the theoretical basis of such analysis.[107] Recently a more rigorous treatment for ideal[108] and regular[109] solutions, has been published but there is still much to be done.

For adsorption by porous solids an approach has been made based on the postulate that adsorption is a process of pore filling, in which the available volume of the pores is the controlling factor.[110,111]

Then

$$^{s}n_1 V_1 + {}^{s}n_2 V_2 = V \tag{2.56}$$

where V_1 and V_2 are the respective partial molar volumes of the adsorbates and V the pore volume per unit weight of adsorbent. For monomolecular adsorption on a plane surface eqn. (2.56) would be written in terms of areas where

$$^{s}n_1 A_1 + {}^{s}n_2 A_2 = A \tag{2.57}$$

where A_1 and A_2 are the partial molar areas occupied at the surface by the two components and A is the specific surface area. However, since it is thought that adsorption is often if not usually as multilayers, no analysis can be attempted until an appropriate relation between $^{s}n_1$ and $^{s}n_2$ has been developed.

Adsorption from partially miscible liquids

The isotherm normally found for adsorption from a partially miscible system of non-electrolytes is usually *S*-shaped, resembling BET type II classification of gaseous isotherms. The adsorption increases rapidly as the miscibility limit is approached, and it seems likely that this usually corresponds to multilayer formation, that is incipient phase separation under the influence of a solid surface.[112] Satisfactory theoretical descriptions of such isotherms are still lacking. The BET equation has been applied because of the analogous shape, with P/P_0 replaced by C/C_0 where C_0 is the concentration at the miscibility limit: there has been no attempt at an adequate theoretical justification of this procedure.

Adsorption of dissolved solids from solution

The majority of isotherms for the adsorption of uncharged dissolved solids from solution are of the Langmuir form. A complete shape classification has been attempted by Giles.[79] The Langmuir equation with pressure replaced by concentration has been remarkably successful in fitting isotherms of adsorption from dilute solution. This is particularly the case where the competition between solute and solvent strongly favours the solute. However, the monolayer capacity evaluated by this means is often open to considerable doubt, and on solids of known specific surface has often been found to correspond to completely unreasonable value of the area occupied by an adsorbed molecule. Therefore the treatment of experimental results has varied widely from system to system and it is difficult to generalise.

Certain isotherms show pronounced steps in their shape, which do not seem to be interpretable in terms of the formation of first and second layers in the majority of cases. These discontinuities have been attributed to phase changes in the adsorbed film, or to changes in orientation of the adsorbed molecules, and it is often impossible on the basis of the isotherm alone to make any statement as to the cause.

The adsorption of polymers

For most systems involving the adsorption of a polymer the adsorption isotherm either reaches some plateau or tends towards a limiting value, and usually multilayer adsorption is not thought to occur. Nevertheless, the amount of polymer adsorbed often exceeds that corresponding to a complete monolayer of monomer units, thus models postulating that polymer molecules always lie flat on the adsorbent surface would seem to be inappropriate, particularly since there is evidence that the adsorbed film is often several monomer units thick. This could be explained if the adsorbed molecules are attached at only a few points to the surface (anchor segments) and that the remainder of the molecule is effectively free from the surface.[113]

Molecules of high molecular weight are thought to exist in the form of random coils which are approximately spherical when the molecular weight is sufficiently high. These coils occupy a larger volume when the polymer is in a good solvent than when it is in a poor solvent. Simha, Frisch and Eirich[114–117] assume that when a flexible polymer is adsorbed from dilute solution, a localised monolayer is formed, but that it is extremely unlikely that each segment of the polymer chain will be attached to the surface. Normally there will only be a few anchor segments per molecule, the remainder of the polymer chain forming loops or bridges extending out into solution.

The adsorption of electrolytes

The adsorption of electrolytes from aqueous solution is the result of electrostatic forces existing between ions in solution and the surface charge on the solid.[118] Adsorption of this type may be considered in terms of electric double layer theory, which is the subject of Chapter 3. If the simple Stern[119] model of the double layer is accepted then the charge density σ_1, in the Stern plane is given by

$$\sigma_1 = \frac{N_1 z_+ e}{1 + \dfrac{N_0}{n_+ M} \exp \dfrac{z_+ e\psi_\delta + \varphi_+}{kT}} - \frac{N_1 z_- e}{1 + \dfrac{N_0}{n_- M} \exp \dfrac{-z_- e\psi_\delta - \varphi_-}{kT}} \tag{2.58}$$

where N_1 is the available number of adsorption sites per cm^2, z_+ and z_- are the valencies of the respective ions having concentrations n_+ and n_-. ψ_δ is the potential in the Stern plane, φ_+ and φ_- the specific adsorption potentials of the ions, M the molecular weight of the solvent, and k the Boltzmann constant. N_0 is the Avogadro number. A diagrammatic representation of the double layer is shown in Fig. 2.14.

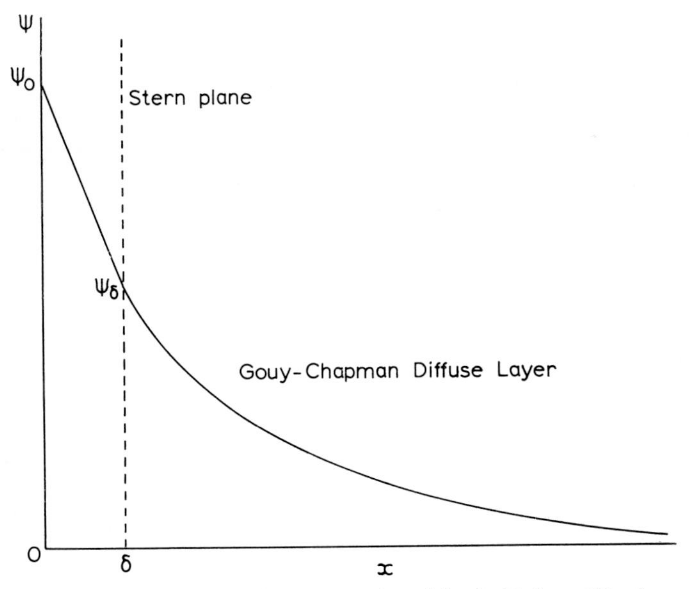

Fig. 2.14. *Diagrammatic representation of the double layer* (*Stern*).

Stern's model is essentially analogous to the Langmuir adsorption isotherm applied to both types of ions. Experimental methods have not reached a state where Stern's theory can be rigorously tested for the solid–liquid interface, although it has been applied a number of times.[120] It should be noted that even if the charge density in the Stern plane has a definite sign, there will normally be a very considerable number of ions of opposite sign also in the plane.

Work carried out on the mercury–solution interface has indicated that the Stern model is too simple, and it has been modified by Grahame.[121] Unfortunately the increased number of parameters involved make it difficult to apply to adsorption at the solid–liquid interface.

In conclusion one might say that the study of the solid–liquid interface represents a study of the forces existing between atoms, molecules and ions. Most of our experimental studies do not penetrate this far being concerned mainly with the observable results of such forces. With simple academic systems the existing theories can be comparatively readily applied, but with far more complex practical systems the application is often more difficult and less obvious.

BIBLIOGRAPHY

A. W. ADAMSON, *Physical Chemistry of Surfaces,* Interscience, New York (1960).
S. J. GREGG and K. S. W. SING, *Adsorption, Surface Area and Porosity*, Academic Press, London (1967).
J. T. DAVIES and E. K. RIDEAL, *Interfacial Phenomena*, Academic Press, 2nd Edition, London (1963).
D. M. YOUNG and A. D. CROWELL, *Physical Adsorption of Gases*, Butterworths, London (1962).
J. J. KIPLING, *Adsorption from Solutions of Non-Electrolytes*, Academic Press, London (1965).
H. R. KRUYT, Editor, *Colloid Science*, Vol. 1, Elsevier, Amsterdam (1952).
Wetting, *Society of Chemical Industry Monograph* No. 25, London (1967).

REFERENCES

1. J. O. Hirschfelder, C. F. Curtiss and R. B. Bird, *Molecular Theory of Gases and Liquids*, Wiley, New York, 1954.
2. E. A. Moelwyn-Hughes, *Physical Chemistry,* Pergamon, Oxford, 2nd ed., 1961.
3. D. M. Young and A. D. Crowell, *Physical Adsorption of Gases*, Butterworths, London, 1962, Chapter 2.
4. G. Mie, *Ann. Physik*, **11** (1903) 657.
5. J. E. Lennard-Jones, Chapter X of R. H. Fowler's *Statistical Mechanics*, Cambridge, 2nd Ed., 1936.
6. J. Willard Gibbs, *The Collected Works of J. Willard Gibbs*, Longmans, Green, London, 1928.
7. E. A. Guggenheim, *Thermodynamics, an Advanced Treatment for Chemists and Physicists*, North-Holland, Amsterdam, 1949, p. 46.
8. T. Young, *Phil. Trans. Roy. Soc.* (London), **95** (1805) 65, 82.
9. J. T. Davies and E. K. Rideal, *Interfacial Phenomena*, Academic Press, New York, 2nd Ed., 1963, Chapter 1.
10. A. W. Adamson, *Physical Chemistry of Surfaces*, Interscience, New York, 1960, Chapter 1.
11. W. D. Harkins, *The Physical Chemistry of Surface Films*, Reinhold, New York, 1952.
12. J. J. Bikerman, *Surface Chemistry*, Academic Press, New York, 2nd Ed., 1958.
13. Ref. 10. Chapter 5.
14. H. Udin, A. J. Shaler and J. Wulff, *J. Metals*, **1** (1949) 186.
15. C. Herring, in *Structure and Properties of Solid Surfaces*, Edited by R. Gomer and C. S. Smith, University of Chicago Press, Chicago, 1953, p. 5.

16. B. Chalmers, R. King and R. Shuttleworth, *Proc. Roy. Soc. (London)*, **A193** (1949) 465.
17. G. C. Kuczynski, *J. Metals*, **1** (1949) 96.
18. J. K. MacKenzie and R. Shuttleworth, *Proc. Roy. Soc. (London)*, **B62** (1949) 833.
19. G. C. Kuczynski, *Acta Met.*, **4** (1956) 58.
20. G. F. Huttig, *Kolloid-Z.*, **98** (1942) 263.
21. Ref. 10, p. 60.
22. H. N. Hersch, *J. Am. Chem. Soc.*, **75** (1953) 1529.
23. G. Cohen and G. C. Kuczynski, *J. Appl. Phys.*, **21** (1950) 1339.
24. R. Gomer, *Advan. Catalysis*, **7** (1955) 93.
25. E. Menzel, *Z. Physik*, **132** (1952) 508.
26. G. Beilby, *Aggregation and Flow in Solids*, Macmillan, New York, 1921.
27. R. C. French, *Proc. Roy. Soc. (London)*, **A140** (1933) 637.
28. H. Raether, *Z. Physik*, **124** (1948) 286.
29. A. R. Bailey, *Rep. Progr. Appl. Chem.*, **34** (1954) 85.
30. T. S. Renzema, *J. Appl. Phys.*, **23** (1952) 1142.
31. W. Cochrane, *Proc. Roy. Soc. (London)*, **A166** (1928) 228.
32. J. J. Trillat, *Compt. Rend.*, **224** (1947) 1102.
33. R. Shuttleworth, *Proc. Phys. Soc. (London)*, **A63** (1950) 444.
34. S. Ross and J. P. Olivier, *On Physical Adsorption*, Interscience, New York, 1964, p. 8.
35. C. Sandford and S. Ross, *J. Phys. Chem.*, **58** (1954) 228.
36. C. Gurney, *Proc. Phys. Soc. (London)*, **A62** (1949) 639.
37. M. M. Nicholson, *Proc. Roy. Soc. (London)*, **A228** (1955) 490.
38. I. F. Guilliatt and N. H. Brett, *Trans. Faraday Soc.*, **65** (1969) 3328.
39. P. J. Anderson and A. Scholtz, *Trans. Faraday Soc.*, **64** (1968) 2972.
40. W. D. Harkins, *J. Chem. Phys.*, **10** (1942) 268.
41. R. Eötvös, *Wied Ann.*, **27** (1886) 456.
42. R. Shuttleworth, *Proc. Phys. Soc. (London)*, **A62** (1949) 167; **A63** (1950) 447.
43. M. Born and W. Heisenberg, *Z. Physik*, **23** (1924) 338.
M. Born and J. E. Mayer, *Z. Physik*, **75** (1932) 1.
44. J. E. Lennard-Jones and P. A. Taylor, *Proc. Roy. Soc. (London)*, **A109** (1925) 476.
45. J. E. Lennard-Jones and B. M. Dent, *Proc. Roy. Soc. (London)*, **A121** (1928) 247.
46. B. M. Dent, *Phil. Mag.*, **8** (1929) 530.
47. M. L. Huggins and J. E. Mayer, *J. Chem. Phys.*, **1** (1933) 643.
48. E. J. W. Verwey, *Rec. Trav. Chim.*, **65** (1946) 521.
49. G. C. Benson, H. P. Schreiber and D. Patterson, *Can. J. Phys.*, **34** (1956) 265.
50. G. C. Benson and R. McIntosh, *Can. J. Chem.*, **33** (1955) 1677.
51. G. C. Benson, P. I. Freeman and E. Dempsey, *Adv. in Chem. Ser.*, **33** (1961) 26.
52. B. M. E. van der Hoff and G. C. Benson, *Can. J. Phys.*, **32** (1954) 475.
53. H. P. Schreiber and G. C. Benson, *Can. J. Phys.*, **33** (1955) 534.
54. Ref. 3, Chapter 6.
55. S. Brunauer, *The Adsorption of Gases and Vapours*, Princeton University Press, Princeton, N.J., 1945.
56. S. Brunauer and P. H. Emmett, *J. Am. Chem. Soc.*, **57** (1935) 1754.
57. P. H. Emmett and S. Brunauer, *J. Am. Chem. Soc.*, **59** (1937) 1553.
58. S. Brunauer, P. H. Emmett and E. Teller, *J. Am. Chem. Soc.*, **60** (1938) 309; for errata see P. H. Emmett and T. W. DeWitt, *Ind. Eng. Chem. (Anal.)*, **13** (1941) 28.

59. G. D. Halsey, *Discussions Faraday Soc.*, **8** (1950) 54.
60. R. A. Beebe, J. B. Beckwith and J. M. Honig, *J. Am. Chem. Soc.*, **67** (1945) 1554.
61. A. C. Zettlemoyer, A. Chand and E. Gamble, *J. Am. Chem. Soc.*, **72** (1950) 2752.
62. R. J. Johansen, P. B. Lorenz, C. G. Dodd, F. D. Pidgeon and J. W. Davis, *J. Phys. Chem.*, **57** (1953) 40.
63. R. T. Davis, T. W. DeWitt and P. H. Emmett, *J. Phys. Chem.*, **51** (1947) 1232.
64. H. L. Pickering and H. C. Eckstrom, *J. Am. Chem. Soc.*, **74** (1952) 4775.
65. R. A. W. Haul, *Angew. Chem.*, **68** (1956) 238.
66. S. Brunauer, L. S. Deming, W. E. Deming and E. Teller, *J. Am. Chem. Soc.*, **62** (1940) 1723.
67. L. G. Joyner, E. B. Weinberger and C. W. Montgomery, *J. Am. Chem. Soc.*, **67** (1945) 2182.
68. C. Orr and J. M. Dalla Valle, *Fine Particle Measurement*, Macmillan, New York, 1959.
69. J. J. Kipling, *Adsorption from Solutions of Non-Electrolytes*, Academic Press, London, 1965, Chapter 17.
70. N. K. Adam, *The Physics and Chemistry of Surfaces*, Oxford University Press, London, 3rd Ed., 1941.
71. W. D. Harkins and D. M. Gans, *J. Am. Chem. Soc.*, **53** (1931) 2804.
72. A. S. Russell and C. N. Cochran, *Ind. Eng. Chem.*, **42** (1950) 1332.
73. C. Orr, H. G. Blacker and S. L. Craig, *J. Metals*, **4** (1952) 657.
74. H. A. Smith and J. F. Fuzek, *J. Am. Chem. Soc.*, **68** (1946) 229.
75. F. Paneth and A. Radu, *Ber.*, **B57** (1924) 1221.
76. A. Clauss, H. P. Boehm and U. Hofman, *Z. Anorg. Allgem. Chem.*, **290** 1957) 35.
77. J. J. Kipling and R. B. Wilson, *J. Appl. Chem.* (*London*), **10** (1960) 109.
78. D. Graham, *J. Phys. Chem.*, **59** (1955) 896.
79. C. H. Giles, T. H. MacEwan, S. N. Nakhwa and D. Smith, *J. Chem. Soc.* (1960) 3973.
80. C. H. Giles and S. N. Nakhwa, *J. Appl. Chem.* (*London*), **12** (1962) 266.
81. H. N. V. Temperley, *Proc. Cambridge Phil. Soc.*, **48** (1952) 683.
82. I. Stranski, *Z. Physik. Chem.*, **136** (1928) 259.
83. M. L. Dundon and E. Mack, Jr., *J. Am. Chem. Soc.*, **45** (1923) 2479.
    M. L. Dundon, *J. Am. Chem. Soc.*, **45** (1922) 2658.
84. M. M. Dubinin, *Chemistry and Physics of Carbon*, P. L. Walker, Editor, **2** (1966) 51.
85. P. H. Emmett, *Catalysis*, **1** (1954) 31.
    H. Mark and E. J. W. Verwey, *Advan. Colloid Sci.*, **3** (1950) 1.
86. H. L. Ritter and L. C. Drake, *Ind. Eng. Chem.* (*Anal.*), **17** (1945) 782, 787.
    L. C. Drake, *Ind. Eng. Chem.*, **41** (1949) 780.
87. J. H. de Boer, B. C. Lippens, B. G. Linsen, J. C. P. Broekhoff, A. van der Heuvel and Th. J. Osinga, *J. Colloid and Interface Sci.*, **21** (1966) 405.
88. K. S. W. Sing, *Proc. Int. Sym. Surface Area Determination,* Butterworths, London, 1970, p. 25.
89. A. Dupré, *Theorie Mechanique de la Chaleur*, Paris, 1869, p. 369.
90. F. M. Fowkes, *Ind. Eng. Chem.*, **56** (1964) 40.
91. A. L. Bailey, *Proc. Intern. Congr. Surface Activity*, 2nd (*London*), **3** (1957) 189.
92. B. R. Ray and F. E. Bartell, *J. Colloid Sci.*, **8** (1953) 214.
93. R. N. Wenzel, *Ind. Eng. Chem.* (*Anal.*), **28** (1936) 988.

94. G. Macdougall and C. Ockrent, *Proc. Roy. Soc.* (*London*), **A180** (1942) 151.
95. B. V. Derjaguin, *Proc. Intern. Congr. Surface Activity*, 2nd (*London*), **3** (1957 446.
96. Ref. 10, p. 270.
97. W. D. Harkins and E. H. Loeser, *J. Chem. Phys.*, **18** (1950) 556.
98. E. W. Washburn and E. A. Anderson, *J. Phys. Chem.*, **50** (1946) 401.
99. W. D. Harkins and H. K. Livingston, *J. Chem. Phys.*, **10** (1942) 342.
100. N. K. Adam and H. K. Livingston, *Nature*, 182 (1958) 128.
101. A. M. Gaudin, *Flotation*, McGraw-Hill, New York, 1957.
102. K. L. Sutherland and I. W. Wark, *Principles of Flotation*, Aus. Inst. M. M., Melbourne, 1955.
103. Ref. 9, Chapter 8.
104. A. B. D. Cassie and S. Baxter, *Trans. Faraday Soc.*, **40** (1944) 546.
105. F. P. Bowden and D. Tabor, *The Friction and Lubrication of Solids*, Oxford, 1954.
106. D. D. Eley, Editor, *Adhesives*, Oxford, 1961.
107. Ref. 66, Chapters 1 to 8.
108. D. H. Everett, *Trans. Faraday Soc.*, **60** (1964) 1803.
109. D. H. Everett, *Trans. Faraday Soc.*, **61** (1965) 2478.
110. R. S. Hansen and R. D. Hansen, *J. Colloid Sci.*, **9** (1954) 1.
111. D. C. Jones and G. S. Mill, *J. Chem. Soc.* (1957) 213.
112. R. S. Hansen, Y. Fu and F. E. Bartell, *J. Phys. Chem.*, **53** (1949) 769.
113. E. Jenckel and B. Rumbach, *Z. Elektrochem.*, **55** (1951) 612.
114. R. Simha, H. L. Frisch and F. R. Eirich, *J. Phys. Chem.*, **57** (1953) 584.
115. H. L. Frisch and R. Simha, *J. Phys. Chem.*, **58** (1954) 507.
116. H. L. Frisch, *J. Phys. Chem.*, **59** (1955) 633.
117. H. L. Frisch and R. Simha, *J. Chem. Phys.*, **27** (1957) 502.
118. H. R. Kruyt, Editor, *Colloid Science*, Elsevier, Amsterdam, 1952, Volume 1, Chapter 4.
119. O. Stern, *Z. Elektrochem.*, **30** (1924) 508.
120. M. B. Abramson, M. J. Jaycock and R. H. Ottewill, *J. Chem. Soc.* (1964) 5034, 5041.
121. D. C. Grahame, *Chem. Rev.*, **41** (1947) 441.

CHAPTER 3

ELECTRICAL PHENOMENA ASSOCIATED WITH THE SOLID–LIQUID INTERFACE

A. L. SMITH

INTRODUCTION

The state of electric charge of the particles of a colloidal dispersion is always an important factor governing the stability to flocculation. In the case of particles which do not greatly influence the dispersion medium in their vicinity and in the absence of adsorbed material of high molecular weight, it is the dominant factor. The correlation between the magnitude of the electrokinetic (zeta) potential and the stability of a lyophobic dispersion has long been recognised and resulted in the approximate working rule[1] that a zeta potential of at least 30 mV is necessary for long term sol stability. While any such universal rule cannot stand up to close scrutiny it illustrates the importance of an understanding of the relationship between interparticle repulsion and the various potentials which it is possible to define, though not necessarily directly measure, at or near the particle surface.

The whole of the region in the vicinity of the solid–solution interface in which any discontinuity or change in mean potentials or electrical properties occurs is usually termed the electric double layer. It will be the purpose of this chapter to discuss the origin and structure of this double layer, how far it is possible to obtain quantitative information about it, and particularly the technique of micro-electrophoresis.

THE ELECTRIC DOUBLE LAYER

Origin of the electric double layer

The redistribution of charge which is implied by the formation of an electric double layer when electrically neutral particles are placed in a solution which is itself (overall) electrically neutral will be governed by the following factors.

(*i*) The dissociation of any ionogenic groups present in the particle surface, *e.g.* –COOH groups which would give the particle a negative charge at high *p*H.
(*ii*) The unequal dissolution of oppositely charged ions of which the particle may be composed. An example of this is the negative charge remaining on silver halides when suspended in water.
(*iii*) The adsorption by the particle of one of the ion-types in the solution in which it is suspended, or alternatively the unequal adsorption of oppositely charged ions. Adsorption must be understood generally here to include negative adsorption.
(*iv*) The adsorption and/or orientation of dipolar molecules at the particle surface. It is evident that such dipoles will not directly contribute to the net charge of the particle but they may have an important effect on the electric double layer as demonstrated below. The dipoles may be the result of the deformation of polarisable molecules in the electric field at the interface.

The disperse particles as normally encountered are more often negatively charged than positively. This is partly because, in aqueous solution, anions are usually less strongly hydrated, and therefore more readily specifically adsorbed, than cations. That this is not the whole explanation, however, is shown by the observation that air bubbles in most aqueous solutions migrate towards the anode when an electric field is applied, despite the fact that surface tension data, interpreted by the Gibbs equation, show that in these solutions there is an overall negative adsorption of electrolyte. Evidently the negative adsorption of cations is more pronounced than that of anions in these cases.

The case of suspended air bubbles illustrates a possible ambiguity in statements about the charge of a particle. The air bubble itself, if by this it is meant to exclude any part of the solution or any ion partly in the solution, is electrically neutral; any observation of a negative electrophoretic mobility indicates that what may be called the electrokinetic unit is negatively charged. The electrokinetic unit will include some solvent and ions close to the surface and may well have a net electric charge. In an extreme case these two interpretations of the charge in the particle may even give rise to opposite signs, *e.g.* AgI particles in a solution containing a slight excess of I^- ions and sufficient cationic surfactant. This should be made clearer by a more detailed discussion of the inner part of the double layer later in this chapter.

The nature and definition of electric potentials in the double layer

In any attempt to discuss the electric potential difference between a point in the solution and a point within a suspended particle the same difficulty arises as is familiar in treatments of electrode potentials. It was pointed out, not first but perhaps most clearly, by Guggenheim[2] that such a potential difference is not accessible to direct measurement. Since disembodied electric charge does not exist, the work done in transporting unit electrical charge from a point in the solution to a point within the particle will necessarily include a contribution from the difference between what may be called the chemical potentials of the charge carrying particle in the two situations. The latter contribution may be looked upon as arising from the change in the short range interactions of the charged particle with its surroundings.

It is not possible to separate unambiguously the electrostatic and chemical contributions to the total work since all the interactions involved are fundamentally electrical in nature. The total work done in the transport of unit electric charge, associated with some body, *e.g.* an electron, from a point in one phase to a point in another is an unambiguous quantity, termed by Guggenheim the *electrochemical potential difference* $\Delta\eta$.

Despite the considerations outlined above it is nevertheless still useful, especially for theoretical purposes, to construct a model in which the components of $\Delta\eta$ are supposed to be separated. The model outlined here is that due to Lange (1933).

In Fig. 3.1 a charged spherical particle, considered initially to be situated in a vacuum, is imagined to be separated into (a) a sphere of

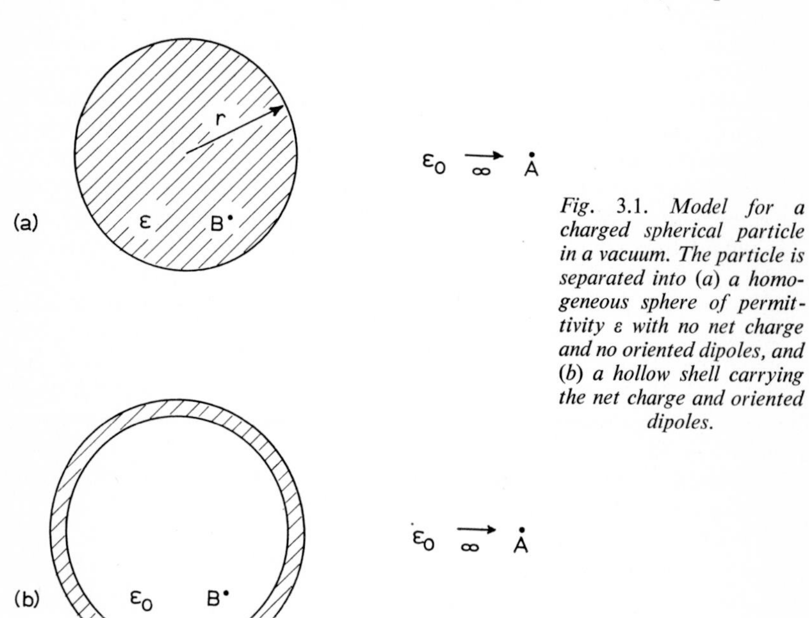

Fig. 3.1. *Model for a charged spherical particle in a vacuum. The particle is separated into* (*a*) *a homogeneous sphere of permittivity* ε *with no net charge and no oriented dipoles, and* (*b*) *a hollow shell carrying the net charge and oriented dipoles.*

dimensions equal to that of the real particle, homogeneous and of permittivity ε but with no net charge and no oriented dipoles in the interface and (b) the excess charge and oriented dipoles concentrated into a thin shell of radius equal to that of the real particle.

In case (a) the work done in taking a species i of charge q_i from a point A distant from the particle to a point B within it is termed μ_i the *chemical* potential. This is not, as might first appear, independent of q_i since if the species i is taken as a sphere of radius a, it will include a term

$$\frac{q_i^2}{2\varepsilon a} - \frac{q_i}{2\varepsilon_0 a} \quad \text{or} \quad \frac{q_i^2}{2a}\left[\frac{1}{\varepsilon} - \frac{1}{\varepsilon_0}\right] \tag{3.1}$$

which vanishes if $q_i = 0$. Thus the term chemical potential for μ_i is not ideal but is, nevertheless, usually used.

In case (b) the work done in taking the charge q_i from A to B is given by $q_i\varphi$ where φ is the *inner* potential (Lange) of the particle phase. Thus

$$\eta_i = \mu_i + q_i\varphi \tag{3.2}$$

The inner potential φ may conveniently be subdivided into a part ψ due to the net charge on the shell and a part χ due to oriented dipoles in the shell. Thus

$$\varphi = \psi + \chi \tag{3.3}$$

Although the φ potential cannot, as explained above, be directly measured, the ψ potential is accessible to experimental measurement (at least in principle) and for a spherical particle of radius r with excess charge q will be given by $q/\varepsilon_0 r$. ψ is termed the *outer* potential (Lange) and may be looked upon as the electric potential just outside the particle surface, not so near that short range interactions with the particle are significant and not so far that the potential has appreciably decayed (say 10^{-4} cm from the surface). The *surface* or *wall* potential which occurs in discussion of colloid stability, etc., is to be identified with this potential.

Although the χ potential was termed by Lange the surface potential, this term is frequently used for ψ and it is perhaps better to refer to the *ki potential.* For a shell consisting of n dipoles per unit area, each of effective moment μ, with positive poles directed inwards, *i.e.* towards the particle, then if potentials are reckoned relative to point A (Fig. 3.1), the dipole contribution to φ will be given by

$$\chi = 4\pi n\mu/\varepsilon_0 \tag{3.4}$$

where ε_0 is the (unrationalised) permittivity of free space.

It must be noticed that for typical dipole layers the effective moment μ

may be very much smaller than the gas phase dipole moment of the molecule concerned. This arises because

(*a*) the dipoles may not be perpendicular to the surface,
(*b*) the dipoles will tend to point in different directions, and
(*c*) imaging in the solid surface may produce a depolarising field.

Of the quantities occurring in eqns. (3.2) and (3.3) it is seen, therefore, that η_i and ψ are accessible to direct measurement whereas μ_i, φ and χ are not. Experiments can, however, give some information on the latter quantities, particularly changes in these quantities, and they can be the subject of a theoretical calculation.

If it is now supposed that the particle is situated not in a vacuum but in some other medium, *e.g.* a solution, eqns. (3.2) and (3.3) become respectively

$$\Delta\eta_i = \Delta\mu_i + q_i\Delta\varphi \tag{3.5}$$

$$\Delta\varphi = \Delta\psi + \Delta\chi. \tag{3.6}$$

$\Delta\varphi$, the difference in inner potentials of the two phases, is termed the Galvani potential difference and $\Delta\psi$, the difference of the outer potentials, is termed the Volta or contact potential difference. It must not be assumed, however, that $\Delta\chi$ is the difference between the two separate χ potentials of the free surfaces (to a vacuum) of particle and solution, and this factor, together with the obvious and universal practice of reckoning potentials of particles relative to bulk solution far from any particle, makes it desirable to write eqns. (3.5) and (3.6) as

$$\eta_i = \mu_i + q_i\varphi$$

and

$$\varphi = \psi + \chi$$

i.e. identical in form to eqns. (3.2) and (3.3) though with slightly adjusted significance. Symbolism such as $\Delta\chi$ can now be used to indicate a change in the χ potential at the particle–solution interface. Some authors prefer to retain χ for the air-solution case and use the symbol g_{dip} for the inter-phase situation. Such nomenclature emphasises the non-additivity of χ values as between different phases.

A more thorough treatment of the matters raised in this section has been given by Parsons.[3]

Potential determining ions and the Nernst equation

Consider a silver iodide particle suspended in a solution containing (inevitably) silver and iodide ions. An exchange will take place between the silver and halide ions in the particle and those in solution until at (a postulated) equilibrium the electrochemical potential η of each ion is the same both in the solid and in solution.

For the Ag^+ ion with a charge equal to e the electronic charge:

$$\eta_{Ag^+}\text{ (solid)} = \eta_{Ag^+}\text{ (solution)}$$

$$\mu_{Ag^+}\text{ (solid)} + e\varphi = \mu^0_{Ag^+}\text{ (solution)} + kT \ln a_{Ag^+}\text{ (solution)}$$

$$[\varphi\text{ (solution)} = 0 \text{ by convention}]$$

so that

$$\varphi = \text{constant} + \frac{kT}{e} \ln a_{Ag^+}$$

which is a form of the Nernst equation.

At 25°C

$$\varphi = \text{constant} - 59{\cdot}2(p\text{Ag})\text{ mV}$$

where pAg is used to indicate $-\log_{10} a_{Ag^+}$

or

$$\Delta\varphi = -59{\cdot}2\Delta(p\text{Ag})\text{ mV} \tag{3.7}$$

The corresponding equation for the iodide ion will be

$$\Delta\varphi = +59{\cdot}2\Delta(p\text{I})\text{ mV}$$

Thus for a 10-fold change in the silver (or iodide) ion activity (which in dilute solution may be taken as equal to the concentration) the φ potential changes in the appropriate sense by 59·2 mV. If the χ potential is taken as constant, and it must be recognised that this is an assumption, then the ψ potential changes by 59·2 mV also. The charge transfer necessary to achieve this will depend on the capacity of the electric double layer which is discussed later in this chapter.

In this and similar cases the ions concerned are termed the *potential determining ions*. It is not appropriate to regard such ions as inhabiting the *Inner Helmholtz Plane* (*see* below) since, after adsorption, they are indistinguishable from the ions which constitute the bulk of the particle.

In the case of silver halide sols the potential determining ions are particularly well defined and this is one reason for their popularity in studies of the electric double layer and its influence on colloid stability.

In the case of metal oxide and hydroxide sols and in sols of biological interest H^+ and OH^- are the most important potential determining ions. For gold sols the Cl^- ion, which forms stable complexes at the particle surface, has been regarded as a potential determining ion.

Applicability of the Nernst equation to surfaces

In the above derivation of the Nernst equation, applied to the AgI-solution case in order to illustrate the concept of potential determining ions, it was assumed that a state of equilibrium exists between the bulk of the solid and the surface. It is by no means self-evident that this is justified even for the apparently straightforward silver halide situation. For the oxide-solution interface and other cases where the H^+ and OH^- ions are seen as potential determining, the difficulty is even more apparent since neither H^+ nor OH^- may exist at all in the bulk solid.

Clearly a more generalised treatment of the Nernst equation is necessary in which the idealised ionic solid, with or without internal equilibrium, is a limiting case. In the following an H^+/OH^- potential determining mechanism in an aqueous system will be used as illustration.

Suppose that the mechanism by which the surface gains a negative charge is

$$AH \rightarrow A^- + H^+$$

and that by which it becomes positively charged is

$$AH + H^+ \rightarrow AH_2^+$$

where A represents some surface site and H^+ ions are present (and at equilibrium) in solution. The negative-going mechanism, represented as the loss of a proton, might equally be the addition of an OH^- ion so far as the following is concerned. In terms of electrochemical potentials the two mechanisms give respectively

$$\mu_{AH}(s) = \eta_{A^-}(s) + \mu_{H^+} \tag{3.8}$$

and

$$\mu_{AH}(s) = \eta_{AH_2^+}(s) - \mu_{H^+} \tag{3.9}$$

where (s) indicates the surface and η has been used in place of μ only where these differ. $\eta_{H^+}(s)$ is clearly equal to μ_{H^+} in bulk solution.

The total, random, number of configurations g available to the three types of site *viz.* positively charged (+), negatively charged (−) and uncharged (0) will be given by

$$g = \frac{N_s!}{n_+!\,n_-!\,n_0!}$$

where N_s is the total number of sites per unit area. Thus the configurational part of the (surface) chemical potential of *e.g.* the positive sites will be given by:

$$-kT\frac{d\ln g}{dn_+} = +kT\ln\frac{\theta_+}{1-\theta_+-\theta_-}$$

where $\theta = n/N_s$.

The remaining part of μ will be taken, in this approximate treatment, as independent of θ and can therefore be written μ with an appropriate implied standard state.

The electrical part of η should involve the micro-potential at the site concerned but here the mean surface (wall) potential ψ_0 will be used giving $e_0\psi_0$ for the positive site and $-e_0\psi_0$ for the negative site.

Elimination of $\mu_{AH}(s)$ between (3.8) and (3.9) then gives

$$\frac{(\mu^0{}_{H_2A^+}(s) - \mu_{A^-}{}^0(s) - 2\mu^0{}_{H^+})}{kT} = 2\ln a_{H^+} - \ln\frac{\theta_+}{\theta_-} - \frac{2e_0\psi_0}{kT} \quad (3.10)$$

where the LHS of eqn. (3.10) has the form of the ratio of two 'equilibrium constants' representing the tendencies of the surface sites to become negatively charged via eqn. (3.8) or positively charged via eqn. (3.9), termed K_- and K_+ respectively. Equation (3.10) now becomes

$$\ln\frac{K_+}{K_-} = 4{\cdot}606\ \text{pH} + \ln\frac{\theta_+}{\theta_-} + \frac{2e_0\psi_0}{kT} \quad (3.11)$$

Elimination of μ_{H^+} between eqns. (3.8) and (3.9) yields

$$K_+K_- = \frac{\theta_+\theta_-}{\theta_0{}^2} \quad (3.12)$$

At the zero point of charge θ_+ and θ_- will take some common value θ_c and, in the absence of specific adsorption of other than H^+ and OH^- ion, ψ_0 will be zero. Equations (3.11) and (3.12) then become

$$\log\frac{K_+}{K_-} = 2\ \text{pH}_0 \quad (3.13)$$

and

$$K_+K_- = \frac{\theta_c{}^2}{(1 - 2\theta_c)^2} \quad (3.14)$$

respectively. At pH $\gg$ pH_0 where θ_+ will be negligible eqn. (3.11) becomes

$$-\ln K_- = 2{\cdot}303\ \text{pH} - \ln\frac{\theta_-}{1 - \theta_-} + \frac{e_0\psi_0}{kT} \quad (3.15)$$

If K_- is known or estimated, from the analysis of potentiometric titrations or otherwise, then K_+ can be obtained from an experimental pH_0 and eqn. (3.13). Equations (3.11) and (3.12) then allow deviations from an ideal Nernst equation (3.7) to be estimated.

As example consider a surface only slightly removed from the zero point with $\theta_+ = \theta_- + \delta$ then

$$\ln\frac{\theta_+}{\theta_-} \simeq \frac{(\theta_+ - \theta_-)}{\theta_-} - \ldots \simeq \frac{(\theta_+ - \theta_-)}{\theta_c}$$

If this term is negligible, *i.e.* $(\theta_+ - \theta_-) \ll \theta_c$ then eqn. (3.11) reduces to an ideal Nernst equation relating ψ_0 to Δ pH. Deviations from the Nernst equation are seen to be most serious when θ_c is very small as would happen when K_+ or K_- or both are small. The zero point is then characterised by the virtual absence of charge rather than by the presence of equal numbers of positively and negatively charged sites.

At the other extreme θ_c takes its maximum value (0·5) for a surface such as the silver halides discussed above, and the Nernst equation is obeyed even at the zero point.

It will be noticed that the potential involved in this (very approximate) treatment is ψ_0 rather than the inner potential φ. The implication of this is that any potential drop within the solid would not be 'seen' from the solution side in a case such as the silver halides whether there was equilibrium between the bulk solid and the surface or not.[4]

In the case of oxide surfaces where the electrical properties are commonly observed to vary slowly with time it seems reasonable to attribute such drifts to solid state effects. Provided such drifts are slow the above treatment can be applied 'instantaneously' and the slow changes seen as drifts of the zero point.

Adsorption of non-potential determining ions and the inner part of the double layer

It is clear that the finite volume of ions places a limit on the closeness with which the centre of an ion in solution can approach a solid surface. To take account of this Stern[5] (1924) proposed a model for the inner part of the double layer in which the ions were taken as point charges in all respects except their inability to approach closer to the solid surface than some distance *d*. Assuming for a moment a flat interface, this plane, later termed the *Stern Plane*, was at first taken to indicate not only the closest distance of approach but also the centres of any ions specifically adsorbed on to the solid surface.

Specific adsorption is often defined as adsorption which takes place (even) at the zero point of charge of the adsorbent surface, but a more general concept is adsorption which depends on the nature, rather than merely the charge, of the ion.

The Stern model for the inner part of the double layer was refined by Grahame[6] who distinguished between the Stern Plane, also called the *Outer Helmholtz Plane* (OHP) to indicate the closest distance of approach of hydrated ions in solution, and an *Inner Helmholtz Plane* (IHP) to indicate the centres of ions (if any) which are specifically adsorbed on to the solid surface.

A distinction between the two planes is in general necessary since even though the specifically adsorbed ions may be the same size as (or the same ions as) the ions governing the position of the OHP they will probably be dehydrated, at least in the direction of the surface, allowing closer approach to the surface than the OHP. Another factor mentioned by Grahame is that the cations and anions in solution will not be the same size. In aqueous solution simple inorganic cations are more strongly hydrated than corresponding anions leading to a larger hydrated radius for the cations and more ready specific adsorption of anions.

It must be remembered that Fig. 3.2 represents only one model, probably as simple as possible, for the inner part of the double layer which will need to be modified in special cases such as the adsorption of large surfactant ions where small counter ions may approach closer to the surface than the specifically adsorbed ions. In the following treatment the model represented by Fig. 3.2 will be used to illustrate typical procedures.

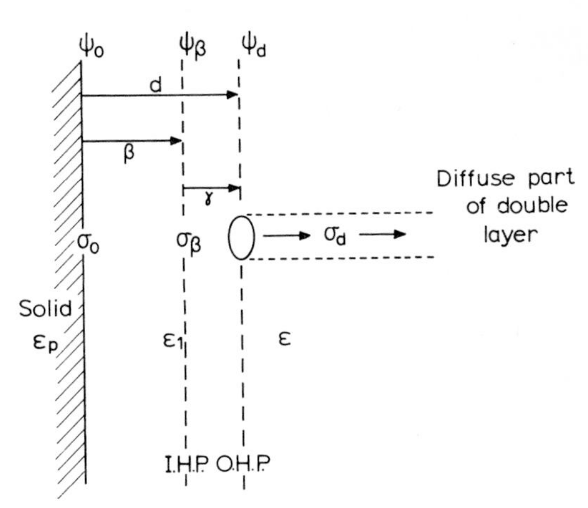

Fig. 3.2. Model for the electric double layer at the solid–solution interface showing potentials ψ and charge densities σ.

The potential referred to as ψ above becomes ψ_0 in Fig. 3.2 so that ψ_0 represents the mean potential just outside the solid surface and will be called the *wall potential.* ψ_β and ψ_d, the mean potentials at the IHP and OHP respectively are, of course, also volta potentials accessible (in principle) to experimental measurement. σ_0 and σ_β are the charges per unit area of the particle surface and IHP respectively; σ_β will be zero in the absence of specific adsorption. σ_d is the net charge contained in a cylinder of unit cross sectional area extending outwards from the OHP (theoretically to infinity).

The effective permittivity ε_1 of the inner region of the double layer has not been assumed equal to that of bulk solution ε. Available evidence shows that in aqueous solution the dielectric constant $(\varepsilon/\varepsilon_0)$ near both the mercury and silver iodide interface at 25°C is only of the order of 10 compared with 78 in bulk solution. This is to be expected in view of the restricted rotation of solvent dipoles near the interface.

The condition of overall electrical neutrality gives

$$\sigma_0 + \sigma_\beta + \sigma_d = 0 \qquad (3.16)$$

The capacity per unit area of the whole of the inner region K is given by

$$K = \frac{\varepsilon_1}{4\pi d} \tag{3.17}$$

so that

$$\psi_0 - \psi_\beta = \frac{\sigma_0 \beta}{Kd} \tag{3.18}$$

and

$$\psi_\beta - \psi_d = \frac{(\sigma_\beta + \sigma_0)\gamma}{Kd} \tag{3.19}$$

If it is not wished to assume that ε_1 is uniform over the inner region then Kd/β can be written K_β and Kd/γ as K_γ so that

$$\frac{1}{K} = \frac{1}{K_\beta} + \frac{1}{K_\gamma} \tag{3.20}$$

and the quantities d, β and γ lose their significance as simple distances.

If the electrolyte in solution around the particles is assumed to be of symmetrical $z{:}z$ valency type with n ions of each type per unit volume, the Gouy–Chapman theory (below) gives

$$\sigma_d = -\frac{\varepsilon k T \kappa}{2\pi z e} \sinh \frac{z e \psi_d}{2kT} \tag{3.21}$$

where κ is the Debye–Huckel parameter defined by

$$\kappa^2 = \frac{4\pi e^2 \sum nz^2}{\varepsilon k T}$$

which, with the assumptions above, becomes

$$\kappa^2 = \frac{8\pi e^2 n z^2}{\varepsilon k T}. \tag{3.22}$$

Combining eqns. (3.18) and (3.19)

$$\psi_0 - \psi_d = \frac{\sigma_0}{K} + \frac{\sigma_\beta \gamma}{Kd} \tag{3.23}$$

If the *zero point of charge* (PZC) is defined as the condition when $\sigma_0 = 0$, as at the mercury solution interface, then at the PZC
from eqn. (3.18)

$$\psi_0 = \psi_\beta$$

from eqn. (3.23)

$$\psi_0 - \psi_d = \frac{\sigma_\beta \gamma}{Kd}$$

and from eqn. (3.16)

$$\sigma_\beta + \sigma_d = 0 \tag{3.24}$$

Figure 3.3 shows the possible situation at the PZC with σ_β negative. Only in the absence of specific adsorption, *i.e.* with $\sigma_\beta = 0$, does it follow that

$$\psi_0 = \psi_\beta = \psi_d = 0$$

The electrokinetic or zeta potential ζ is presumably to be identified either with ψ_d or the potential some distance further out into the diffuse part of the double layer. At low potentials $\zeta \approx \psi_d$ and certainly when

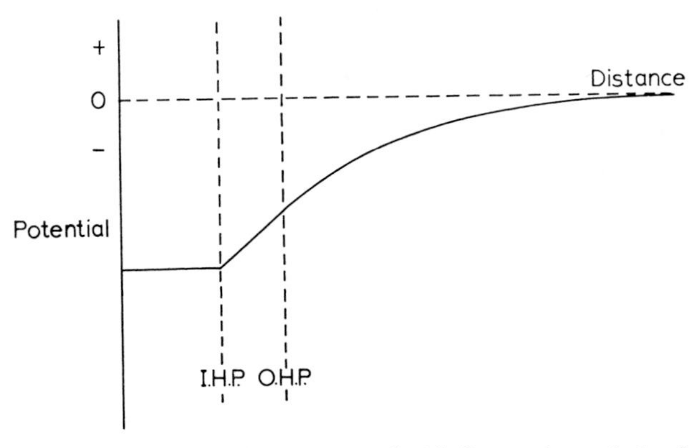

Fig. 3.3. *Idealized variation of electric potential with distance into solution for a surface at its PZC but with specific adsorption of anions into the Inner Helmholtz Plane.*

$\zeta = 0$ then $\psi_d = 0$ so that the zero point of electrophoretic mobility corresponds to $\psi_d = 0$, not to the PZC as defined above, though the two conditions are sometimes confused. When $\psi_d = 0$ then from eqn. (3.21) $\sigma_d = 0$ so that $\sigma_0 = -\sigma_\beta$; from eqn. (3.19) $\psi_\beta = \psi_d\,(=0)$ and from eqn. (3.18)

$$\psi_0 = \frac{\sigma_0 \beta}{Kd}$$

Figure 3.4 shows a possible situation at the zero point of electrophoretic mobility with σ_β negative.

The zero point of electrophoretic mobility sometimes termed the iso-electric point, does, of course, correspond to the zero point of total charge within the OHP and, being a directly measurable quantity, is perhaps a more convenient parameter than the PZC defined as $\sigma_0 = 0$, especially for systems without well defined potential determining ions. In the absence of specific adsorption the two zero points coincide.

Adsorption of dipolar molecules

It might be thought that since an adsorbed neutral dipolar molecule cannot contribute to the net charge of a particle it will have no effect on the electric double layer, except perhaps to displace it to slighter greater distances from the surface, and no effect on observed quantities such as the electrophoretic mobility. This, however, is not necessarily the case.

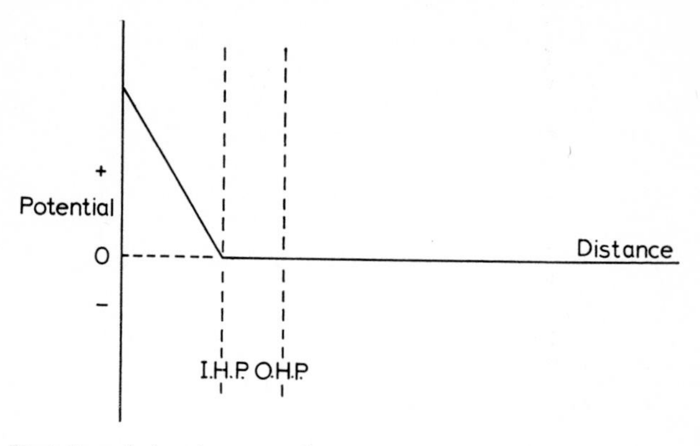

Fig. 3.4. *Variation of electric potential with distance into solution for a surface at the zero point of electrokinetic potential but with specific adsorption of anions.*

Consider a particle suspended in a certain (constant) concentration of potential determining ions. Thus, according to the Nernst equation, the φ potential is fixed and

$$\varphi = \psi_0 + \chi$$

where χ is due to the orientation of dipoles in the particle surface. Suppose now that neutral dipolar molecules are introduced into the solution and adsorbed at the particle–solution interface with their positive poles towards the solid. This introduces a positive contribution to the χ potential which will be written $\Delta\chi$. If it is assumed that the activity of the potential determining ions is unchanged at the (perhaps small) concentration of neutral molecules used then φ is unaltered and

$$\Delta\psi_0 = -\Delta\chi$$

Thus ψ_0 will move to more negative values and with it, provided that $|\psi_0|$ was not already large, ψ_d and the electrokinetic potential. If the particles were at the PZC (specific adsorption of ions may be temporarily ignored) before the dipolar adsorption, then to restore the PZC after this adsorption a higher concentration of positivity charged potential determining ions will be needed. In a similar way, at the mercury solution interface the same adsorption would require more positive polarisation of the mercury to restore the electro-capillary maximum.

These considerations indicate that *e.g.* non-ionic stabilising or sensitising agents do not necessarily act solely by non-ionic mechanisms. In any attempt to relate $\Delta\chi$ to the orientation and adsorbed density of dipoles it must be remembered that in addition to the factors influencing the effective dipole moment in any adsorbed layer, the dipole moment and orientation of any displaced solvent layer must be considered. The latter could even predominate.

As has been noted above, it is often assumed that the χ potential remains constant during variations of surface charge and potential. That this is not universally justifiable is seen by considering the case where dipolar molecules are adsorbed on a surface at the ZPC with the dipoles oriented parallel to the surface. As the surface charge is increased (say positively) it is likely that the adsorbed dipole direction will rotate somewhat to give a negative $\Delta\chi$.

The Stern adsorption isotherm and the discreteness of charge effect

The charge in the IHP is not smeared out but consists of discrete ionic charges. When an ion from solution is adsorbed into the IHP the discrete charges will be rearranged and this must be taken into account when writing the electrostatic work of adsorption. The rearrangement will impose a *self-atmosphere* potential on the adsorbed ion which is the two-dimensional analogue of the three-dimensional self-atmosphere potential occurring in the Debye–Hückel theory of electrolytes. This effect, which has been termed the *discreteness of charge* effect or *discrete ion* effect was introduced by Esin and Shikov[7] after an original suggestion by Frumkin,[8] to explain the Esin–Markov[9] effect at the mercury–solution interface.

The importance of the effect in colloid stability theory has been stressed by Levine, Bell and collaborators and a review[10] has appeared. The main consequence of the theory to colloid stability is that it predicts that under suitable conditions the OHP potential $|\psi_d|$ can reach a maximum as $|\psi_0|$ is increased so that further increase in $|\psi_0|$ leads to a decrease in $|\psi_d|$ and therefore to a decrease in interparticle repulsion.

For the adsorption into the IHP of a cation of charge ze, the electrostatic work of adsorption will be not $ze\psi_\beta$ but

$$ze\psi_{\mathrm{A}} = ze(\psi_\beta + \varphi_\beta) \tag{3.25}$$

where φ_β is the self atmosphere potential referred to above and ψ_{A} is the *micro-potential* in the *hole* which re-arrangement of other adsorbed ions has provided. In the treatment of Ershler[11] the adsorbed ions are assumed to form a hexagonal array, but the more convenient method of Grahame[12] will be followed here in which an adsorbed ion is regarded as at the centre of a circular area in the IHP of radius r_0 such that $\pi r^2 \sigma_\beta = ze$ from which the uniform charge density has been removed. φ_β is now the potential at the centre of this disc due to a uniform charge density of $-\sigma_\beta$ over its whole area. The calculation of φ_β on this model is in general a matter of some complexity[13] involving multiple electrical images of the disc in the particle wall and OHP, but can be simplified if certain assumptions are acceptable.

If $\varepsilon \gg \varepsilon_1$, and $\varepsilon_p \gg \varepsilon_1$ as is a reasonable approximation at the mercury–solution interface, and if $r_0 \gg d$, *i.e.* low charge density in the IHP, then φ_β becomes the potential of an infinite plane of charge density $-\sigma_\beta$ between two earthed conducting plates, and simple electrostatics gives

$$\varphi_\beta = -\frac{\beta\gamma\sigma_\beta}{Kd^2} \tag{3.26}$$

If $\varepsilon \gg \varepsilon_1$ and $\varepsilon_p = \varepsilon_1$, a reasonable approximation at the interface between a dielectric and aqueous solution, and as before $r_0 \gg d$ then

$$\varphi_\beta = -\frac{\gamma\sigma_\beta}{Kd} \tag{3.27}$$

Following Levine and Bell[14] eqns. (3.26) and (3.27) may be combined in

$$\varphi_\beta = -\frac{\beta\gamma\sigma_\beta}{Kd^2} g \tag{3.28}$$

where the factor g is d/β in the conditions leading to eqn. (3.27), though this treatment is greatly oversimplified. If $n_\beta(=\sigma_\beta/ze)$ is the number of adsorbed cations per unit area of the IHP, each occupying p of the N_s adsorption sites available per unit area, then equating the electrochemical potentials for an adsorbed cation and a cation in bulk electrolyte

$$kT \ln \frac{n_\beta}{(N_s - pn_\beta)^p} + ze\psi_A = \text{constant} + kT \ln nf \tag{3.29}$$

where a is the activity of cations in bulk solution and the constant term includes the chemical energy of adsorption.

If $p = 1$ and a is written as the mole function then eqn. (3.29) becomes the Stern adsorption isotherm which can be written in the form

$$\frac{N_s}{n_\beta} = 1 + \frac{n_0}{n} \exp\left[\frac{ze\psi_\beta + \Phi}{kT}\right] \tag{3.30}$$

where n_0 is the number of solvent molecules per unit volume, and Φ is the specific adsorption potential.

If eqns. (3.17), (3.19), (3.20), (3.25), (3.28) and (3.29) are combined the following equations result[14]

$$r - \ln r + p \ln(1 - pbr) = \frac{ze\psi_d}{kT} + \frac{2\upsilon\gamma}{d} \sinh\left(\frac{ze\psi_d}{2kT}\right) - \ln x + \text{constant} \quad (3.31)$$

and

$$\frac{e\psi_0}{kT} = \frac{e\psi_d}{kT} + \frac{2\upsilon}{z} \sinh\left(\frac{ze\psi_d}{2kT}\right) - \frac{rd}{\gamma g z} \quad (3.32)$$

from which it follows that if ψ_0 is varied at constant electrolyte concentration, $|\psi_d|$ will have a maximum at a value of r given, if κ is assumed constant, by

$$1 - \frac{1}{r} - \frac{p^2 b}{1 - prb} = 0 \quad (3.33)$$

where

$$r = \frac{\sigma_\beta}{\sigma_\beta^0}, \qquad \sigma_\beta^0 = \frac{kTKd^2}{ze\beta\gamma g}$$

$$b = \frac{kTd^2K}{(ze)^2\beta\gamma g N_s}, \qquad \upsilon = \frac{\kappa\varepsilon}{4\pi K}.$$

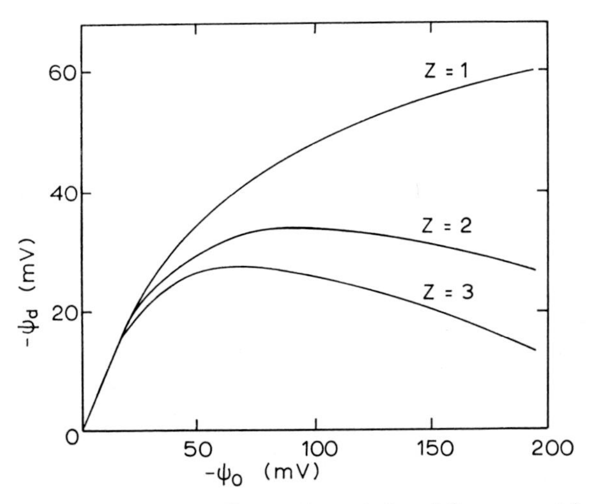

Fig. 3.5. Influence of discrete-ion effect on the variation of Stern potential ψ_d with wall potential ψ_0 at a negatively charged surface, in the presence of 1:1 (p = 1), 2:2 (p = 3) *and* 3:3 (p = 5) *valent electrolytes at their respective flocculation concentrations.* K = 40 *μF/cm*², $N_s = 5 \times 10^{14}$, $\gamma/d = 1/4$ *and* g = 4/3.

This maximum in $|\psi_d|$ does not occur if the self atmosphere potential φ_β is omitted. Figure 3.5 shows[14] the variation of ψ_d with ψ_0 for a negatively charged surface at concentrations of 1:1, 2:2 and 3:3 electrolytes corresponding to their flocculation concentrations, using the above equations with reasonable values for the parameters concerned. The discreteness of charge effect has been able to give a satisfactory explanation for results at the mercury–solution interface,[10,15] and at the silver iodide–solution interface has been used to explain maxima in the co-ion contribution to the capacitance,[16] flocculation kinetic results, and maxima in electrokinetic potentials.[17]

Correction of the Stern adsorption isotherm to include the self-atmosphere potential has been shown[17] to reduce the specific adsorption only $-$ 0·5kT which agrees with the observation that potassium ions have only an insignificant effect on the PZC of silver iodide.

The diffuse region of the double layer in solution

In terms of the model for the double layer adopted in Fig. 3.2 the part extending outwards from the OHP or Stern plane is termed the *diffuse* region. Here the thermal motion of the ions in solution results in an approximately exponential decay in potential and in the classical treatment of the diffuse layer the concentration of ions is taken to be governed by a Boltzmann distribution in the form

$$n'_{\pm} = n_{\pm} \exp\left[\mp \frac{ze\psi}{kT}\right] \tag{3.34}$$

where $n'_{\pm}$ is the number of ions of either charge per unit volume at a point in the diffuse layer where the mean potential is ψ, and $n_{\pm}$ is the corresponding concentration in bulk solution.

If this distribution law is combined with the Poisson equation

$$\nabla^2\psi = -\frac{4\pi}{\varepsilon}\rho \tag{3.35}$$

where ρ is the charge density $\left(\sum n'_+ z_+ e - \sum n'_- z_- e\right)$, ε the permittivity of the solution (taken as independent of ψ) and the differential operator ∇^2 is given in Cartesian co-ordinates by

$$\left[\frac{\partial^2}{\partial x^2} + \frac{\partial^2}{\partial y^2} + \frac{\partial^2}{\partial z^2}\right]$$

then the Poisson–Boltzmann equation results, namely

$$\nabla^2\psi = -\frac{4\pi}{\varepsilon}\left[\sum z_+ e n_+ \exp\left(-\frac{z_+ e\psi}{kT}\right) - \sum z_- e n_- \exp\left(+\frac{z_- e\psi}{kT}\right)\right]. \tag{3.36}$$

For a plane interface ∇^2 becomes $\partial^2/\partial x^2$ and for a single symmetrical $z:z$ electrolyte in solution, eqn. (3.36) simplifies to

$$\frac{d^2\psi}{dx^2} = +\frac{8\pi}{\varepsilon} zen \sinh\left(\frac{ze\psi}{kT}\right) \tag{3.37}$$

which is readily solved (Gouy 1910), with the boundary conditions $\psi = \psi_d$ at $x = 0$ and $\psi = d\psi/dx = 0$ at $x = \infty$ to give

$$\psi = \frac{2kT}{ze} \ln\left[\frac{1 + \gamma \exp(-\kappa x)}{1 - \gamma \exp(-\kappa x)}\right] \tag{3.38}$$

and

$$\sigma_d = -\left[\frac{2n\varepsilon kT}{\pi}\right]^{\frac{1}{2}} \sinh\left(\frac{ze\psi_d}{2kT}\right) \tag{3.39}$$

where

$$\gamma = \frac{\exp\left[\frac{ze\psi_d}{2kT}\right] - 1}{\exp\left[\frac{ze\psi_d}{2kT}\right] + 1} \tag{3.40}$$

The assumption of a symmetrical electrolyte in the above solution is less restricting than might at first be supposed since it is well known in colloid phenomena that the co-ion (*i.e.* ion of the same charge as the surface) is of little importance. This is indeed shown by splitting σ_a into σ_{d+} the contribution due to an excess of cations and σ_{d-} the contribution due to the excess of anions, excess being understood algebraically.

Then

$$\sigma_{d+} = ze \int_0^\infty (n'_+ - n)\, dx$$

$$= \left[\frac{\varepsilon nkT}{2\pi}\right]^{\frac{1}{2}} \left[\exp\left(-\frac{ze\psi_d}{2kT}\right) - 1\right] \tag{3.41}$$

and

$$\sigma_{d-} = -ze \int_0^\infty (n'_- - n)\, dx$$

$$= \left[\frac{\varepsilon nkT}{2\pi}\right]^{\frac{1}{2}} \left[1 - \exp\left(\frac{ze\psi_d}{2kT}\right)\right] \tag{3.42}$$

It can be seen that if ψ_d is *e.g.* large and positive, then the major part of σ_d is σ_{d-} and the nature of the positively charged co-ion is relatively unimportant.

At a spherical interface eqn. (3.36) cannot be integrated analytically unless some approximation to the exponential term is made. If the linear approximation is justified, *i.e.* $ze\psi/kT < 1$ which implies $z\psi < 25$ mV, the treatment becomes that of Debye and Hückel (1923). The equation corresponding to eqn. (3.37) is then

$$\frac{1}{r^2}\frac{d}{dr}\left(r^2\frac{d\psi}{dr}\right) = \kappa^2\psi$$

which, with the boundary conditions $\psi = \psi_d$ at $r = a$ and $\psi = d\psi/dr = 0$ as $r = \infty$, leads to

$$\psi = \psi_d \frac{a}{r} \exp\left[-\kappa(r-a)\right] \tag{3.43}$$

This equation gives the potential ψ at a distance r from the centre of the sphere, where a is the radius at the inner boundary of the diffuse layer.

If the Debye–Hückel linear approximation is applied to eqn. (3.37) the solution, valid only at low potentials, is

$$\psi = \psi_d \exp\left(-\kappa x\right) \tag{3.44}$$

a result which can be obtained from eqn. (3.38) with ψ_d small. The same approximation applied to eqns. (3.41) and (3.42) leads to the result that the net charge in the diffuse layer is due equally to an excess of counter ions and deficit of co-ions. It is perhaps here that the deficiencies of the Debye–Hückel approximation in the treatment of colloid problems are most apparent.

More exact solution of the Poisson–Boltzmann equation for the spherical case have been obtained by Müller (1928) graphically: by Gronwall, La Mer and Sandved (1928) analytically, retaining terms up to the fifth order in the exponential expansion for symmetrical electrolytes; and more recently by Loeb, Overbeek and Wiersema[18] who have published a very full compilation using numerical methods and an electronic computer.

The validity of all these treatments, however, including that of the plane diffuse layer, depends upon the applicability of the Poisson–Boltzmann equation in the form of eqn. (3.36). It is beyond the scope of this discussion to review the very considerable literature in which corrections to this equation have been treated but the most recent review at the time of writing is that of Levine and Bell[19] who conclude that, as a result of some cancellation of effects, the use of the uncorrected Poisson–Boltzmann equation probably gives rise to little error, in the case of 1:1 electrolytes, even at concentrations of 0·1M for potentials up to 75 mV.

Capacity of the diffuse region of the double layer

From the way in which σ_d has been defined it is evident that as ψ_d becomes more positive σ_d becomes more negative, so that the capacity per unit area of the diffuse region of the double layer is given by $-\sigma_d/\psi_d$ and the differential capacity by $-d\sigma_d/d\psi_d$.

Differentiating eqn. (3.39) for a plane double layer

$$C_d = -\frac{d\sigma_d}{d\psi_d} = \frac{\kappa\varepsilon}{4\pi}\cosh\left(\frac{ze\psi_d}{2kT}\right) \tag{3.45}$$

At low potentials, where cosh $(ze\psi_d/2kT)$ tends to unity

$$C_d = \frac{\kappa\varepsilon}{4\pi} \tag{3.46}$$

and since this expression is independent of ψ it is also, under these conditions, the integral capacity $-\sigma_d/\psi_d$. For a 1 mM aqueous solution of a 1:1 electrolyte at 25°C this becomes 7·22 $\mu F/cm^2$.

The capacity of the diffuse part of the double layer is not accessible to direct measurement and it is indeed only in certain cases that the capacity of the whole double layer σ_0/ψ_0 which consists of an inner capacity in series with the diffuse capacity, can be measured. At the mercury–solution interface, as a result of the mercury being conducting and polarisable, the capacity may be measured by a direct AC bridge method and such measurements, after allowing for the capacity of the diffuse region calculated as above, have given valuable information[15] on the inner region which may be applied, with suitable reservations, to the inner regions at solid particle–solution interfaces.

Capacity measurements at the silver halide–solution and oxide–solution interfaces have also been measured by AC methods but most estimations have been made by potentiometric titration[20,21] in which, using silver halides as example, $\Delta\sigma_0$ and $\Delta\varphi$ can be obtained from the change in emf of a silver halide–calomel electrode pair when a known quantity of potassium halide or silver nitrate is introduced into the silver halide dispersion of known surface area. If the χ potential is assumed constant $\Delta\varphi$ becomes $\Delta\psi_0$ and, knowing the ionic concentrations corresponding to the PZC, σ_0 can be plotted against ψ_0 to give integral (σ_0/ψ_0) and differenerential $(d\sigma_0/d\psi_0)$ capacities of the double layer.

Some results of Lyklema[22] are shown in Fig. 3.6 in which the differential capacity $d\sigma_0/d\psi_0$ is shown as a function of ψ_0 for aqueous silver iodide sols in the presence of various nitrates at an ionic strength of 1 mM. The form of these curves is similar to results obtained at the mercury–solution interface though there is greater specificity of cation effect on silver iodide. This specificity provides an explanation for the widely observed sequence of flocculation concentrations for the alkali metal cations on negatively charged sols, *viz.* $Li^+ > Na^+ > K^+$. The Li^+ ion, being more strongly hydrated, is less specifically adsorbed than the succeeding members of the series so that $|\psi_d|$ remains higher and with it the flocculation concentration. Some later experiments of Lyklema[21] indicate that at temperatures in excess of 50°C the specificity largely diappears both from the capacity curves and flocculation concentrations.

The limited value of the capacity at elevated negative potentials as shown in Fig. 3.6 is largely due to the effect of the capacity of the inner part of the double layer which is effectively in series with that of the diffuse part. If the Gouy–Chapman expressions are used to give the capacity of the diffuse region, then the measured total capacities indicate a value for the integral capacity of the inner part of the double layer varying from $\sim 30\ \mu F/cm^2$ near the zero point to $\sim 10\ \mu F/cm^2$ at elevated negative charges. These figures are similar to those found for the mercury–aqueous

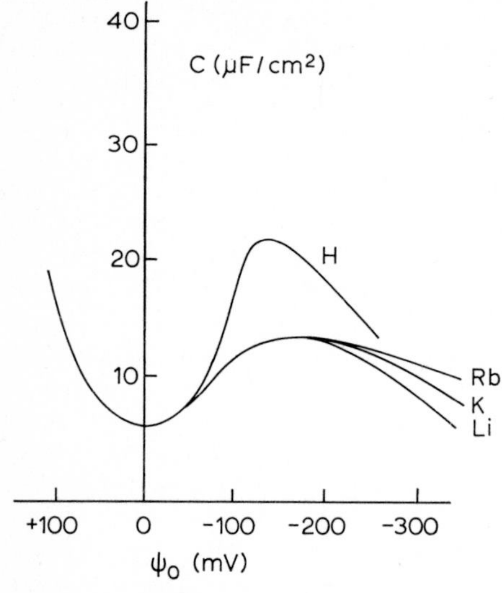

Fig. 3.6. Total differential capacities at the AgI–solution interface in aqueous solutions of various nitrates at $10^{-3}M$. *(after Lyklema*[22]*).*

solution interface and for any reasonable thickness of the inner layer indicate a substantial lowering of the effective dielectric constant of water near the interface. At both interfaces the capacities rise at positive polarisations, an effect which is probably partly due to the weaker hydration and therefore greater adsorbability of anions compared to cations, and partly due to water dipole re-orientation. There has been considerable, and largely unresolved discussion of the origin of a maximum in the differential capacity of mercury at positive polarisations which persists through many electrolytes and many non-aqueous solvents (of sufficient dielectric constant to dissociate electrolytes).

At the oxide–aqueous solution interface a rather different capacity pattern emerges. There is virtual symmetry about the zero point and the capacities seem markedly higher than for Hg or AgI. It is difficult, however, to derive capacities at the zero point from experimental data because of the extra factor θ_c (eqn. (3.14)) which occurs. It seems reasonable to attribute the higher capacities to a higher effective dielectric constant of water near oxide interfaces, which will be in a different state to that at the hydrophobic Hg or AgI. The symmetry about the zero point could well arise from considerably less dehydration of adsorbed ions which would certainly diminish differences between anions and cations.

Double layer on the solid side of the interface

In the case of a conductor such as mercury or a solid metal there can be no potential gradient (*i.e.* electric double layer) within the conductor except possibly very near the interface. In such a case the whole of any electric double layer is on the solution side of the interface. In the case of an ionic solid, however, in which the presence of lattice defects allows the migration of charge it will be quite possible to have a substantial fraction of the total potential drop from bulk solution to bulk solid within the solid. A general treatment of the distribution of an electric double layer between two phases has been given by Verwey and Niessen.[23]

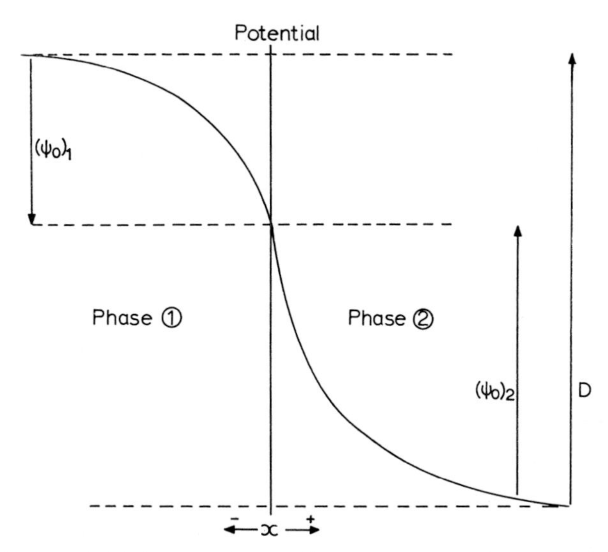

Fig. 3.7. Electric double layer on the two sides of a phase boundary, potential and inner parts of the double layers omitted.

The situation in Fig. 3.7 has been simplified by ignoring the inner parts of the electric double layers near the interfaces, *i.e.* the Gouy diffuse layers have been assumed to continue right up to the interface. The χ potential has been omitted and $(\psi_0)_1 + (\psi_0)_2$ written as D. The condition of overall electrical neutrality gives

$$\int_{-\infty}^{0} \rho_1 dx + \int_{0}^{\infty} \rho_2 dx = 0 \tag{3.47}$$

The Poisson equation may be applied on each side of the interface to give

$$\rho = -\frac{\varepsilon}{4\pi}\frac{d^2\psi}{dx^2}$$

and substitution in eqn. (3.47) yields

$$\varepsilon_1 \int_{-\infty}^{0} \left(\frac{d^2\psi}{dx^2}\right)_1 dx + \varepsilon_2 \int_{0}^{\infty} \left(\frac{d^2\psi}{dx^2}\right)_2 dx = 0$$

With the boundary conditions $(d\psi/dx)_1 = 0$ as $x = -\infty$ and $(d\psi/dx)_2 = 0$ as $x = \infty$ we therefore have, at $x = 0$

$$\varepsilon_1 \left(\frac{d\psi}{dx}\right)_1 = \varepsilon_2 \left(\frac{d\psi}{dx}\right)_2 \tag{3.48}$$

showing that there is a discontinuity in potential gradient at the interface.

Equation (3.37) gives for $d\psi/dx$ at any point in the diffuse layer

$$\frac{d\psi}{dx} = -\left[\frac{8\pi nkT}{\varepsilon}\right]^{\frac{1}{2}} 2 \sinh\left(\frac{ze\psi}{2kT}\right)$$

and substitution in eqn. (3.48) gives

$$(n_1\varepsilon_1)^{\frac{1}{2}} \sinh\left[\frac{ze(\psi_0)_1}{2kT}\right] = (n_2\varepsilon_2)^{\frac{1}{2}} \sinh\left[\frac{ze(\psi_0)_2}{2kT}\right] \tag{3.49}$$

Thus the division of the total potential drop D into its components $(\psi_0)_1$ and $(\psi_0)_2$ is governed by the ratio

$$\alpha = \left(\frac{n_1\varepsilon_1}{n_2\varepsilon_2}\right)^{\frac{1}{2}} = \frac{\kappa_1\varepsilon_1}{\kappa_2\varepsilon_2}. \tag{3.50}$$

In most cases the medium with the lower dielectric constant will also have the lower concentration of ions so that the greater part of the potential drop D is in the medium of lower dielectric constant. It has been suggested that this explains the instability of oil-in-water emulsions in the absence of stabilising agents.

Ottewill and Woodbridge[24] and Honig[25] have applied these ideas to the double layer at the silver halide–solution interface using figures of order 10^{19} for the number of interstitial silver ions per cm^3 and in both cases finding a significant part of the double layer to be within the solid. Honig questions the usefulness of a Stern layer in the description of double layers but Levine, Levine and Smith[4] have used Honig's figures and concepts to predict capacities at the AgI–solution interface, which are a more sensitive test of model than are charge densities, and find that agreement cannot be obtained with experimental data. It is nevertheless possible that potential drop within the solid considered with a Stern layer of suitable capacity, will give such agreement. However, the later authors point out that it is possible that the double layer within the solid is 'hidden' from the solution side (*see* page 94) except for its effect on the position of the zero point of charge.

Experimental values for $d\zeta/dp\mathrm{Ag}$ measured at $\zeta \to 0$ for AgI particles have been held[24] to indicate the presence of a double layer within the solid in so far as the numerical value for this quantity falls short of the Nernst value $2{\cdot}303kT/e_0$, *i.e.* 59 mV at 25°C. The interpretation then rests on

$$\left[\frac{d\zeta}{d(p\mathrm{Ag})}\right]_{\zeta\to 0} = \left[\frac{d(\psi_0)_2}{d(p\mathrm{Ag})}\right]_{(\psi_0)_2\to 0} = -59\left[1+\frac{1}{\alpha}\right]^{-1} \mathrm{mV} \quad (3.51)$$

However, the ζ potential is more correctly related to ψ_d rather than ψ_0 and in this case, assuming no specific adsorption, ζ will fall short of ψ_0 even with no double layer in the solid, to the extent

$$\left[\frac{d\zeta}{d(p\mathrm{Ag})}\right]_{\zeta\to 0} = -59\left[\frac{K^0}{K^0 + C_d^{\,0}}\right] \mathrm{mV} \quad (3.52)$$

where K^0 is the inner capacity at the zero point and $C_d^{\,0} = (\kappa\varepsilon/4\pi)$ is the diffuse layer capacity at the same point. In 0·1 M of 1:1 electrolyte $C_d^{\,0} \simeq 72\ \mu\mathrm{F/cm}^2$ and if K^0 is taken as $30\ \mu\mathrm{F/cm}^2$ then $(d\zeta/dp\mathrm{Ag})_{\zeta\to 0}$ becomes ~18 mV, which is precisely the sort of magnitude observed. Of course the value for K^0 may include a contribution from a capacity within the solid but a value for $[d\zeta/dp\mathrm{Ag}]_{\zeta\to 0}$ lower than 59 mV is not in itself evidence for double layer within the solid. Comparison with mercury where there can be no capacity in the metal suggests that any capacity in solid AgI would have to be at least $\sim 70\ \mu\mathrm{F/cm}^2$ which, as a diffuse capacity, would involve a fraction of defects in the solid as high as 0·4%, which seems much too large.

To summarise, there seems to be no experimental evidence for dispersions of solids in liquids in which the surface electrical properties are influenced by any double layer within the solid. Since some such solids definitely contain finite concentrations of mobile charge carriers (*e.g.* silver halides) it becomes attractive to suppose that the surface screening mechanism discussed above in fact applies.

Free energy of the diffuse double layer

The electric double layer forms spontaneously around a charged particle or interface in solution so that the free energy of formation of the double layer must be negative. When two such particle or interfaces approach each other so that the respective double layers overlap it is evident that a part of the previously existing double layers has been destroyed. This will give a positive contribution to the free energy which is manifested as a repulsion between the particles.

It is perhaps not immediately apparent how the free energy of the double layer comes to be negative. In this connection it is useful to consider an analogous case—the charging of a simple parallel plate condenser. In Fig. 3.8 when the switch S is closed the condenser C will charge from the

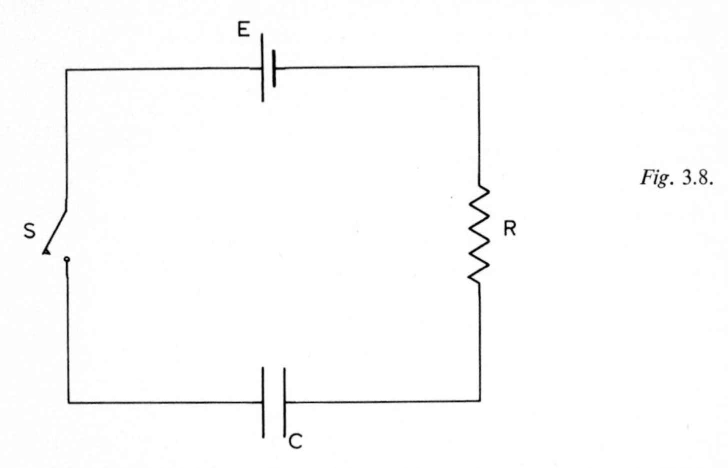

Fig. 3.8.

cell of emf E through the resistance R. If an increment of charge dq flows into the condenser when the potential difference between its plates is V then the work done is Vdq and for the whole of the charging process the work will be

$$\int_{q=O}^{q=Q} Vdq. \tag{3.53}$$

Writing $V = q/C$ and taking C to be constant this evaluates to $\frac{1}{2}Q^2/C$, *i.e.* $\frac{1}{2}QE$. This is the familiar expression for the potential energy of a charged condenser and it will be noted that it is positive. To obtain the total free energy change in the charging process, however, it is necessary to consider the state of the cell which supplied the charge to the condenser. Since the emf remains constant at E while a total charge Q is extracted, the decrease in free energy of the cell (as a result of the chemical reaction proceeding with it) is QE. Thus the total increase of free energy on closing S is

$$\tfrac{1}{2}QE - QE = -\tfrac{1}{2}QE. \tag{3.54}$$

The contribution $+\frac{1}{2}QE$ may be looked upon as the electrical part of the free energy change and $-QE$ as the chemical part.

In the case of the electric double layer, where it must be noted that the capacitance is not constant but depends upon the wall potential and electrolyte concentration, the free energy of formation per unit area of (plane) double layer will be, by analogy with eqns. (3.53) and (3.54)

$$\int_0^{\sigma} \psi' d\sigma' - \psi\sigma = -\int_0^{\psi} \sigma' d\psi' \tag{3.55}$$

where the primes indicate values of σ and ψ during the charging process and eqn. (3.55) gives the free energy (of formation) of a double layer, or part of a double layer, extending from a potential ψ out into bulk solution ($\psi = 0$). It can be seen from eqn. (3.55) that the free energy will always be negative as the argument at the beginning of this section indicated.

For a diffuse double layer of inner Gouy potential ψ_d, $\sigma = -\sigma_d$ and substitution of eqn. (3.39) into eqn. (3.55) gives for the free energy

$$-\left[\frac{2n\varepsilon kT}{\pi}\right]^{\frac{1}{2}} \int_0^{\psi_d} \sinh\left(\frac{ze\psi'}{2kT}\right) d\psi'$$

$$= -\frac{8nkT}{\kappa}\left[\cosh\left(\frac{ze\psi_d}{2kT}\right) - 1\right] \tag{3.56}$$

When particles approach to such distances that the double layers overlap, some electric double layer is destroyed, so that there will be a positive contribution to the total free energy which is manifested as a repulsion between the particles. This question is considered more fully in Chapter 1.

ELECTROKINETIC PROPERTIES OF DISPERSIONS

Classification

Electrokinetic properties include all phenomena in which there is relative tangential motion between charged phases. This may be the result of an applied electric field or, conversely, may be mechanically achieved and produce an electric field. A convenient and usual subdivision of these phenomena is into

(i) electrophoresis,
(ii) electro-osmosis,
(iii) streaming potential and
(iv) sedimentation potential (Dorn effect).

In electrophoresis experiments an electric field is applied to a dispersion and the velocity of the particles measured, either directly as in micro-electrophoresis or as a moving boundary between the sol and pure dispersion medium. Electro-osmosis differs only in that the solid phase is kept stationary as a porous plug or single capillary while the velocity of the liquid is measured. As an alternative to a direct measurement of the liquid velocity the volume of liquid transported may be measured or the reverse pressure necessary to prevent electro-osmotic flow.

(*iii*) and (*iv*) represent the reverse situation where the motion is achieved mechanically and the resulting potential difference is measured between suitable electrodes. In the case of streaming potentials the solution is streamed through a porous plug of the solid phase. If the solid phase is particulate it is convenient to contain the plug by means of the electrodes themselves which may be of blacked platinum or of the reversible type. An electrometer should be used to record the resulting potentials to minimise the current passed between the electrodes. In the case of sedimentation potentials particles are allowed to move through the dispersion medium in a gravitational or centrifugal field with suitably placed electrodes and an electrometer to record the resulting potential difference. The sedimentation potential technique presents the greatest experimental difficulties of the four alternatives listed above and would not normally be used to obtain information on dispersions. It will not therefore be discussed further.

The zeta potential

The potential yielded by any of the electrokinetic effects is referred to non-commitally as the zeta (ζ) potential and will be expected to be the potential at the surface of shear between the phases in relative motion. The concept of such a surface of shear is possibly complicated by any variation of viscosity with distance from the surface. If electrokinetic experiments are interpreted in terms of charge then this will be the total charge within the effective surface of shear. In terms of Fig. 3.2 the ζ potential will not be expected to refer to any surface nearer to the particle than ψ_d and could well in some cases be the potential further out into the diffuse layer if the particle immobilises a considerable amount of solvent.

When it is further borne in mind that electrophoretic mobilities are not easy to measure with accuracy and that the conversion of the mobilities into zeta potentials is still open to some doubt, especially at high mobilities, it is perhaps understandable that zeta potentials have been described as 'difficult to measure and impossible to interpret'. This is, however, an over pessimistic assessment of the situation and considerable information can be obtained from electrokinetic measurements which is relevant to the problem of colloid stability.

If silver halides are typical of lyophobic surfaces then certainly for such surfaces ζ is equal to ψ_d within normal experimental accuracy when both are small, and there is evidence that the two potentials do not differ by more than about 2 or 3 mV up to 85 mV in 0·01 M 1:1 electrolyte and up to 30 mV in 0·1 M 1:1 electrolyte. This range covers that most often of experimental interest.

In order to investigate the relationship between ζ and ψ_d suppose that ζ refers to a plane distant Δ from the Stern plane out into the diffuse layer (Fig. 3.2). Then in the limiting case of low potentials

$$\zeta = \psi_d \exp(-\kappa\Delta)$$

Taking silver halides as example, since here there are well established potential determining ions, then at a given κ, *i.e.* at a given electrolyte concentration

$$\left[\frac{d\zeta}{d(pAg)}\right]_{\zeta\to 0} = S = \left[\frac{d\zeta}{d\psi_d}\frac{d\psi_d}{d\psi_0}\frac{d\psi_0}{d(pAg)}\right]_{\zeta\to 0} \tag{3.57}$$

On the right-hand side of eqn. (3.57) the factor

$$\left[\frac{d\psi_0}{d(pAg)}\right]$$

is the Nernst factor which can be treated as experimentally determinable and termed N. The factor $(d\psi_d/d\psi_0)$ can be simplified, in the absence of specific adsorption to $K^0/(K^0 + C_d{}^0)$ where K^0 and $C_d{}^0$ are the inner layer and diffuse layer capacities respectively at the zero point (eqn. 3.52), so that

$$S = N\left[\frac{K^0}{K^0 + C_d{}^0}\right]\exp(-\kappa\Delta) \tag{3.58}$$

or

$$S^{-1} = N^{-1}\left[1 + \frac{C_d{}^0}{K^0}\right]\exp(+\kappa\Delta) \tag{3.59}$$

Figure 3.9 shows an experimental plot of S^{-1} versus $C_d{}^0$ (*i.e.* $\kappa\varepsilon/4\pi$) for AgI at 25°C in KNO^3 (aq).

The intercept at $C_d{}^0 = 0$ (*i.e.* $K = 0$) is seen to be 1/59 mV^{-1} as expected and the plot is linear within experimental error. This indicates that Δ is small and the slope gives $K^0 = 30$ $\mu F/cm^2$. Since this latter figure is that which results from the now extensive potentiometric titration and directly measured capacity data, it can be taken as reinforcing the conclusion from the linearity of Fig. 3.9 that Δ is small. Any specific adsorption, which could be allowed for formally in eqn. (3.59), will clearly have the effect of reducing S, and therefore increasing S^{-1}, at large $C_d{}^0$. Since this is in the same sense as $\Delta > 0$ the conclusion of small

Δ is again reinforced. At 100 mM KNO_3, with $C_d{}^0 = 72{\cdot}2\ \mu F/cm^2$, a value of Δ as small as 1 Å would increase S^{-1} by 10%, which seems the limit of experimental error.

This experimental indication that $\zeta = \psi_d$ even at high electrolyte concentration (0·1 M), is restricted to the limiting case of low potentials. However, if Δ is assumed to remain zero as the potentials are increased, a good fit is obtained between experimental ζ values and those calculated by

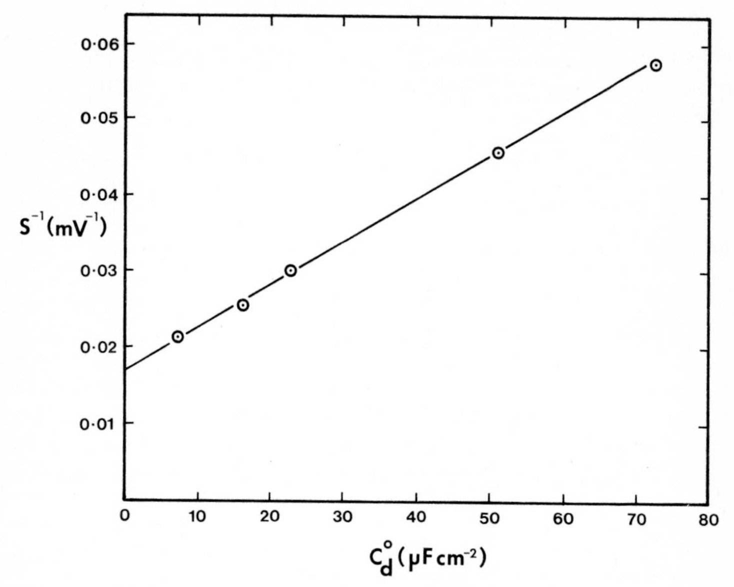

Fig. 3.9. Experimental plot of S^{-1} against $C_d{}^0$ (= $\kappa\varepsilon/4\pi$) for AgI in KNO_3 (aq) at 25°C. At $C_d{}^0 = 0$, $S = 59$ mV; slope yields $K° = 30\ \mu F/cm^2$.

assuming that specific adsorption follows a Stern isotherm (eqn. (3.30)), up to potentials around 85 mV. To achieve such potentials it is, however, necessary to reduce the 1:1 electrolyte concentration to 0·01 M.

The interpretation of S^{-1} plots for oxide and similar surfaces is complicated by the factor θ_c (eqn. 3.14).

Experimental determination of electrophoretic mobilities

Since electrophoresis is the electrokinetic technique of widest application to dispersion stability problems, experimental details will be discussed at some length. The first choice to be made is between the moving-boundary method and microelectrophoresis in which the motion of individual particles is followed with ultra-microscopic illumination.

The microelectrophoretic technique may be said to have the following advantages.

(*i*) The particles are observed while in their normal sol environment.
(*ii*) Very dilute sols can be studied so that flocculation rates, even at high electrolyte concentration or near the ZPC are negligible.
(*iii*) The high magnification of the ultra-microscope system leads to very short observation times and high sensitivity.
(*iv*) In a polydisperse sol the particles in a chosen (though necessarily wide) size range can be observed while others are ignored.

These advantages are so extensive that the moving-boundary method will probably only be used in cases where it is desired to study colloidal dispersions in a concentrated state or the particles are too small or too similar in refractive index to that of the dispersion medium to be visible. In this connection it may be noted that polystyrene latex particles of radius 400 Å dispersed in water are readily visible in suitable illumination. Some workers[26] have preferred the moving-boundary method where the effect of surfactants on a dispersion is to be studied, on the grounds that non-uniform adsorption of the surfactant on the walls of the cell makes the interpretation of microelectrophoretic experiments difficult or impossible. While this is true it must be noted that any such non-uniform adsorption is readily detected and does not seem to give trouble in practice if the cells are kept scrupulously clean. Acidified sodium fluoride solution (*i.e.* dilute HF aq.) has been recommended[27] for cleaning electrophoresis cells where there is a danger of interference from silicate ions from the glass.

Stationary levels in microelectrophoresis cells

Before discussing the factors involved in the choice of type of microelectrophoresis cell it is necessary to realise that when an electric field is applied between the ends of a microelectrophoresis cell, of whatever cross section, the solution contained in the cell will, in general, experience electro-osmotic flow on account of surface charge on the cell walls. Only where this electro-osmotic flow velocity is zero, the so called *stationary levels*, can the electrophoretic mobility of a particle be observed directly. Occasionally the electro-osmotic flow is taken into account in the simultaneous determination of the zeta potentials of the particles and cell walls.

Figure 3.10 shows the situation in a microelectrophoresis cell which is assumed, in the first instance, to be of circular cross-section. The electro-osmotic effect alone will give rise to a solution velocity v_{eo} uniform across the cell cross-section, towards the electrode of the same polarity as the charge on the cell wall. If the cell is closed it is evident that a reverse flow will be set up so that there is no net transport of liquid. This will be true even if the cell is not closed once the necessary hydrostatic reverse pressure is built up. Since the reverse flow will follow Poiseuille's law, the condition of no overall liquid transport gives

$$\int_{r=0}^{r=a} 2\pi v r dr = 0 \tag{3.60}$$

with

$$v = v_{eo} - C(a^2 - r^2) \tag{3.61}$$

where v is the liquid velocity at a distance r from the centre of the tube of radius a, and C is a constant.

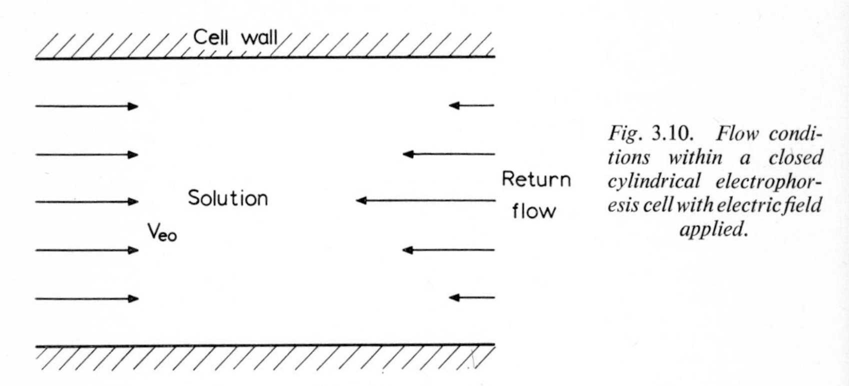

Fig. 3.10. *Flow conditions within a closed cylindrical electrophoresis cell with electric field applied.*

Equations (3.60) and (3.61) solve to give $C = 2v_{eo}/a^2$ so that

$$v = v_{eo}\left[2\frac{r^2}{a^2} - 1\right]. \tag{3.62}$$

At the cell wall $r = a$ and $v = v_{eo}$ as expected; at the centre $r = 0$ and $v = -v_{eo}$ so that the velocity of liquid flow is equal in magnitude but opposite in direction to that at the wall. The condition for zero liquid velocity is given by eqn. (3.62) as $r = a(2/2)^{\frac{1}{2}}$. Thus the stationary level, which in this case will be cylindrical, is 0·707 of the radius from the centre, or as more usually expressed, 0·146 of the (internal) diameter from the wall. Looking into a cylindrical cell along a diameter there will, therefore, be two such levels at which particles show their electrophoretic velocity.

For cells with rectangular cross-section the situation is more complex.[28,29] Using the nomenclature illustrated by Fig. 3.11

$$\frac{v(x = 0)}{v_{eo}} = 1 - \frac{3}{2}\left[1 - \frac{y^2}{b^2}\right]\left[1 - \frac{192b}{\pi^5 a}\right]^{-1}. \tag{3.63}$$

If the ratio a/b is taken as infinite then $v(x = 0)$ is zero when $y/b = 0·5774$

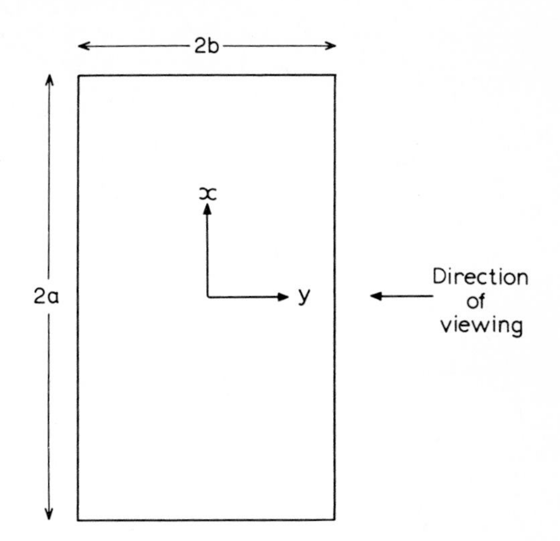

Fig. 3.11.

so that the stationary levels are 0·211 of the cell thickness (2*b*) from both the front and back walls. The position of the stationary levels for other values of *a*/*b* are shown in Table 3.1.

TABLE 3.1

POSITION OF STATIONARY LEVELS IN ELECTROPHORESIS CELL OF RECTANGULAR CROSS SECTION

Nomenclature shown in Fig. 3.11, $x = 0$ in all cases

a/b	$(b-y)/2b$
∞	0·211
50	0·208
20	0·202
10	0·194

Choice of microelectrophoresis cell

Single channel microelectrophoresis cells may be of rectangular or circular cross-section and the latter may be thick or thin-walled. Each type has its advantages. Those of the flat cell may be listed as follows.

(*i*) There is no optical correction to apply (as would be the case with curved walls) in the sense that fractional depths in the cell are not affected by the refractive index of the solution within the cell or of the cell wall or of the surrounding liquid (if any).

(*ii*) When the observing microscope is focussed on a particular level within the cell the whole of the field of view is at this level and velocity measurements are valid even at the extremities of the field of view. This can be quite important when observing very dilute sols.

(*iii*) Provided that the cell is used within the dimension *a* (Fig. 3.11) vertical, any particles falling under the influence of gravity will remain in the stationary level while in view, probably for long enough to measure their velocity. The particles will, furthermore, fall to a part of the cell where their presence is relatively unimportant both with regard to interference with illumination and to distortion of the electro-osmotic flow.

(*iv*) The cell is readily constructed and silica can be used as easily as glass.

However, the large cross-sectional area of the flat cell compared with cylindrical cells will make the use of reversible electrodes necessary in most aqueous solutions if polarisation is to be avoided and these are much less convenient in use than blacked platinum electrodes. Furthermore, when used with the dimension *a* vertical in order to take advantage of (*iii*) above, convection currents can be troublesome, especially with intense illumination. The advantages of the cylindrical cell are as follows.

(*i*) The small cross-sectional area leads to small electric currents for a given field strength so that large area blacked platinum electrodes can be used with negligible polarisation errors and the more troublesome reversible electrodes are avoided.

(*ii*) The volume of sol required to fill the cell can be made very small.

(*iii*) The cell can be illuminated from a direction perpendicular to that of observation in the traditional ultra-microscope configuration which is much more satisfactory when very small particles have to be observed than the dark ground condensers used with flat cells.

(*iv*) Thermostatting can be made very effective, especially with thin walled cells, leading to the virtual elimination of convection currents and a great reduction in the self-heating effect resulting from the current passed.

The situation may be summarised by saying that for electrophoresis in aqueous or other highly conducting solutions the cylindrical cell is to be preferred, but for most non-aqueous (low conductance) electrophoresis the flat cell, constructed of silica, is usually better.

Cylindrical cells may be thick-walled as with the Mattson[30] cell or thin-walled as the van Gils[29] cell. Thick-walled capillary tubing of uniform bore is readily purchased and forms a mechanically strong cell, but it is necessary to grind a flat on the cell for observation (and usually another for illumination) and even with this flat and the tube surrounded by thermostatting fluid, the optical correction which must be considered for apparent positions within the cell is large, and distortion makes it difficult to observe particles beyond the centre of the cell. The latter point is of some importance since reliable mobility measurements should include observations at both stationary levels.

Considerable skill is required both to draw thin-walled capillary tube of accurately circular cross-section and to join such tubing to more substantial end tubing. The resulting cell is also rather fragile but apart from this it has every advantage over the thick-walled cell particularly with regard to efficient thermostatting and negligible optical corrections at the very small thicknesses which become possible (*e.g.* 20 μ).

An interesting cell is that due to Smith and Lissie[31] in which the electrode compartments are connected by two separate capillaries. The advantage of this cell is that, provided the capillaries are of identical material and that their lengths (l) and radii (a) satisfy the relationship

$$\frac{l_2}{l_1} = \left(\frac{a_2}{a_1}\right)^2 \left[\left(\frac{a_2}{a_1}\right)^2 - 2\right] \tag{3.64}$$

the stationary level is at the centre of the tube of smaller cross section where the electro-osmotic velocity gradient is zero.

That this cell has been so little used is probably due to the difficulty of satisfying the necessary condition given above. The only satisfactory test would presumably be to check it against a more conventional cell.

Experimental detail

All types of microelectrophoresis cell are best immersed in a thermostat bath since the temperature coefficient of particle mobility, as for ion conductance, is around 2% per °C. For best observation conditions the microscope objective may also be immersed and possibly also the illuminating lens. It is convenient to build the apparatus around a standard microscope assembly. Such a microscope should have the coarse and fine focusing controls operating on the microscope tube so that the rather heavy thermostatting tank can be safely rested on the stage and should have sufficient space above and around the stage to accommodate such a tank. The fine focusing control must be accurately calibrated at 1 μ or 2 μ intervals and binocular optics are an advantage.

Such a microscope is the Vickers Patholux and Fig. 3.12 shows some details of an assembly using this microscope with side illumination of a van Gils type cell. The side illumination tube which carries an adjustable slit and immersed $\times 10$ objective can be moved in all senses so that the cell can be fixed rigidly within the thermostat tank. The 100 watt quartz-iodine illumination unit used in the microscope quoted is removable and makes a suitable light source for most work. For particles of very low visibility a laser can be an advantage—as an example polystyrene latex particles of 400 Å radius are readily detected in the focused beam of a 5 milliwatt helium-neon laser. There must, however, be some compromise between focusing the beam for maximum intensity of illumination and having a sufficient area illuminated to time particles over a reasonable distance. The observing microscope should have an overall magnification of around 200 and a long working distance objective is needed (suitable objectives are manufactured by Vickers Instruments and by Nikon) both to ensure that the bottom of the cell is visible and to avoid working with the objective very close to the fragile cell. When calibrating the eyepiece graticule (which is conveniently engraved in 2 mm squares) it must be remembered the objective is immersed when in use.

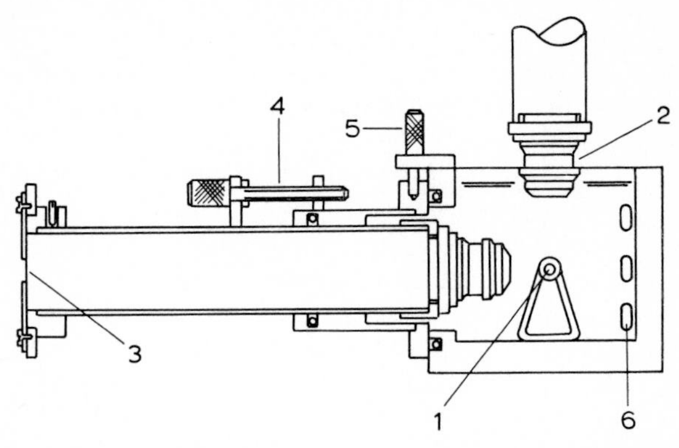

Fig. 3.12. *Thermostat tank and illumination arrangements for a van Gils electrophoresis cell mounted on the stage of a microscope.* (1) *Cell,* (2) *microscope objective,* (3) *adjustable slit width,* (4) *light focus adjustment,* (5) *vertical adjustment,* (6) *temperature control coils.* (*By courtesy of Mr. J. C. Barnett, College of Technology, Liverpool.*)

Figure 3.13 shows a cell of rectangular cross-section which is particularly convenient for non-aqueous work. It consists of a 1 mm path length silica spectrophotometer cell on to which side arms and electrode compartments have been fused. Taps on the side arms reduce the bulk flow of liquid arising from bench vibrations and conveniently have PTFE barrels. Figure 3.14 shows a commercially available instrument with which both flat and cylindrical cells can be used.

When using the flat cell for non-aqueous electrophoresis or the cylindrical cell for aqueous work it is not usually necessary to use reversible electrodes, a few square centimetres of well blacked platinum foil being sufficient to avoid polarisation effects. The advantage of platinum electrodes is that the effective length of the cell l remains constant so that the field strength at the point of observation is given by V/l, where V is the applied emf, and it is not necessary to measure the conductance of each suspension investigated or to include an ammeter in the circuit (except to

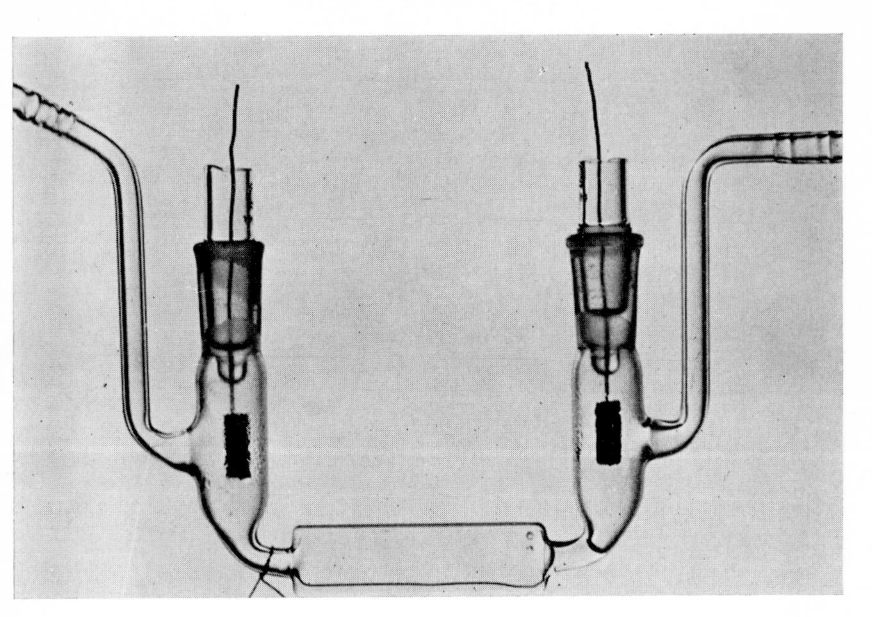

Fig. 3.13. *Flat electrophoresis cell constructed from a* 1 *mm spectrophotometer cell.*

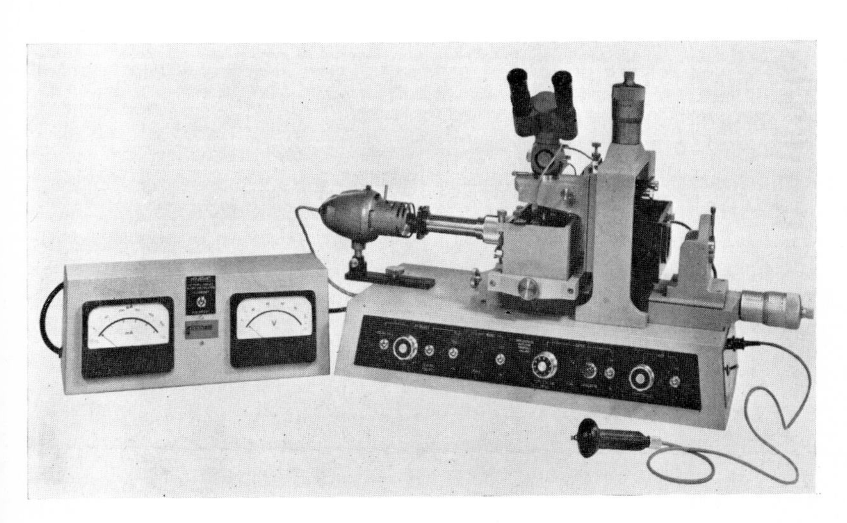

Fig. 3.14. *Instrument manufactured by Rank Bros., Bottisham, Cambridge, with which both cylindrical and flat cells can be used.*

verify that polarisation is absent). The effective length l is given by $R\kappa A$ where R is the resistance, conveniently measured by a conductance bridge, between the electrodes of the cell when it contains a solution of known specific conductance κ, and A is the cross-sectional area of the cell at the point of observation. It is not necessary that the cross-sectional area of the cell should be uniform over its whole length and it is indeed desirable that the electrode compartments should be enlarged, both to accommodate large area electrodes and to ensure that slight irreproducibility in the position of the electrodes has a negligible effect on l.

Polarisation is minimised by the technique, which is in any case desirable, of measuring the transit time for particles across the eyepiece graticule successively in opposite directions.

For a new cell it should be verified that eqn. (3.62) or (3.63) is obeyed. In the case of the cylindrical cell it is most convenient to plot observed particle velocity against $(r/a)^2$ which should result in symmetrical straight lines on each side of the centre ($r/a = 0$). The symmetry of the electro-osmotic flow should be checked each time a sol is investigated, either by taking measurements at both stationary levels or, with greater sensitivity and convenience, by checking that the velocity of particles just within the nearer wall is the same as that of particles just within the further wall.

It is usual to measure about 20 transit times (10 particles in each direction) to obtain a reasonable average.

Choice of electrodes

Blacked platinum electrodes are convenient to use in that they need little preparation or servicing and can be used in the '(V/l) mode' (above). However, if care is not taken to switch off the current when not in use, or in any case if the electrolyte concentration is above (say) 0·01 M, gassing can take place and a polarisation emf is set up. A small amount of gassing is not of itself important and indeed the polarisation emf will never exceed about 2 V which might be acceptably small for a long cell. The polarisation can be significantly delayed by using palladium electrodes which have been pre-charged with hydrogen before use. The pre-charging can conveniently be achieved by making each electrode in turn the cathode, using a platinum anode, while passing 10–50 mA for 10–30 min.

However, both palladium and platinum electrodes have the disadvantage of producing H^+/OH^- ions in most electrolytes which can surprisingly quickly diffuse to the capillary portion of the cell and cause the current to increase appreciably, in addition to any increase caused by self-heating in badly thermostatted or thick-walled cells.

Reversible electrodes of the type Ag/AgCl/KCl aq. 1·0 M/porous plug or $Cu/CuSO_4$ 1·0 M/porous plug do not have this disadvantage, but as usually used require that the electrical conductivity of each solution must be independently measured in order to calculate the field strength at the point of viewing from $I/\kappa A$ where I is the observed current passed. The reason for operation in this mode is that inconsistencies in diffusion through the porous plug can change the effective cell length l so that the field cannot be reliably calculated from V/l.

Perhaps the best compromise for an electrode system at high electrolyte concentration is to allow the current to enter and leave the cell via reversible electrodes but to have also small potential sensing electrodes nearer to the capillary portion which, when connected to a high input impedance voltmeter, allow the cell to be operated in the convenient '(V/l) mode'. The potential sensing electrodes can be very small indeed, of blacked platinum or, better, of Ag/AgCl type.

It is, of course, impossible to avoid self-heating effect in the cell by any choice of electrodes and at high electrolyte concentration this will be a limiting feature. A very thin-walled cell ($<60\ \mu$m) and well circulating thermostatting liquid are of great advantage in this respect. The use of a constant current supply rather than constant voltage is also helpful in that the effect of increased mobility but decreasing field strength, as the mean cell temperature increases, gives an automatic compensation.

Optical corrections

In the case of flat-walled cells there is no optical correction to apply in the sense that apparent fractional depths within the cell as measured by the observing microscope correspond to the true fractional depths. For the Mattson cell with, in effect, a plano-concave cylindrical lens on top of the observation tube it is necessary to apply an appreciable correction to the apparent depths for at least one of the points of focus of an inevitably distorted image as seems first to have been pointed out by Henry.[32] The correction for thin-walled cylindrical cells is much less serious than for Mattson cells and can be made insignificant.

Figure 3.15 shows a cross-section of a cylindrical cell with wall of thickness δ and refractive index n_2 containing liquid of refractive index n_1 and immersed in a thermostatting liquid of refractive index n_3. Assuming a vanishingly small aperture for the observing objective then for a point such as P at a distance u from the top inside face, successive refractions at the inner and outer wall respectively yield the relations

$$\frac{n_1}{u} - \frac{n_2}{v'} = -\frac{(n_2-n_1)}{a} \tag{3.65}$$

$$\frac{n_2}{v'+\delta} - \frac{n_3}{v+\delta} = \frac{(n_2-n_3)}{(a+\delta)} \tag{3.66}$$

where all quantities have their numerical values (*i.e.* are positive). After elimination of v' these relations give v the apparent distance of P from the top inside face as seen by the observing microscope.

A point at the centre of the capillary O gives (obviously) $v = u$ and is unshifted. If for simplicity it is assumed that $n_1 = n_3$ ($= n$) then points in the top inside edge ($u = 0$) are shifted upwards by a distance t given by

$$t = \delta a(n_2 - n)/(n_2 a + n\delta) \tag{3.67}$$

while points in the lower inside edge are shifted upwards by $at/(a + 2t)$ which is very nearly equal to t for a typical thin-walled cell. Since the inside face of the capillary is used as a reference point for the observing microscope this means that, in effect, the upper and lower inside faces are unshifted

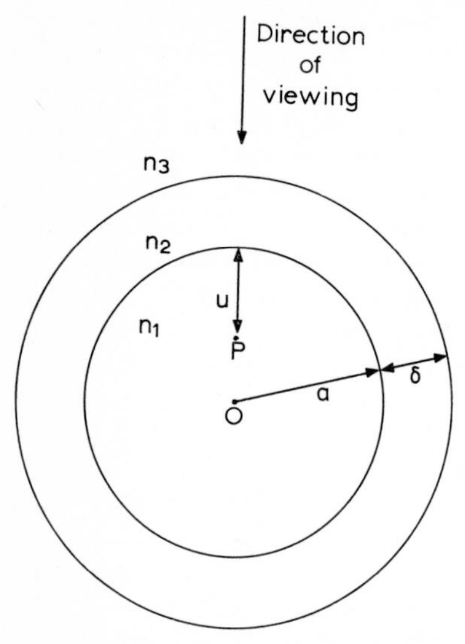

Fig. 3.15. Cross-section of a cylindrical microelectrophoresis cell.

(*i.e.* the apparent inside diameter is correct) and the centre shifted (relatively) downwards by t. The stationary levels will be shifted by intermediate amounts. Since the cell wall is cylindrical rather than spherical the correction to the stationary level position will be zero in the axial direction; thus if the eye is assumed to choose the 'circle of least confusion' the corrections to stationary level positions as calculated above should be halved. For a well designed thin walled cell with $\delta < 40\ \mu$ the correction is in any case negligible.

Accuracy of electrophoretic mobilities

The chief sources of error, not already discussed, in a well designed and calibrated cell are probably (a) the depth of focus of the observing microscope and (b) Brownian motion of the particles.

The gradient of electro-osmotic liquid velocity with distance along a diameter of a cylindrical cell is given from eqn. (3.62) by

$$dv/dr = 4v_{eo}\,(r/a^2)$$

so that at the stationary level

$$\frac{dv}{dr} = \frac{2\sqrt{2}}{a} v_{eo}.$$

For a typical arrangement with $a \approx 2$ mm and particles with mobility $\approx v_{eo}$ then an observing microscope with a depth of focus of 30 μ implies a possible error in a single particle mobility measurement of $\approx 2\%$.

The error due to Brownian motion of the particles will depend upon the size of particle and also upon the distance over which the particles are timed. The latter is commonly around 250 μ with a convenient transit time of ~ 10 sec achieved typically for particles with mobility 2 to 4×10^{-4} cm/sec per volt/cm by applying a potential gradient of 5 to 10 volt/cm. Particles of radius 0·1 μ will have a diffusion coefficient of $2{\cdot}5 \times 10^{-8}$ cm^2/sec in water at 25°C giving Brownian displacements of the order of 5 μ in 10 seconds again leading to errors in single mobility measurements of the order of 2%.

While the error in mobility measurements can be minimised by careful attention to the factors discussed above and by averaging over large numbers of particles, it seems reasonable to estimate that the probable error in electrophoretic mobility measurement is seldom less than 2%.

Conversion of mobilities to zeta potentials

Historically the first equation relating the observed mobility of a colloid particle to the ζ potential was due to Helmholtz and Smoluchowski and is usually referred to as the Smoluchowski equation

$$v_e = \frac{E\varepsilon\zeta}{4\pi\eta} \qquad (3.68)$$

where v_e is the velocity of the particle under the influence of an electric field of intensity E in a solution of viscosity η and permittivity ε. The ratio (v_e/E) is usually termed the electrophoretic mobility and has magnitudes typically in the range 0 to 6×10^{-4} cm^2 volt^{-1} sec^{-1} (*cf.* $7{\cdot}6 \times 10^{-4}$ cm^2 sec^{-1} per volt/cm for the ion K^+). It is therefore convenient to record mobilities in μ/sec per volt/cm.

If the permittivity of the dispersion medium ε is put equal to the dielectric constant (*e.g.* 78·35 for water at 25°C), this implies that $\varepsilon_0 = 1$ and the use of corresponding esu/cgs units for other quantities, so that if E is to be understood as in volt/cm and ζ in volts a numerical factor must be introduced and

$$v_e = \frac{E\varepsilon\zeta}{4\pi\eta} \times \frac{1}{299{\cdot}8^2}$$

where η is in poise. For aqueous sols at 25°C, therefore,

$$\zeta\ (\text{mV}) = 12{\cdot}83 \times (\text{mobility in } \mu/\text{sec per volt/cm}).$$

The use of MKS units makes it possible to use volts directly for both E and ζ but then of course η and ε will take different numerical values.

The Smoluchowski equation is often derived by a method of doubtful validity which assumes a parallel-plate condenser model for the electric double layer at the particle–solution interface. A more satisfactory derivation equates the force $E\rho dx$ exerted by the external field on an element in the double layer of unit cross-sectional area and length dx perpendicular to the (plane) particle surface, with the frictional force on the element due to the velocity gradient across it

$$E\rho dx = \frac{d}{dx}\left(\eta \frac{dv}{dx}\right) dx.$$

Substituting for ρ from the Poisson equation in one dimension

$$\rho = -\frac{1}{4\pi}\frac{d}{dx}\left(\varepsilon \frac{d\psi}{dx}\right)$$

gives

$$\frac{d}{dx}\left(\varepsilon \frac{d\psi}{dx}\right) = -\frac{4\pi}{E}\frac{d}{dx}\left(\eta \frac{dv}{dx}\right).$$

Integrating once with $d\psi/dx = dv/dx = 0$ at $x = \infty$

$$\frac{d\psi}{dx} = -\frac{4\pi}{E}\left(\frac{\eta}{\varepsilon}\right)\frac{dv}{dx}. \tag{3.69}$$

If η and ε are taken to be constant (*i.e.* independent of position outside the shear plane), eqn. (3.69) may be integrated again with $\psi = \zeta$ at the shear plane where $v = 0$, and $\psi = 0$ at $x = \infty$ where the solution velocity is equal to the electrophoretic velocity v_e (strictly the electro-osmotic velocity on this model of stationary solid and moving solution) to give

$$v_e = \frac{E\varepsilon\zeta}{4\pi\eta}.$$

This derivation makes it clear that the Smoluchowski equation only applies to plane double layers or to those in which the extent of the diffuse double layer (measured qualitatively by κ^{-1}) is small compared with the radius of curvature a of the particle, *i.e.* κa is large (say >500). Subject to this limitation, which should apply to all parts of the particle, the Smoluchowski equation is valid for all particle shapes.

For spherical particles (only) with small values of κa (say $<0{\cdot}05$) the Hückel equation may be used

$$v_e = \frac{E\varepsilon\zeta}{6\pi\eta}. \tag{3.70}$$

This equation is readily derived from the Debye–Hückel expression for the potential ψ at a distance r from an ion of radius a and charge q

$$\psi = \frac{q}{\varepsilon r} \frac{1}{1 + \kappa a} \exp\left[-\kappa(r - a)\right]. \tag{3.71}$$

In an applied field E the force Eq will be balanced in the steady state by the frictional force $6\pi\eta v$ if the particle is spherical and Stokes law applies. Substituting for q from eqn. (3.71) with $\psi = \zeta$ at $r = a$ and $\kappa a \ll 1$ the Hückel equation results.

In aqueous or other solutions in which electrolytes are largely dissociated the use of the Hückel equation is seldom justified since, even for particles as small as 200 Å radius, the electrolyte concentration will have to be only $\sim 10^{-6}$ M to give $\kappa a < 0{\cdot}1$ and such low ionic concentrations are difficult to control. For non-ionising solvents the values of κ will be extremely small (despite the direct effect on κ of the low ε) and the use of the Hückel equation may be justified.

Henry[33] showed that the Hückel and Smoluchowski equations were limiting forms, at low and high values of κa respectively, of the equation

$$v_e = \frac{E\varepsilon\zeta}{6\pi\eta}\left[1 + \lambda f(\kappa a)\right] \tag{3.72}$$

in which

$$\lambda = \frac{\sigma_0 - \sigma}{2\sigma_0 - \sigma}$$

where σ_0 is the electrical conductivity of the electrolyte and σ that of the particle.

In principle the effect of particle conductivity is to allow movement of the charge with respect to the particle under the influence of the applied field, an effect which should be negligible for small particles with extended double layers (κa small) as is borne out in the Henry treatment. Even where a correction is indicated by eqn. (3.72) for particle conductivity (*i.e.* $f(\kappa a) \neq 0$) it is probably best to omit it since the particle will quickly become polarised in the applied field, preventing any further charge transport so that the particle behaves as if it were non-conducting.

Henry[33] gave values for $f(\kappa a)$ for particles of spherical and other shapes, the limiting values of which are summarised in Table 3.2. Table 3.3 shows values of the correction factor for spherical particles over a range of κa values. For the reasons given above λ has been taken as $\frac{1}{2}$ in these tables.

The Henry equation, while taking into account the *electrophoretic effect*, *i.e.* the retardation caused by the movement of counter ions in the opposite direction to that of the particle under the influence of the applied field, does not include the *relaxation effect* familiar in the theory of electrolyte conductance. This is again a retardation and results from the deformation of the otherwise symmetrical double layer around a charged particle when it is moving in the applied field. It is evident that the relaxation retardation will be most serious when the counter ions in solution have low mobilities and when ζ is high.

TABLE 3.2

LIMITING VALUES FOR $[1 + \frac{1}{2}f(\kappa a)]$ IN THE HENRY EQUATION (eqn. (3.72))

	Low κa	*High* κa
Sphere	1·0	1·5
Cylinder parallel to field	1·5	1·5
Cylinder perpendicular to field	0·75	1·5

The relaxation correction is most serious in the range $0{\cdot}2 < \kappa a > 50$ which is unfortunately that commonly encountered in colloid problems. The analytical treatments of Overbeek[34] and Booth,[35] especially the former, have been widely used to interpret mobilities in this range, both yielding power series in ζ but restricted in the number of terms by the mathematical difficulties. In both treatments exponential terms have been linearised with the implication $e\zeta/kT < 1$, *i.e.* $\zeta < 25$ mV though, as in other applications of the Poisson–Boltzmann equation, the actual range of validity seems to be considerably greater than this.

TABLE 3.3

VALUES FOR $[1 + \frac{1}{2}f(\kappa a)]$ IN THE HENRY EQUATION FOR A RANGE OF κa VALUES

κa	0·01	0·1	0·3	1	3	5	10	20	50	100	1000
$[1+\frac{1}{2}f(\kappa a)]$	1·000	1·001	1·004	1·027	1·100	1·160	1·239	1·340	1·424	1·458	1·495

The equations of Overbeek and Booth have been superseded both in range of validity and convenience of use by the later work of Wiersema, Loeb and Overbeek[36] where approximations have been avoided by using numerical methods and an electronic computer. Before discussing the use of these calculations it is opportune to list the main assumptions on which they rest.

(i) The particles are not sufficiently close to influence each other.
(ii) The particles are rigid non-conducting spheres with the charge uniformly distributed over the surface.

(iii) The dielectric constant of the particle and the dielectric constant and viscosity in the solution are not functions of position.

(iv) Only one type of positive ion and one type of negative ion are present in the ionic atmosphere of the particle.

The Henry equation can be looked upon as the limiting form of this treatment at low ζ, the Smoluchowski equation at high κa and the Hückel equation at low κa. In these extreme cases the conditions of applicability are as already discussed. It may be noted that the effect of surface conductance, at least outside the surface of shear, is automatically included in the results of this work.

TABLE 3.4

PARTICLE MOBILITIES IN μ/SEC PER VOLT/CM AS A FUNCTION OF THE ZETA POTENTIAL AT VARIOUS VALUES OF κa FOR PARTICLES SUSPENDED IN AQUEOUS SOLUTION AT 25°C CONTAINING ONLY 1:1 ELECTROLYTE WITH IONIC MOBILITIES CORRESPONDING TO $\lambda = 70$ ohm^{-1} cm^2/eq[1]

κa / $\zeta(mV)$	0	0·05	0·1	0·5	1·0	2·0	5·0	10	20	50	100	200	500	∞
25·7	1·33	1·33	1·33	1·32	1·35	1·39	1·51	1·63	1·77	1·89	1·93	1·96	1·99	2·00
51·4	2·67	2·65	2·63	2·56	2·55	2·57	2·78	3·03	3·35	3·68	3·83	3·91	3·96	4·00
77·1	4·00	3·95	3·87	3·62	3·50	3·47	3·66	4·08	4·57	5·23	5·60	5·79	5·94	6·00
102·8	5·34	5·20	5·04	4·48	4·17	3·99	4·10	4·58	5·29	6·39	7·18	7·58	7·87	8·00
128·5	6·67	6·40	6·12	5·07	4·58	4·27	4·23	4·63	5·47	7·00	8·30	9·18	9·78	10·00
154·1	8·00	7·55	7·18	5·48	4·84	4·36	4·11	4·46	5·25	7·03	8·86	10·54	11·65	12·00

Table 3.4 shows a selection of the results given by Wiersema, Loeb and Overbeek[36] for the particular case of an aqueous solution at 25°C containing only 1:1 electrolyte with ionic mobilities corresponding to $\lambda = 70$ ohm^{-2} cm^2/eq. These figures show that the Overbeek and Booth equations overestimate the relaxation correction, that the implication of these two equations that there are two values for ζ corresponding to each mobility at high ζ is a result of the approximations inherent in the analytical method, and that the Henry equation very slightly underestimates the electrophoretic retardation. The slight maximum in mobility at around $\zeta = 130$ mV shown in Table 3.4 for $\kappa a = 5$ to 20 is almost certainly without physical significance.

Although there will be regions of ζ and κa for which the Overbeek and Booth equations estimate the relaxation correction with the required accuracy, it may no longer be worth using these rather cumbersome equations in the routine interpretation of observed mobilities, especially where, as is frequently the case, the data all refer to the same value of κa and a graphical interpolation of the Wiersema computations can easily be made. The Henry equation and its limiting cases the Hückel and Smoluchowski equations will still find applications at low ζ and extreme κa values.

Wiersema, Loeb and Overbeek[36] have published a few computations for 2:1, 2:2 and 3:1 electrolytes. As predicted by the earlier equations the valence of the counter-ion has a much greater effect on electrophoretic mobility than that of the co-ion, multivalent co-ions giving the possibility of a negative relaxation effect (*i.e.* an augmentation of mobility). The Overbeek equation is seen to break down at lower values of ζ in the case of multivalent counter-ions than in the univalent case and, rather curiously, the Henry equation becomes a better approximation than the Overbeek equation where the co-ions are tri-valent.

Where the ions constituting the atmosphere around the colloid particle have λ values differing appreciably from 70 ohm^{-1} cm^2/eq a correction may be necessary to the mobilities given in Table 3.4. So far as co-ions are concerned this correction is almost negligible, being of the order of $+1\%$ for H^+ at $\zeta \approx 75$ mV and -3% for ions with $\lambda \approx 22$ such as dodecyl sulphate. Where ions such as these are counter-ions, which is not likely in the case of the large surfactant ions, the correction is rather more serious being *e.g.* $+4\%$ for H^+ at $\zeta \approx 75$ mV. The necessary corrections, which are additive where both co- and counter-ions are concerned, may be taken from the rather limited computations of Wiersema or estimated from the appropriate terms in the Overbeek equation.

In all of the treatments so far discussed it has been assumed that ε and η are constants. Just outside the surface of shear, however, the electric field strength may be sufficiently high to increase η and/or decrease ε significantly, both effects reducing the particle mobility for a given ζ potential.

Lyklema and Overbeek[37] examined this problem and concluded that though the effect of field strength on ε could probably be neglected, the effect on η may be more serious. The variation of η with field strength may be expressed by the equation

$$\eta = \eta_0 \left[1 + f_0 \left(\frac{d\psi}{dx}\right)^2\right]$$

where η_0 is the viscosity in zero field strength. Unfortunately the value of the viscoelectric constant f_0 in aqueous solution is not known with any certainty. Lyklema and Overbeek used $f_0 = 10^{-11}$ cm^2/volt^2, but it now appears that this may be an overestimate by a factor between 50 and 100.

Using the Booth equation

$$\varepsilon = \varepsilon_0 \left[1 - B \left(\frac{d\psi}{dx}\right)^2\right]$$

for the variation of permittivity with field strength, with B between 1 and

12×10^{-14} cm^2/volt^2 for water, Hunter[38] has recently concluded that the variations of e and η with field strength are of similar significance and that their combined effect is probably small in most cases.

REFERENCES

1. F. Powis, *Z. Phys. Chem.*, **89** (1915) 186.
2. E. A. Guggenheim, *J. Phys. Chem.*, **33** (1929) 842.
3. R. Parsons, *Modern Aspects of Electrochemistry* (Ed. Bockris), Butterworths, London, 1954, p. 103.
4. P. L. Levine, S. Levine and A. L. Smith, *J. Colloid and Int. Sci.*, **34** (1970) 549.
5. O. Stern, *Z. Elektrochem.*, **30** (1924) 508.
6. D. C. Grahame, *Chem. Revs.*, **41** (1947) 441.
7. O. A. Esin and V. M. Shikov, *Zh. Fiz. Khim.*, **17** (1943) 236.
8. A. N. Frumkin, *Phys. Z. Sovjetunion*, **4** (1933) 256.
9. O. A. Esin and B. F. Markov, *Zh. Fiz. Khim.*, **13** (1939) 318.
10. S. Levine, J. Mingins and G. M. Bell, *J. Electroanalyt. Chem.*, **13** (1967) 280.
11. B. V. Ershler, *Zh. Fiz. Khim.*, **20** (1946) 679.
12. D. C. Grahame, *Z. Electrochem.*, **62** (1958) 264.
13. S. Levine, G. M. Bell and D. Calvert, *Canad. J. Chem.*, **40** (1962) 518.
14. S. Levine and G. M. Bell, *J. Colloid Sci.*, **17** (1962) 838.
15. D. C. Grahame and B. A. Soderberg, *J. Chem. Phys.*, **22** (1954) 449.
16. S. Levine, A. L. Smith, J. Mingins and G. M. Bell, *Proc. IVth Int. Cong. Surface Active Substances, Brussels* (1964).
17. S. Levine and A. L. Smith, *Disc. Faraday Soc.*, **42** (1966) 97.
18. A. L. Loeb, J. Th. G. Overbeek and P. H. Wiersema, *The Electric Double Layer around a Spherical Colloid Particle*, M.I.T. Press, Cambridge, Mass., USA, 1960.
19. S. Levine and G. Bell, *Disc. Faraday Soc.*, **42** (1966) 69.
20. E. L. Mackor, *Rec. Trav. Chim.*, **70** (1951) 763.
21. J. Lyklema, *Disc. Faraday Soc.*, **42** (1966) 81.
22. J. Lyklema, *Kolloid Z.*, **175** (1961) 129.
23. E. J. W. Verwey and K. F. Niessen, *Phil. Mag.*, **28** (1939) 435.
24. R. H. Ottewill and R. F. Woodbridge, *J. Colloid Sci.*, **19** (1964) 606.
25. E. P. Honig, *Trans. Faraday Soc.*, **65** (1969) 2248.
26. C. J. van Oss and J. M. Singer, *J. Colloid and Int. Sci.*, **21** (1966) 117.
27. K. N. Davies and A. K. Holliday, *Trans. Faraday Soc.*, **48** (1952) 1061.
28. S. Komagata, *J. Electrochem. Soc. Japan*, **1** (1933) 97.
    J. Allen, *Phil. Mag.*, **18** (1934) 488.
29. G. E. van Gils and H. R. Kruyt, *Kolloid-Beih.*, **45** (1936) 60.
30. S. Mattson, *J. Phys. Chem.*, **37** (1933) 223.
31. M. E. Smith and M. W. Lissie, *J. Phys. Chem.*, **40** (1936) 339.
32. D. C. Henry, *J. Chem. Soc.* (1938) 997.
33. D. C. Henry, *Proc. Roy. Soc.*, **A133** (1931) 106.
34. J. Th. G. Overbeek, *Kolloid-Beih*, **54** (1943) 287 and *Adv. Coll. Sci.*, **3** (1950) 97.
35. F. Booth, *Proc. Roy. Soc.*, **A203** (1950) 514.
36. P. H. Wiersema, A. L. Loeb and J. Th. G. Overbeek, *J. Colloid Sci.*, **22** (1966) 78.
37. J. Lyklema and J. Th. G. Overbeek, *J. Colloid Sci.*, **16** (1961) 501.
38. R. J. Hunter, *J. Colloid Sci.*, **22** (1966) 231.

CHAPTER 4

SURFACE-ACTIVE COMPOUNDS AND THEIR ROLE IN PIGMENT DISPERSION

W. BLACK

INTRODUCTION

Surface-active compounds (substances which alter the conditions prevailing at interfaces) have been obtained from natural products by extraction or modification since prehistoric times. The cave dwellers in their drawings and paintings used egg albumen and other natural gums to disperse their pigments in liquid media and to give some degree of preservation to their works of art.

The early pigments were natural earths, charcoal or occasionally naturally occurring pigments and in the main egg albumen was used as a binder. It was not until the Renaissance that naturally occurring varnish extracts and rosins came into general use. For several hundred years only minor modifications were made in the use of pigments, binders and paint making but with the advent of the industrial revolution greater demands were made on the properties of paint and other film forming binders.

It is clear that many of the pigments used up to sixty years ago were not well dispersed as we would interpret that word today, and that the dispersion properties depended to a large extent on the naturally occurring surface-active components in the film-forming binders. These compounds were among the earlier macromolecular dispersing agents although this was not realised at that time. The presence of these surface-active compounds was the probable explanation of the rather limited and very specific success of certain natural agents such as tallow, stearine, pitches and waxes which were added in small amounts in attempts to improve the working properties of the paints or inks.

The introduction of new pigments, such as the synthetic organic pigments which seem to be inherently more difficult to disperse in many media, led to a greater interest in dispersion techniques. The new film-forming binders and the problems associated with improved application and drying times in addition to their use in motor-car and other production lines, placed even greater demands on dispersion and led to an increasing interest in surface behaviour and surface-active compounds.

Synthetic surface-active agents, and by this term we understand substances which have been specially synthesised to obtain surface-active effects, represent a fairly modern development, which stemmed from demands on the technology required in modern industrial nations. The application of these surface-active materials is very widespread in everyday life and examples which at once come to mind are the washing and other applications in textile materials, the preparation of dispersions and emulsions, the application of agricultural sprays and other special uses the number of which increases steadily year by year. Many books and articles have been written on the use of such substances in recent years but the main theme of this chapter will be on the use of such materials in the dispersion of pigments and their application in the pigment industry.

The author considers surface-active compounds in this chapter under three main headings (*i*) surface activity, the physical requirements, general properties and selection of surface-active agents; (*ii*) the use of such agents in aqueous media and (*iii*) their use in non-aqueous media. In this review considerable reference is made to patents which have been treated purely as *literature*, and no regard is paid to their validity as patents nor to their value as monopoly instruments. The term surface-active compound is interpreted perhaps more loosely than many purists would desire. This is done deliberately, since it is the author's belief that restriction of this term to conventional surface-active agents could not adequately cover the surface-chemical effects, such as adsorption, especially with regard to dispersion of pigments in non-aqueous media. Accordingly there are several places in the text where reference is made to altering the adsorption behaviour of the surface of pigments by various coating techniques which deposit a layer or layers of a substance which may be completely insoluble in the medium into which the pigment is to be dispersed. If this coating alters the adsorption of any species in the media to enhance dispersion stability then the coating agent can also be considered to be a surface-active or perhaps better a surface-activating compound and as such reference to it in this chapter is considered justified.

SURFACE ACTIVITY

The general requirement for surface activity is, of course, the adsorption of a solute, usually from a liquid phase, at one or more interfaces of the system under consideration. There are several ways in which surface-active agents can be adsorbed but the most general manner and the way which can be most generally predicted lies in the *amphipathic* behaviour of certain organic molecules.

This amphipathic behaviour occurs most frequently in aqueous solutions and is exhibited by a wide variety of organic types and can be most easily understood by reference to some examples of amphipathic surface-active agents such as

$C_{17}H_{35}COO^- Na^+$	(sodium stearate)
$C_{16}H_{33}OSO_3^- Na^+$	(sodium cetyl sulphate)
$C_{16}H_{33}N^+(CH_3)_3Br^-$	(cetyl trimethyl ammonium bromide)
$C_{16}H_{33}(OCH_2CH_2)_{20}OH$	(cetyl alcohol condensed with 20 molecular proportions of ethylene oxide)
$(C_3H_7)_2C_{10}H_5SO_3^- Na^+$	(sodium di*iso*propyl naphthalene sulphonate)
$C_8H_{17}OOC.CH.SO_3^- Na^+$ / $C_8H_{17}OOC.CH_2$	(sodium dioctyl sulphosuccinate)

All these compounds have amphipathic character, *i.e.* they all contain a hydrocarbon group which is expelled by water, and a polar group which is water-liking and which tends to seek to remain in the water. This gives rise to amphipathic adsorption in which the hydrophobic groups are oriented away from the water and the polar groups towards it.

The nature of the solubilising group on the surface-active species gives rise to a convenient classification of the various agents into anionic, cationic and non-ionic types. Surface charge is a very important property of colloid systems as shown in earlier chapters and can lead to very important technical effects. An example of this is *specific adsorption* which occurs when strong attractions are set up between the polar groups of a surface-active agent and specific groups in the surface of a solid phase, then adsorption with reversed orientation can occur, in which the hydrophobic groups are forced to orientate towards the aqueous phase. This is a highly specific phenomenon which is more difficult to predict; it occurs, for instance, when cetyl trimethyl ammonium bromide is adsorbed from an aqueous medium onto Prussian Blue and is really a case of chemisorption. This type of technique has, of course, been used in the pigment industry to render pigments more oleophilic, in flushing processes for example. In non-aqueous media specific adsorption tends to play a very important role in surface treatments and this makes the selection of surface-active agents for particular technological improvements more difficult.

The distinction between amphipathic and specific adsorption is that the former is determined mainly by the general tendency of the medium to control the adsorption process by an expulsion of one part of the molecule or ion whereas the adsorption process in specific adsorption occurs by interactions between the surface and the agent molecule.

The adsorption behaviour of polymeric surface-active agents does not fall conveniently into either of the two types of adsorption, *i.e.* amphipathic and specific, which we have already examined. Certain water soluble surface-active polymers such as sulphonated polystyrene, hydrolysed styrene/maleic-anhydride copolymers and condensation products of naphthalene-2-sulphonic acid and formaldehyde, might be considered as having a hydrophobic *backbone* and hydrophobic side groups, but it seems unlikely that such a simple concept can adequately interpret the surface-activity of such complicated macromolecules. Specific interactions between certain sites on the solid phase and groups in the polymeric surface-active agent certainly occur but, because of the cumulative effect of several attachment points between the surface and one macromolecule, the specific bonds do not require to be as strong as with the simpler surface-active species which are not polymeric.

The adsorption of polymeric materials is discussed in Chapter 2 and it seems likely that because of the solution properties of such materials that a contributory factor in their surface-active behaviour may be related to a sort of configurational entropy, the most probable configuration depending on both solution properties and behaviour at the surface.

Because of the difference in behaviour it would seem desirable in addition to anionic, non-ionic and cationic types to consider macromolecular surface-active agents in another classification.

In addition to this it is possible to obtain surface-active materials which belong to more than one class, *e.g.* they may contain both cationic and anionic groups (as in the betaine structures)

$$C_{12}H_{25}N^{+}\begin{matrix} \diagup (CH_3)_2 \\ \diagdown CH_2COO^- \end{matrix} \qquad \text{(dodecyl betaine)}$$

or both ionising and non-ionic hydrophilic groups as in structures of the type

$$R-(OCH_2CH_2)_nOSO_3{}^-Na^+$$

or

$$R-N\begin{matrix} \diagup (CH_2CH_2O)_nH \\ \diagdown (CH_2CH_2O)_mH \end{matrix} \qquad \text{(in strongly acid media).}$$

Ampholytic surface-active agents which can be anionic or cationic depending on the *p*H of the medium, have recently been introduced.[1] An example of such a surface-active agent in its various forms is

$C_{12}H_{25}NHCH_2CH_2COO^-Na^+$
sodium dodecylaminopropionate

$C_{12}H_{25}N^+H_2CH_2CH_2COOHCl^-$
dodecylaminopropionic acid hydrochloride

and $C_{12}H_{25}N^+H_2CH_2CH_2COO^-$
dodecylaminopropionic acid.

PROPERTIES OF SURFACE-ACTIVE AGENTS

It is only within the scope of this chapter to discuss briefly the properties of the various types of surface-active agent, and more extensively treatments can be found elsewhere as indicated in the bibliography at the end of the chapter. The theoretical development in earlier chapters has indicated the importance of surface tension, the thermodynamics of adsorption processes and wetting, and the problems associated with the forces between colloid particles. All these can be affected by surface-active agents as indeed can precipitate formation and possibly one or two examples of how such effects arise might be advantageous. Before examining these aspects, however, it seems desirable to discuss the broad chemical characteristics of surface-active compounds.

Surface-active agents are organic compounds and are subject to the same factors which affect the stability of other non-surface-active organic molecules. This is an important factor to be taken into consideration when selecting surface-acting compounds for use in various technological applications. For example, although sulphated alcohols are stable in alkaline or neutral aqueous solution, they are hydrolysed under acid conditions to the free alcohols and sulphuric acid, and because of this, can only be used in acid conditions if they have completed their purpose before or very shortly after being exposed to the acid. Sulphonic acids do not suffer this hydrolytic attack and are better in this respect and in their thermal stability. Cationic agents can be used in acid solution but quaternary ammonium compounds, especially the pyridinium types are liable to decompose in alkaline solution. In general it is preferable to use anionics under neutral or alkaline conditions and cationics neutral or slightly acid. The non-ionic agents are relatively insensitive to acids or alkalis except the esters which are subject to hydrolysis under alkaline conditions.

Chemical reactions are possible between surface-active agents and other species in the formulations. An example of this is the specific adsorption of cetyl pyridinium bromide on Prussian Blue which we have already mentioned. Anionic agents are generally incompatible with cationic agents in aqueous media but non-ionic agents are compatible with both these types.

Polymeric surface-active compounds usually exist in various molecular weight ranges and these have different adsorption and solubility characteristics. This makes it difficult to predict the effects of other chemicals, both organic and inorganic, incorporated into a common solvent. Several papers have been published[2a–c] concerning interactions between ionic surface-active agents and polymers in aqueous media, and possibilities of interactions of this kind must be considered in emulsion paint formulations.

All surface-active agents tend to form oriented molecular aggregates or micelles at certain concentrations. The formation of micelles is a result of the tendency of the solvent to expel certain parts of the surface-active molecule, and occurs in both aqueous and non-aqueous media. In fairly dilute solutions the micelles are probably spherical although at higher concentrations they may exist in rod shaped aggregates. In aqueous media the expulsion of the non-polar part of the molecule gives rise to the interior of the micelle having an oily nature and the polar groups orientated towards the water. The reverse is the case in non-aqueous media. The formation of micelles reduces the total free energy of the system and is therefore based on firm thermodynamic grounds and indeed the thermodynamics of micelle formation has been the subject of many papers in the literature.

One of the important properties of micellisation is that of solubilisation. Solutions of surface-active agents, at or above the critical micelle concentration (C.M.C.), have the property of dissolving considerable amounts of added chemicals which have only very limited solubility in the solvent. For example, certain simple azo compounds are considerably more soluble in aqueous solutions of surface-active agents above the C.M.C. than in water, and the solubility of stryrene is considerably increased in the presence of micelles, this being the basis of emulsion polymerisation which is examined in more detail later. In non-aqueous media the presence of micelles can lead to the solubilisation of water and this is of considerable importance in the application and performance of paints.

The concentration at which micelles form depends on the polarity and size of both radicals which can be considered as forming the surface-active agent, and these are the factors which determine the efficiency of the surface-active material. The concept of the balance of polar groups and non-polar groups being related to performance has always played a part in the selection of particular agents and the introduction of a measurement of this quantity, the hydrophilic-lipophilic balance (HLB) which is considered in more detail later, has considerably improved the selection of surface-active agents for particular technological requirements.

Adsorption and wetting are dealt with in Chapters 1 and 2 but it is perhaps useful here to summarise the conclusions especially as they relate to the application of surface-active agents.

The cohesive nature of liquids and solids arises from the existence of attractive forces such as van der Waals forces. The forces are non-symmetrical on the molecules in the surface layer and this non-uniformity of forces at the surface gives rise to a boundary tension acting parallel to the surface. The tension is numerically equal to the work done to increase the interface reversibly by unit area, at constant temperature, pressure, composition, and areas of all other interfaces, and can be defined as

$$\gamma_i = \left(\frac{\partial G}{\partial A_i}\right)_{p,\mathrm{T},\mathrm{n},\mathrm{A_j},\ldots} \tag{4.1}$$

where γ_i is the interfacial tension, A_i the area of the i-th interface and G the total Gibbs free energy of the system. When a surface-active compound is reversibly adsorbed γ_i is reduced according to the Gibbs equation

$$\Gamma_r = -\frac{d\gamma_i}{d\mu_r} = -\frac{a_r}{RT}\cdot\frac{d\gamma_i}{da_r} \tag{4.2}$$

where μ_r is the chemical potential, a_r the activity of a surface-active species r, and Γ_r the surface excess relative to the solvent, which in the majority of cases can be taken as equal to the adsorption at the i-th interface. From eqn. (4.2) it follows that positive adsorption leads to a reduction in the interfacial tension. If r is an ionised material or there are mixtures of micelle forming surface-active materials, special considerations have to be applied but this statement is still qualitatively valid.

Both eqns. (4.1) and (4.2) assume that equilibrium is established but in many processes, especially in the adsorption of polymeric material in viscous media, this can take a considerable time. In certain wetting processes also, the interfacial tension is lowered, but not to the equilibrium value in the time available, so that *dynamic* values are the important ones in many practical systems.

In the pigment industry wetting is a pre-requisite for all dispersion processes in both aqueous and non-aqueous media, and is also important in paint and other applications with reference to the final gloss, texture and adhesion of the product.

Young's equation

$$\gamma_{13} = \gamma_{23} + \gamma_{12}\cos\theta \tag{4.3}$$

is usually applied to describe wetting processes where the γ's are the relevant interfacial tensions and θ the contact angle provided $180° < \theta > 0°$. If $\gamma_{13} > \gamma_{23} + \gamma_{12}$ then phase (2) spreads completely over the surface and wetting is complete. In this case there is no equilibrium contact angle and the driving force behind the process is expressed by the spreading coefficient S, given by

$$S = \gamma_{13} - \gamma_{23} - \gamma_{12}. \tag{4.4}$$

Surface-active compounds will therefore be of assistance in processes described by eqns. (4.3) and (4.4) if they can lower θ or increase S.

Special considerations apply when we consider the wetting of porous masses such as a large quantity of pigment powder. Here two different situations can arise. The first is when the mass of pigment is thrown on the surface of the liquid. In this case the liquid has to penetrate into the fine capillaries contained between the pigment particles and displace the air. The equations for capillary rise predict that the force driving the liquid into the system of capillaries is proportional to $\gamma_{12} \cos \theta$, where γ_{12} is the boundary tension between the liquid entering the capillaries and the air which is hoped will be expelled from the capillaries. This means that γ_{12} should be as *high* as possible, consistent with a small value of θ. These two variables γ_{12} and θ are not independent variables and it is $\gamma_{12} \cos \theta$ which has to be maximised for the best result. If, on the other hand, the liquid is poured over the mass of powder, air has to escape from the capillaries through the liquid. This case therefore demands a *low* value of γ_{12}.

These requirements can be used to select agents to improve wetting but rather than measure contact angles (a very difficult and laborious process) and interfacial tensions, it is invariably easier, especially recalling that the kinetics of such processes may be involved, to devise a simple test procedure based on the particular type of wetting which is required.

The use of surface-active agents for the dispersion of pigments in a variety of media is becoming of increasing importance. In addition to deflocculation which takes place by surface-active agents which are reversibly adsorbed, many surface-coating treatments, which are irreversible, may be said to be the application of a surface-active compound in that a considerable modification of the pigment surface occurs and this can lead to improved dispersion in certain organic binders. In all these cases, however, for any improvement adsorption of some component has to occur in either a reversible, or in certain coatings, an irreversible manner. The prime requirement for dispersion of solids in liquids is adsorption, and unless this occurs dispersion cannot be improved. Although adsorption is a necessary condition it is not a sufficient condition as certain other factors involving the forces between the particles have also to be satisfied. These forces and the effect on dispersion stability are discussed in Chapter 1. Surface-active agents modify the electric charge on surfaces, they can introduce steric barriers which increase the stability and they can alter the adsorption characteristics of the surface to increase the adsorption of polymeric materials in paint and other pigment systems thus enhancing the barriers to flocculation.

In aqueous systems the stability of a suspension is due to an energy barrier which is set up by one or more of three complimentary mechanisms, *viz*., an electric double layer (*see* Chapter 3), a steric protective layer and a solvation energy effect. In many cases it is impossible to separate these factors, since the electric double layer contributes to the solvation effect in water, and to the possible hydration sheath.[3] Steric protection might be considered to be exactly the same as a hydrated sheath, and here again it is difficult to separate the various factors involved, but the importance of the steric effect can be discerned in the preparation of redispersible powders, which will re-disperse into water to give highly disperse, stable colloidal dispersions.

Since the above mechanisms are independently operative, to a certain extent, it follows that there are several possible types of disperse systems which are stabilised by (*i*) an electric double layer only, (*ii*) a steric mechanism, (*iii*) a solvated sheath, and (*iv*) a combination of two or more of these effects. Systems which are stabilised by combinations of some of these mechanisms tend to be the most stable. These concepts of dispersion stability are most useful when we consider very fine dispersions where Brownian motion is sufficiently powerful to overcome any tendency towards sedimentation. The great majority of technically useful dispersions contain some material which is coarser than this size range, and this fact has two main consequences; (*i*) the adhesion between flocculated particles is more easily broken down mechanically, and (*ii*) the dispersed particles will settle even if they are completely deflocculated. The latter consequence leads to an effect which is rather paradoxical at first sight. If such particles are deflocculated or only very slightly flocculated, either in aqueous or non-aqueous media, their mutual approaches during sedimentation will only very occasionally result in adhesion, consequently the particles will glide past each other during sedimentation and the final volume of a given mass of sediment will be small. If the disperse phase is a solid, an extremely stiff clay-like mass may be formed which is very difficult to stir into the supernatant liquid. If, however, the suspended particles are brought into a more flocculated condition, so that an appreciable number of particles are adhering as a result of collisions during settling, the final sediment tends to be more voluminous and is more easily broken up and stirred into the supernatant liquid. Such flocculated suspensions tend to exhibit rather anomalous viscosity behaviour and may even be thixotropic *i.e.* they may form gel structures which break down on stirring and re-form when they are left undisturbed.

In aqueous media the adsorption of ionogenic surface-active agents leads to increases in the capacity of the electric double layer provided that the adsorption process does not destroy the ionisation. Anionic agents will tend to make the stabilising potential more negative, and cationic agents will make the potential more positive. In view of the tendency for ions to flocculate dispersions of opposite ionic charge it is not surprising to find that anionic agents are more effective deflocculators in alkaline solutions, or in solutions containing multivalent anions, while the opposite is true of cationic surface-active agents.

The electric double layer as an explanation of stability is satisfactory for fairly dilute aqueous systems, but in most technological dispersions the effect of concentration and presence of inorganic electrolytes is such that it seems unlikely to play a predominant role in stabilising such dispersions. It cannot account for the enhancement of stability shown by many non-ionic surface-active compounds in aqueous media, nor can it account for the very much greater than expected stability of certain ionic surface-active agents to added electrolytes. It seems likely that steric and/or solvation barriers have to be involved to explain such behaviour and that these barriers are even more operative in non-aqueous media, in the relatively concentrated suspensions encountered in technological systems.

In addition to affecting dispersions of solids in liquids discussed rather briefly above, surface-active agents play a part in other dispersion processes such as liquids in liquids (emulsions), and gases in liquids (foams). The theory behind these other dispersion processes is basically the same, in that the stabilising mechanisms already discussed operate in these systems also. Certain different conditions do apply, however, in that instead of flocculation through minor contacts as with solids, the final state in an emulsion is the coalescence of two drops to form a larger drop which can lead to complete separation of the two phases, and in a foam complete collapse through continued ruptures of the various lamellae constituting such a foam. We shall return to these items for further discussion later when we examine emulsion paint systems.

SELECTION OF SURFACE-ACTIVE AGENTS

The bewildering variety of types and numbers of surface-active materials which are available from the many manufacturers of such chemicals can be a disheartening experience to anyone considering using such aids for the first time. In this section we examine the general ideas behind the selection and examination of these agents and indicate, where possible, the aids to selection which can be put forward with the present state of knowledge in this field. More detailed applications and uses of specific surface-active compounds are described in later sections.

In most technological problems concerning the dispersion of solids in liquids there is an obvious or more frequently less obvious connection with surface properties, and hence with surface activity. It is important, in the first instance, to consider what possible relations exist between effect and cause. The effect can all too obviously be seen but the cause is often more difficult to determine. This is often due to the fact that many of the factors involved are interconnected and it is difficult to separate the variables. In such cases it is often advisable to carry out some simple tests, even if they are not apparently commercially or technically feasible, in order to define the problem. An application of the basic theories of surface chemistry with this problem in mind can often result in a solution or partial solution to the technological difficulty.

In aqueous systems it is usually fairly easy to decide the requirements for using an anionic, cationic or non-ionic surface-active agent, and the general rule that adsorption must occur at the solid–liquid interface enables us to carry out some simple tests to select a short list of possible agents. In our testing procedures, of course, we must either know whether the problem can be solved by either improving the wetting of the pigment for example, or increasing the stability of the final suspension under our particular application conditions. At this stage it is sometimes desirable to carry out further tests, perhaps of increasing severity, in order to select the preferred agent for our purpose. Economic considerations have also to be examined as many problems can be solved to the required extent by more than one type of surface-active agent. It is very difficult indeed to generalise on the cost of surface-active materials as so much depends on the ready availability of raw materials but, as a result of the large market in the detergent field and simpler processing from petroleum products, the anionic agents tend to be the cheapest, the non-ionics generally are of the order of twice as expensive and the cationics four times the cost of the anionics, but considerable variations exist within this rough guide. Among the polymeric materials the β-naphthalenesulphonic acid/formaldehyde condensates and sulphite cellulose lye liquors in many modifications are widely used because of their low cost and very wide applicability to dispersing a wide variety of solids from calcium carbonate to highly priced sophisticated organic pigments.

In aqueous media amphipathic adsorption is the most frequent type of adsorption process encountered and, as this is by far the most predictable behaviour, this considerably aids us in the selection of surface-active compounds for particular uses.

The widespread use of surface-active agents in aqueous media such as detergency, textile processing and the like has created a considerable demand for these processing aids and because of this and the understanding of the requirements for amphipathic adsorption, which is most general in aqueous media, there is a very large selection of water soluble surface-active agents available for examination. In most cases a satisfactory solution to most technological problems in aqueous media can be obtained from the large selection of commercially available surface-active compounds.

When we come to consider dispersions in non-aqueous media considerable differences are observed. In the first place the great majority of surface-active agents used as dispersing agents in non-aqueous media are highly specific to the systems involved. There do not appear to be any agents of comparable ubiquity to the sulphite cellulose lye liquors or naphthalenesulphonic acid/formaldehyde condensates, for example, in aqueous media. Secondly, as has been mentioned previously, many non-aqueous systems such as paint vehicles, some printing ink media and rubber, have fairly powerful deflocculating properties in themselves (due to surface-active constituents arising from their manufacture), so that the addition of a special surface-active compound is not necessary. An important corollary of this is that in certain cases where such a surface-active component is required, considerable difficulties can be experienced because of the very severe competition (for the surface of the solid) between the added surface-active material and the usually much larger amounts of surface-active components in the media.

The problem of selecting surface-active agents for use in non-aqueous media is made more difficult by this specificity and usually it is often advantageous to exercise our chemical knowledge to select agents which might be chemisorbed on the surface of the solid or to modify, by chemical means, the surface of the pigment in order that adsorption of surface-active materials, already present in our medium, may be increased. In such cases, in order to differentiate between several agents, a combination of a practical test as devised by an experienced technologist and a simple test for wetting or adsorption should be applied. Tests for adsorption can be very simple; a row of glass vials containing (*i*) a solvent or dilute solution of the medium concerned, (*ii*) a certain weight of the solid to be dispersed, and (*iii*) various surface-active agents at a convenient concentration can, by examination of the relative rates and final volume of the sediment, provide considerable information easily and fairly rapidly. Similar tests can be devised for wetting, flow and other technical problems with simple apparatus and some ingenuity.

The desirability of having a convenient yard-stick to measure the surface-activity of surface-active materials and the relationship between this measurement and technological performance has, no doubt, occupied the minds of many surface chemists at some time in their careers. The development of the concept of the HLB number in the past ten years is, by far, the most interesting concept of this kind.

The surface-activity of a compound depends on the size and polarity of the hydrophilic and lipophilic groups in the molecule. This is the basis of the hydrophile-lipophile balance or HLB system of classification which was initially introduced by Griffin[4] based on emulsion data and originally of an empirical nature.

Griffin[5] found that in addition to a very laborious experimental determination of HLB numbers he could calculate the numbers approximately from the formula of the surface-active agent, for example, for polyhydric alcohol fatty acid esters the following formula applied

$$\mathrm{HLB} = 20\,(1 - S/A) \text{ where } S = \text{saponification number of the ester group and}$$
$$A = \text{acid number of the acid.}$$

For the fatty acid esters, which do not give good saponification values, the calculation was based on the formula

$$HLB = \frac{E + P}{5} \text{ where } E = \text{percent by weight of oxyethylene and}$$
$$P = \text{percent by weight of the polyol.}$$

Griffin pointed out in this paper that this approach still lacked exactness in that different agents could have the same HLB value but be of a different chemical type. This difference in chemical type can give considerably different technological performances and as an example Griffin cites the possibility of esters of a particular fatty acid providing better emulsification in a given system than any other fatty acid esters. He attributes such behaviour to the differences in the attraction of the lipophilic group in the agent for the lipophilic surface.

The same paper lists the *required* HLB values for the preparation of water in oil emulsions (4–8), oil in water emulsions (9–17) and for the solubilisation of hydrophobic compounds in water (14–17). Davies[6] has extended the calculation of HLB values by devising a system of group numbers and has arrived at the formula

$$HLB = \Sigma\,(\text{hydrophilic group numbers}) - n(\text{group number per } CH_2 \text{ group}) + 7.$$

Davies[6] has also shown that, at least under certain conditions, the experimental HLB values can be correlated with a theory based on the relative rates of coalescence of droplets of oil in water and water in oil.

The application of the HLB system in the selection of surface-active agents has been extended to pigment coloration by Pascal and Reig[7] and more recently by Weidner.[8] The former authors found that the required HLB values for a number of organic and inorganic pigments were constant in both aqueous media (for emulsion paints) and in non-aqueous paint systems. The values found using three series of non-ionic surface-active agents of different HLB value established that peak performance (colour value) occurred with the same HLB number in each series but that the different lipophilic groups exerted an influence on the level of the maximum colour effect. This will of course have a considerable use in the selection of surface-active agents for use in universal tinters as is pointed out by Pascal and Reig. It is interesting to note that the required HLB values of the organic pigments range from 8–11 for the relatively easily dispersed toluidine pigments to 14–16 for the very much more difficult phthalocyanine blues. Inorganic pigments excluding lamp-black, which is considerably less polar than the others on the list, has required HLB values of 13 to over 20. The degree of strength of the derived paints varied with the composition of latex, thickener and other additives but the required HLB for the pigment remained constant.

The use of HLB values has been examined further with regard to the compatibility of titanium dioxide with phthalocyanine green in an aqueous emulsion paint system[9] and for the selection of emulsifiers for emulsion polymerisation. Greth and Wilson[10] found that while certain emulsifiers and emulsifier blends affected the physical properties of polystyrene latices such as particle size, viscosity, and stability, the maximum stability and most rapid conversions were obtained with HLB values of 13 to 16. Testa and Vianello,[11] on the other hand, conclude, from an examination of the emulsion polymerisation of vinyl chloride, that the HLB value of the emulsifiers does not correlate with their performance and suggest that this is due to the relatively low solubility of the monomer in the polymer. Bondy[12] has given some broad generalisations on the required HLB values for emulsifiers for emulsion polymerisation systems. He points out that, not surprisingly, the more polar the monomer the higher is the required HLB; the more hydrophobic the disperse phase the more sensitive it is to deviations from the optimum HLB value; with systems containing two or more monomers the required value is intermediate to those required for the pure components. More interestingly, he observes that blends of chemically dissimilar types of agents give broader stability peaks than do blends of similar chemical types. This latter observation suggests that interactions between the different chemical types of agent in the surface-layer may be contributing to the stability of the emulsion drops.

It seems likely that further correlations will be found between HLB values and the performance of surface-active agents, but as noted above, it must always be kept in mind that specific interactions can occur between surface-active agents, solid surfaces and non-surface-active constituents and that such interactions could swamp the concept of the HLB number, especially in non-aqueous systems.

DISPERSION OF PIGMENTS IN AQUEOUS MEDIA

The dispersion of solids in aqueous media is widespread in many industrial processes from the washing of clothes to the preparation of dispersions of expensive pharmaceutical products. A vast accumulation of knowledge as to the factors affecting dispersion and the large number of special surface-active materials which have been synthesised, indicates the tremendous range of interest in this branch of surface chemistry. In many processes where dispersed pigments are used such as in the coloration of emulsion paints, and the mass-pigmentation of regenerated cellulose and synthetic fibres, it is important for the manufacturer to understand the surface chemistry involved in the manufacturing process in which the pigment may be only a minor constituent.

In viscose pigmentation an aqueous dispersion of the pigment is stirred into cellulose xanthate in dilute caustic soda which is then de-aerated, to prevent the formation of small air bubbles which could cause weaknesses and unattractive defects in the finished yarn. This solution is then passed through a spinnerette, containing many very tiny holes, into a spinning bath containing sulphuric acid, water, zinc and possibly nickel sulphate and some dispersing agent. There are several important surface chemical effects involved in this process. Firstly, the pigment dispersion must be fine enough initially not to block the spinnerette holes, and it must be stable enough not to form flocculates which could block the holes. The initial pigment dispersion must be wetted out thoroughly in order to avoid the occurrence of bubbles in the finished fibre and this high degree of wetting must be preserved throughout the process in order to avoid the nucleation of gases (at the pigment–viscose interface) during processing. In general, dispersing agents of sodium salts of β-sulphonic acid/formaldehyde condensates have found considerable application in such systems. One of the advantages of these types of agent is that although they are well adsorbed at the pigment–water interface they do not decrease the surface tension at the air–water interface to the same extent as the conventional amphipathic surface-active agents and this leads to better milling behaviour in that foaming and frothing are less likely to occur. Non-ionic agents are used as additives in the spinning bath to prevent *crater* formation (build up of inorganic salts) on the spinnerettes and for producing special skin effects. Alkyl-phenol/ethylene-oxide condensates and amine/ethylene-oxide condensates are used for this purpose and some pigment pastes using alkyl-phenol/ethylene-oxide condensates as dispersing agents have also been manufactured. These non-ionic agents are probably efficacious in that their dispersing and wetting powers are still operating to a great extent even in the very strong electrolyte solutions in the viscose spinning bath.

An interesting use of quaternary surface-active agents has been patented by Hoechst,[13] who claims that by using pigments containing a feebly acidic group or groups which can become acidic in the presence of a strong base, adding a quaternary ammonium or phosphonium agent, a polyglycol, water and a heavy metal salt and heating to 80°C, the pigment goes into solution and remains in solution on cooling. These soluble products are said to be especially useful in viscose pigmentation but the rather large quantities of cationic surface-active agent used would appear to be a financial difficulty compared to the more conventional surface-active agents used at present.

The advantages of water-based paint systems over solvent based systems are freedom from obnoxious vapour, very much reduced fire hazard in storage of large quantities of paint, and for the user greater ease in cleaning brushes, paint kettles and mopping up spillages. The latter advantages are very attractive for the do-it-yourself market which has become considerable in the last ten to fifteen years. Despite the disadvantages quoted above solvent based paints have been developed over the years to give excellent coverage, adhesion and protection to a variety of substrates. The main problems associated with the introduction of water-based systems are how to achieve a sufficiently high performance/cost relationship compared to conventional solvent paints with regard to adhesion, protection and durability. Water based emulsion paints containing fine latices of a variety of synthetic co-polymers, pigments, thickeners and other additives were the first type to achieve commercial success, especially for interior decoration, and more recently paints containing water-soluble alkyd resins have also had a considerable impact on industrial finishes. These products, together with improved application methods such as the electro-coating technique, have had a considerable impact on surface coating technology and, with the increasing understanding of the surface chemistry and physics of these processes, seem certain to continue to show improvements.

In the field of emulsion paints this success is largely due to surface-active agents. The essence of all emulsion paints is the formation of a fine dispersion or latex of water insoluble organic polymer which, after evaporation of the water, will coalesce to form a film-forming binder. Such latices are formed by the emulsion polymerisation of a certain poly-merisable monomer (or monomers) which are insoluble in water. The presence of surface-active agents in the aqueous polymerisation of water-soluble monomers (such as methyl methacrylate) increases the rate of polymerisation and allows it to go almost to completion.[14]

We have seen that amphipathic substances in aqueous solution exist as single molecules (or ions) and, above the critical concentration range, as aggregates or micelles. These micelles, because of their essentially hydro-carbon interior, are able to solubilise considerable quantities of hydro-carbons, so that the solubility of a hydrocarbon in a micelle-containing solution may be many times its solubility in water. The amount solubilised depends on the relative solubility of the hydrocarbon in water and in the micelle and on the capacity for the amphipathic compound to form micelles.

When a monomer is emulsified in an aqueous solution of an amphipathic compound, it is distributed in three loci; (*i*) in the emulsion droplets which comprises by far the largest fraction of the total monomer, (*ii*) in true solution in the aqueous medium and (*iii*) solubilised in the micelles. In emulsion polymerisation processes polymerisation takes place *via* a free radical mechanism and water soluble initiators are used mainly. Polymerisation, according to the general theory as first propounded by Harkins,[15] takes place primarily almost entirely in the micelles where the solubilised polymer is converted to polymer. At a certain stage the polymer particle can no longer be said to be in the micelle. This is due to the monomer diffusing through the aqueous phase, swelling the polymer and continuing the polymerisation process. At this stage the polymer exists in the aqueous phase as a discrete particle stabilised by an adsorbed layer of the amphipathic emulsifying agent.

The monomer emulsion droplets then supply monomer to the aqueous solution to keep the dissolved monomer concentration constant and thus slowly disappear. The emulsifying agent in solution also decreases as the polymerisation progresses as this is used to stabilise the growing polymer dispersion, and the number of micelles therefore rapidly decreases until with their disappearance the formation of primary polymer particles virtually ceases. The polymerisation reaction continues to completion in the polymer–monomer particle, but since the initiation of new polymer particles has stopped, it is evident that the amount of emulsifying agent is a factor in determining the molecular weight of the final polymer. In practice, modifiers (chain transfer agents, inhibitors, and retarders) are also used to control the polymerisation. It has been suggested by several authors that micelles do not act as the locus of initiation of polymerisation. Roe,[16] in a paper containing several interesting calculations on particle generation and stabilisation, has argued that particle generation occurs always in solution and that the growing particle or oligomer is stabilised by the adsorption of surface-active agent. It would appear that while this may be a true picture and would certainly explain several anomalies described by Roe in the above paper, in normal emulsion polymerisation recipes where the micellar concentration is relatively high, the rate of radical generation of about 10^{13}/sec/ml would suggest that collision of the radical or growing low molecular weight oligomer with a micelle is far more probable than the chances of mutual termination. As such the most likely situation is that within a very short time of initiation the radical is solubilised within a micelle and this, together with the capacity for the micelles to solubilise the monomer, gives rise to the ideal conditions for rapid polymerisation. The very large number of micelles found at concentrations not much above the C.M.C. thus accounts for the large number of particles formed in emulsion polymerisations carried out in such surface-active agent solutions.

McCoy[17] has reviewed the role of surface-active agents in emulsion polymerisation and, in addition to the classical case outlined above, examines two other cases, (*i*) where the monomer is soluble in water but the polymer is insoluble and (*ii*) where the polymer is insoluble in and only slightly swollen by the monomer. In the first case, exemplified by vinyl acetate, colloidal macromolecules such as polyvinyl alcohol, carboxymethyl cellulose and high molecular weight non-ionic surface-active agents are used as stabilisers. These compounds do not form micelles and the main polymerisation route occurs initially in solution until the polymer becomes water insoluble and precipitates. The water soluble macromolecules stabilise this precipitating polymer and allow the polymerisation to continue to an extent governed by concentration of stabiliser and other modifiers. The nature of the emulsifier or protective colloid in this case can affect the final particle size of the latex. Micellar surface-active agents can also be used to polymerise vinyl acetate but here we can encounter complications resulting from polymerisation taking place both in solution and in the micelles, so we get two routes to polymer particles.

The other case, where the polymer is insoluble in, and only slightly swollen by, its monomer, has been examined by Evans *et al.*[14] using vinylidene chloride as the monomer. A micellar surface-active agent, sodium lauryl sulphate was used as stabiliser. Initially polymerisation takes place in the micelles as in the case of styrene, etc., but, because the monomer cannot swell the polymer particles, there is a less rapid transfer of monomer from the emulsified monomer droplets than in the case of styrene polymerisation. This, coupled with the reduction in surface-active agent available to stabilise these monomer droplets, leads to coalescence of monomer droplets and to polymer–monomer droplet collisions which overcome the slow-down in reaction rate governed by the diffusion of monomer through the solution.

More recently van der Hoff[18] has given a comprehensive review of emulsion polymerisation in which the role of surface-active agents is examined with respect to initiation, particle formation and stability.

The stability of latices when polymerisation is completed depends on the amount and type of surface-active agent present. The behaviour of the latex is most important in application and stability to mechanical shear, freezing and the effect of the stabilisers on flow behaviour have all to be considered. At an early stage in the study of such systems it was discovered that the usable non-ionic surface-active materials were freer from objectionable foaming, and latices using such stabilisers were exceptionally stable, even to freezing and thawing. This is an important factor in the stability of aqueous paint systems but non-ionic agents, when used as the sole emulsifier for monomers which are essentially water soluble, give latices of increased particle size. Combinations of non-ionic and anionic agents are used in practice as this enables greater control to be obtained over particle size and molecular weight.[17,17a] Weidner[8] has examined some aspects of the interactions of anionic and non-ionic surface-active agents in latex paints and finds that there are considerable differences in rheological behaviour depending on the anionic agent used, and that the hiding power of the paint dropped with increasing concentration of non-ionic agent in the paint. It seems likely that many more complex interacting factors will arise when pigments are added to the latex system and when this pigmented material is formulated as an emulsion paint, and indeed this is the case.

In addition to pigments and the latex suspension other additives are present in an emulsion paint. Plasticisers to promote coagulation and give improved paint films, thickeners to improve can stability, evaporation, rheology, etc., and possibly defoamers are also present in the final formulated paint. These additions are made to improve scrubbing resistance, etc., but they can have possible adverse effects on pigment dispersion, and indeed their incorporation can present considerable difficulties depending on the adsorption–desorption characteristics and relative concentration of surface-active agent in both the latex and the concentrated aqueous pigment paste.

Many of these effects have been described in an excellent paper by O'Neill[19] on problems associated with polyvinyl acetate emulsion paints. O'Neill points out that although flocculation of pigment can occur on mixing a white base paint with a coloured pigment leading to a reduction in tinting strength, not all losses of tinting strength can be ascribed to failures in the surface-active agent stabilising mechanism. The slow loss in tinting strength of certain benzidene-yellow pigments was shown to be due to a crystallisation of the yellow pigment in plasticised latex particles, and that this could be inhibited by the presence of certain colloid macromolecules and certain surface-active agents which presumably acted by hindering contact of the pigment particles and plasticised latex particles. Such products are probably effective in preventing crystallisation because they form a more coherent and less labile adsorption layer around the pigment particles.

The adverse effect of excess surface-active agents on wet-scrubbing resistance of the paint film and water vapour permeability has been emphasised by many authors. Excess surface-active agent does not improve pigment or latex dispersion stability, and indeed can markedly alter the rheological behaviour of certain compositions in an adverse way. The adverse effects of surface-active agents can be overcome by either using a latex formed by co-polymerising the vinyl compound with a small quantity of a water soluble vinyl derivative (such as acrylic acid), which on neutralisation imparts a charge stabilisation to the latex particle, and/or by a careful selection of surface-active agent. Another possible way of overcoming the problem of excess surface-active agent affecting the resistance of the paint film to wet-scrubbing has been claimed by Hoechst[19a] who, by using an ethylene oxide condensate of parahydroxybenzophenone as the emulsifier, found that exposure to light has the effect of causing the emulsifier to act as a cross-linking agent for the resin in the paint film. Other surface-active agents have been examined in an attempt to overcome this difficulty and the following examples illustrate the use of high molecular weight non-ionic agents which are claimed to give products of low water sensitivity. Plastic milling of pigments with polyvinyl-alcohol and polymeric N-vinylpyrrolidone[20] of formula

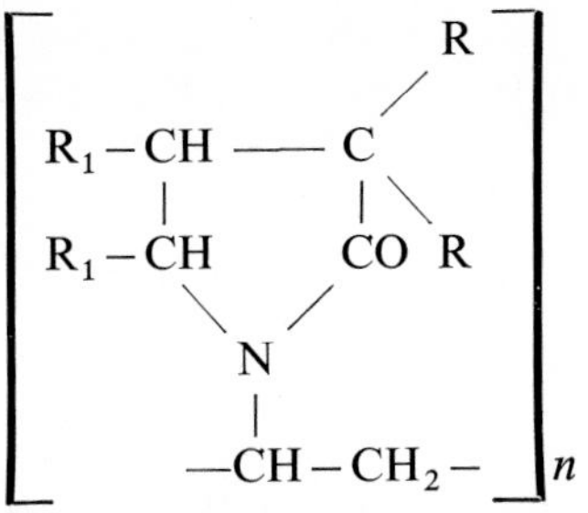

$$\left[\begin{array}{ccc} & & R \\ & & / \\ R_1-CH & — & C \\ | & & | \diagdown \\ R_1-CH & & CO\ \ R \\ & \diagdown\ \ / & \\ & N & \\ & | & \\ & —CH-CH_2— & \end{array}\right]_n$$

where R is hydrogen or methyl, R_1 is hydrogen, methyl or ethyl and the molecular weight is between 500 and 200 000, and mixing with styrene-butadiene latices is said to give paints of low water sensitivity, while Bowyer[21] claims that excellent gloss and water resistant vinyl ester paints can be obtained by using mixtures of an anionic (sodium salt of sulphated methyl oleate) and a non-ionic block co-polymer of propylene oxide and ethylene oxide. Such block co-polymers are surface-active due to the fact that propylene oxide polymers are essentially hydrophobic. Propylene oxide can be condensed with alcohols, glycols or amines to give an almost infinitely adjustable hydrophobic *block* which can be then further condensed with ethylene oxide to incorporate one or more hydrophilic entities into the molecule.

Combinations of non-ionic and anionic surface-active agents of relatively high HLB values have been suggested for stabilising pigments and latices for emulsion paint systems. Ethylene oxide condensates of the readily available alcohols such as cetyl and lauryl alcohols and alkyl phenols with alkyl aryl sulphonates have been used. Mixtures of alcohols condensed with ethylene oxide and carboxylic acid salts of the type

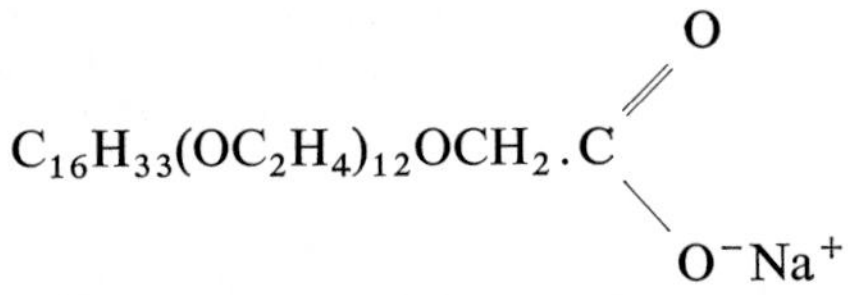

$$C_{16}H_{33}(OC_2H_4)_{12}OCH_2.C\begin{array}{l} \nearrow\!\!\!\!/\ O \\ \searrow\ O^-Na^+ \end{array}$$

have also been suggested[22] and these probably owe their effectiveness as dispersing agents to the close packing which might be expected at the pigment–water and latex–water interfaces. We have previously mentioned the use of polymeric surface-active components as pigment stabilisers and in this connection it is interesting to note that the effect of lecithin on the dispersion of titanium dioxide, which has been discussed by Kronstein,[23] might be ascribed to its polymeric nature.

Williams[24] also advocates the use of polymeric material for giving good aqueous dispersions of pigments with substantial shelf-life to settling, etc. He claims that olefin/maleic-anhydride co-polymers reacted with molar proportions of substituted amines to form the half-amide, when used as the salts give good dispersion of many pigments in water and these dispersions can be used for tinting aqueous latex paints.

Considerable efforts have been made to develop water-based air-drying gloss paints in the hope that the advantages of easy application and perhaps cheapness would appeal to the ever increasing do-it-yourself market. Hunt[25] has recently considered this subject and has drawn several important conclusions from his investigations. In water-based gloss paints the film-forming binder is present in the aqueous phase either as an emulsion or as a water-soluble resin. In the first case the emulsified resin must wet and flow out over the pigment particles so that a non-chalky film is obtained. The emulsified particles are rather viscous, and to improve the flow plasticising components are added. Hunt reports that he attempted to incorporate the pigment into such plasticised emulsified resin bases using aqueous pigment dispersions, by triple-roll milling dry pigment into the alkyd then emulsifying and in every case while the gloss obtained was good, wet-edge time and brushing behaviour satisfactory, the rheological properties of the brushed film were such that brush-marks did not flow out.

Several problems are also met using water soluble resins. Normal ester type alkyds tend to hydrolyse in the aqueous alkaline conditions required to keep them in solution and when this is overcome by the use of special hydrolysis resistant resins, the better resins are solubilised by the rather expensive method of using ionic carboxylic acid amine salts. The rheological properties of these moderately high molecular weight polyelectrolytes in water are such that, in order to obtain viscosities similar to those of good solvent based paints, certain glycol ester water soluble solvents must be added to reduce the viscosity. This, together with the necessity to adjust the rate of evaporation of the solvent system to achieve a good wet-edge retention, levelling and drying time, means that fairly substantial quantities of fairly expensive solvents have to be used. Hunt concludes that it is unlikely that water-based systems will ever be cheaper than organic solvent-based gloss paints because of the increased sophistication of the binders required, the necessity of using expensive glycol ether solvents, the use of amines to solubilise the resins and the possibility that more expensive surface treated pigments may be required. While it seems possible that suitably surface-treated pigments might be obtained at little or no increase in cost over conventional products, the other difficulties are unlikely to be remedied easily.

Foaming

The problem of foam production which can arise due to incorporation of a surface-active agent to improve pigment dispersion in aqueous media has been mentioned previously. Foam formation in liquids arises because of the mechanical properties of the surface layer and, as large enough differences in properties between the surface layer and the bulk liquid can only be brought about by adsorption at the air–liquid interface, it is easy to see that the persistent foams are encountered generally in aqueous systems where amphipathic adsorption is greatest. The theoretical background to foams and foaming has been comprehensively reviewed by Cooper and Kitchener[26] and several monographs (listed in the bibliography) have been devoted to this subject.

Stabilisation of foams depends on the properties of the surfaces of the thin liquid films which separate the dispersed gas. The main factors have been known for almost a hundred years in that Gibbs showed that when a film containing an adsorbed surface-active agent is stretched locally, the disturbance of the adsorption equilibrium leads to a decrease in the amount of surface-active agent adsorbed per unit area. This gives rise to an *elasticity* effect, since the surface tension immediately rises and tends to counteract further expansion. Gibbs' arguments apply to static conditions and Marangoni showed that under dynamic conditions the effect of *surface elasticity* is greater than Gibbs' equations indicate. The other main factor in foam stability is surface viscosity, first postulated by Plateau, and now completely substantiated experimentally by many workers. The surface viscosity can be several orders of magnitude higher than the bulk viscosity of the liquid due to the concentration of surface-active species at the interface and can be altered quite dramatically by specific interactions between two or more species giving considerable synergistic effects.

Our main interest is in the destruction of foams and the technical literature contains hundreds of examples of compositions which are claimed to inhibit or destroy foams. These substances must in general possess properties which are the opposite of those needed for foam stabilisation, *i.e.* they must eliminate or considerably diminish surface elasticity, increase the speed of drainage in the lamellae, decrease the stability of the interfacial films, wet out solid stabilisers into the liquid, and so on. Anti-foaming agents can therefore work in one or more ways, by displacing the foaming agent from the interface and thus reducing surface elasticity. In order to do this effectively the agent must have a low intrinsic surface tension, and must spread over the foam lamellae. A low solubility is also an obvious advantage as this increases the effectiveness by enabling small quantities of agent to be used. These factors are all present with the silicone anti-foamers and are also likely to be important in their effect on controlling *flooding* of pigments in non-aqueous paints. Care must always be exercised in the application of anti-foam agents, however, as their powerful surface-active effects can lead to other problems if they are used incorrectly or at too high a concentration. In addition to anti-foam agents which lower the surface elasticity other agents are known which act only by increasing the rate of drainage, *i.e.* lowering the surface viscosity. Many cases are also known in which foams are stabilised by finely divided, partly wetted solids and addition of wetting agents which allow the liquid to completely wet the solid will break such a foam and usually hinder the formation of fresh foam.

Water-based stoving paints

In recent years there has been a considerable upsurge in interest in water reducible coatings for industrial applications. The reduction in fire and fume hazards and the resulting lowering of insurance costs are the obvious advantages here, but consumer reaction is governed by performance and cost compared to the well established paints using organic solvent based coatings and most of the development work in this field has been concentrated on reconciling these requirements.

An excellent review of the present position of water soluble resins has been given by Tasker and Taylor[27] with special reference to electrodeposition which offers considerable advantages in the use of such systems. To date such paints have been mainly used as primer stoving finishes and are based on either emulsion systems somewhat similar to those used in emulsion paints, or on solutions of water soluble modified alkyd, urea-formaldehyde, phenolic and acrylic resins. After stoving it is important that the film contains as small a concentration of water soluble impurities as possible otherwise the painted article shows poor water resistance. This obviously could create problems in the dispersion of pigments for such finishes, as most conventional surface-active agents would lower the water resistance of the final film to a considerable degree. Fortunately the solubilised resins behave as dispersing agents and milling in conventional milling equipment gives satisfactory dispersions. Fry and Bunker[28] have examined the problem of wetting of various substrates and conclude that the paint has to have a surface-tension below that of the lowest energy surface concerned in order to obtain satisfactory wetting. In many cases the use of alcohols, to improve the solubility characteristics of the resin, has a beneficial action in this respect but they suggest that the use of surface-active compounds which could be rendered hydrophobic on stoving might be beneficial in wetting. At the moment the continued improvement in water soluble resins has increased the latitude for water soluble impurities in pigments[29] and in addition to the use of alcohols, white spirit and other solvents small quantities of non-ionic surface-active agents have been used as foam inhibitors in such systems.

Pigment dispersion does not seem to be regarded as too great a problem in electrodeposition systems at the moment although this does seem to have received some attention by Landon and Ashton[30] who find that the addition of Calgon S in small quantities of the order of 0·15% on paint gives increased gloss in a titanium-dioxide/acrylic-resin system. The use of ammonium polyacrylate dispersing agents which give increased gloss and thicker films, is also mentioned by these authors. Tasker and Taylor[31] have examined the dispersion of a tinted titanium dioxide in two different water soluble acrylic systems and find that flocculation is not markedly different from that obtained in the usually well deflocculated sprayed films. There is still some difficulty in relating adequate gloss and opacity to film thickness and here control of the pigment volume concentration may be important. The relationship between the pigment volume concentration in the bath and that in the deposited film has been examined by Robinson and Tear[32] who found, not unexpectedly, that this depended on the relative electrophoretic mobilities of the water-soluble resin and the pigment particles. As electrophoretic mobility is affected by bath temperature, *p*H, solids content and applied voltage it is quite easy to appreciate that very good control of these factors is necessary in order to obtain deposited films of adequate gloss and appearance. Improvements in the control of adsorption on the pigment particles, possibly by a carefully selected surface-active agent, could permit greater latitude in the bath conditions particularly if the surface charge density on the pigment could be brought to an optimum value for that of the resin system being employed.

Zinc oxide pigment has been used in paints, especially exterior paints, to improve resistance to ultra-violet light, controlled chalking, improved tint retention, increased durability and mildew resistance. Zinc oxide is a reactive pigment and forms zinc soaps with carboxylic acids. Morley-Smith[33] has examined the formation of such soaps using several fatty acids in a suitable solvent system and has reported on the viscosity behaviour with several zinc oxides including surface modified grades. More recently, Princen[34] has examined interactions between zinc oxide and titanium dioxide in aqueous media. The behaviour here depends on the sign of the ionic charges on the surface of the pigments; if opposite signs exist at certain *p*H regions the pigments mutually flocculate. As might be expected there is a complex relationship between exchange of ions between the surface and solution and surface area, particle size and particle volume relationships. Princen points out that non-ionic surface-active agents are unable to overcome this problem because such agents are only physically adsorbed on the surface and are competing with the chemisorption process. The addition of phosphates is said to completely overcome this problem by chemisorption of the phosphate. Conventional anionic agents of the naphthalene-β-sulphonic acid/formaldehyde condensate type decreased the initial viscosity of the paints but increased the rate of interaction between the pigments.

An interesting solution to this problem and to the effect of surface-active agents on water resistance has been suggested by Kubie *et al.*[35] who formulated emulsion paints with and without zinc oxide present using linseed fatty acids or esters reacted with sorbitol or sorbitol/ethylene-oxide condensates. These agents are capable of reacting with film forming binders in the paint and polymerising, thus increasing the resistance to water and in addition give paints, containing zinc oxide pigment, which maintain a stable viscosity for over two years.

Miscellaneous aqueous dispersions

In addition to the use of aqueous pigment dispersions in emulsion paints and viscose pigmentation they are used for colouring paper both for newsprint and wall-paper. The difficulties associated with the dispersion of Prussian Blue are well-known in the printing industry and we examine these later, but even in aqueous media it is found that large quantities of dispersing agents are required to obtain stable dispersions. Rogers and Todd[36] have found that the use of oxalic acid together with the sodium salt of a naphthalene-β-sulphonic-acid/formaldehyde condensate enabled them to obtain satisfactory dispersions with low (0·3%) anionic surface-active agent concentrations. Another interesting method of treating pigments with non-ionic agents is the subject of a patent by Bayer,[37] who claim that aqueous dispersible pigment powders can be made by treating with alkylphenol/ethylene-oxide condensates and the like which are salted out by electrolyte on heating to 70°C. The product obtained is then filtered off, dried and pulverised. These powders can be used easily in aqueous media provided they are allowed to swell in contact with water before being stirred in.

DISPERSION OF PIGMENTS IN NON-AQUEOUS MEDIA

The use of surface-active compounds to facilitate dispersion of pigments in non-aqueous media presents many problems to any reviewer. We have mentioned some of these difficulties in the section dealing with the selection of surface-active agents but the greatest difficulty by far is the lack of any predictive theory of dispersion stability in non-aqueous systems. The lack of understanding of all the factors involved in the dispersion of pigments in such media means that the worker in this field must rely to a very large extent on empirical sorting tests when seeking a dispersing agent for a particular system. In recent years, however, a number of useful papers have been published containing summaries of *ad hoc* information and these, together with several more fundamental papers on the nature of interactions at the pigment–vehicle interface, can form a useful starting point in this field.

Ten years ago the number of fundamental papers on dispersion stability in non-aqueous media was small and the work tended to be rather fragmentary but since then several groups of workers and many individuals have made important contributions to this field as can be seen by reference to Chapter 1. Notable among the centres where such work has been carried out are the National Printing Ink Research Institute at Lehigh University and the Paint Research Station of the Research Association of British Paint, Colour and Varnish Manufacturers. As a result of this and other work our understanding of interactions at the vehicle–solid interface has improved and, as adsorption is a pre-requisite for both wetting and dispersion, this has considerably improved our chances of being able to successfully apply surface-active compounds to solve dispersion problems in non-aqueous media.

Various stabilising mechanisms have been suggested for the stability of dispersions in non-aqueous media. A very useful summary of the situation in this field has been given by Chessick[38] who concludes that the repulsive forces set up by the interaction of electrical double layers round the particles confer the greatest degree of stabilisation. This may well be so with some systems, especially when the solvent is fairly polar, but in most commercial non-aqueous dispersions complete stabilisation is extremely rare and often undesirable, as it could lead to hard compact claying on storage, and with commercial non-aqueous dispersions control of the degree of flocculation tends to be the important factor. Controlled stabilisation by electrical repulsion only tends to be rather difficult to apply over the variety of conditions which the suspension might be subjected to in use and for this reason the author considers that steric barriers are more likely to offer practical solutions to dispersion stabilisation in non-aqueous media. Recently there has been a considerable interest in attempts to quantify the physical barriers set up by adsorbed polymeric molecules and considerable progress has been made (*see* Chapter 3).

One of the major difficulties in working in non-aqueous media is in finding a dispersing agent which is adsorbed on the surface of the pigment. As we have mentioned previously, in many non-aqueous systems there are already present surface-active materials which can compete for the adsorption sites on the pigment surface. A possible solution to this difficulty is to coat the pigment with the surface-active compound before attempting to disperse it in the medium and many successes have been achieved by this means. In the case of polymeric surface-active agents such treatment probably owes its success to the occurrence of adsorption hysteresis. This arises because in many systems involving polymer adsorption, equilibrium is achieved very slowly, due to the large number of attachment points of the adsorbed molecule, and the polymer which is adsorbed initially is therefore likely to remain on the surface.

The surface-active compound can be regarded as forming a bridge between the solid and the liquid in which it is dispersed. This is especially true when the polarity of the solid and the medium differs. An interesting paper by Trudgeon and Prihoda[39] illustrates this point among others in that they obtained some evidence that a pigment–solvent series could be set up which indicated that in the sequence of hydrophilic to organophilic pigments the best dispersing solvents were in the order of the hydrogen bonding parameters of the solvents as defined by Lieberman.[40] The former authors also show that a similar resin–pigment series can be set up and point out that according to their measurements, the largely used industrial solvents such as xylene and white spirit flocculate all the pigments examined.

The work discussed above was concerned with single solvents and obviously to extend such a series to mixed solvents and resins requires considerably more information. By the nature of the manufacturing process alkyd resins are mixtures of molecules of differing molecular weight and acid value and some work on the relative adsorption of these compounds has shown that this polydispersity could explain several problems encountered in pigmented systems. At the moment not enough is known of the relative adsorption from mixed solvents or from resin-containing mixed solvents to give a complete understanding of the pigment–resin interface and it is in this context that we further complicate the behaviour when we add surface-active compounds either as an insoluble or slow to equilibrate surface coating treatment, or as a soluble surface-active additive.

Titanium dioxide is a widely used white pigment and there are numerous grades with different coatings and/or isolation techniques for particular purposes. It would be expected that these coatings materially affect adsorption behaviour at the pigment–solvent interface and thus these coatings can be regarded to a certain extent as surface-active compounds. It is not our intention to examine this problem in any detail here as it is dealt with in Chapter 8, but it is interesting to note[41] that treatment of titanium dioxide (and other pigments) with fairly simple organic compounds such as acetylacetone or phthalimide and drying at 120°C is said to give improved brilliance to the final paint and more rapid dispersion in an alkyd-resin/white-spirit medium.

Several papers of a review nature which contain a considerable amount of *ad hoc* information on the dispersing of pigments in non-aqueous media and which might repay closer study are the papers by Carr,[42] Florus and Hamann[43] and Ozols.[44] Of these authors Ozols appreciates the difficulties of selection of agents through the lack of information on adsorption behaviour and lack of general theory, Florus and Hamann attempt to explain all their data by electrical effects, which for reasons we have already discussed we would consider rather unlikely, and Carr's opinion is that if lecithin does not improve dispersion no other additive is likely to have much effect. In a later paper,[45] Carr takes the view, which the writer would support, that the addition of surface-active agents to the medium is unlikely to improve dispersion stability, and that better results are more likely to be obtained by surface treatments at the pigmentation stage in colour manufacture. This does not mean, however, that the addition of simple surface-active agents to the medium will have no effect in paint manufacture. The addition of cationic surface-active agents is said to aid the wetting of pigments probably in some cases by forming bonds of a chemisorption type[46] and this property of chemisorption probably plays a part in the improved adhesion of paints to damp surfaces which agents of this type are said to encourage. With agents which do not form bonds of this type it is possible because of their smaller molecules that their rate of adsorption is high relative to the polymeric dispersing fraction of the resin, and although in time the simpler molecules are replaced on the surface by the polymeric material, the improved wetting obtained initially by the use of simple surface-active agents increases the rate of attainment of fine particle size and thus lowers the grinding time in the paint mill. Kresse[47] encountered a wetting problem which resulted in the deposition of an iron oxide pigment at the juncture of the air, paint, container interface. Cationic surface-active agents based on diamine oleates successfully overcame this difficulty. Another effect which is claimed to be explained by wetting is the subject of a patent by van Loo and Bitter,[48] who claim that dialkyl sulphosuccinates improve the gloss finish when paint is applied in humid conditions by maintaining a low (wetting) contact angle throughout the drying cycle and thus preserving the gloss.

The discovery of copper phthalocyanine as a potential blue pigment over thirty years ago, and its development to the present day where it is the acknowledged most versatile and universal blue pigment has required the solution to several problems associated with its physical form and surface characteristics. One of the largest uses for this pigment is in the surface coating field, in paints based on alkyd, nitrocellulose and stoving lacquers, etc. In spite of the many good qualities of copper phthalocyanine such as its shade and intensity three major defects have been associated with its use in paints, namely crystallisation, flocculation and the associated poor rheological properties of the pigmented mill bases.

The high tinctorial strength of copper phthalocyanine leads to a considerable quantity of paints being prepared as reduced shades with white pigments such as titanium dioxide and it is in such reduced shades that the flocculation effect can be most easily seen. The white and blue particles flocculate to different extents depending on the shear rate at which the surface-coating is applied and on the rate of drying of the solvent. The differences in polarity between the pigments have been suggested as the reason for such differing flocculation behaviour and certainly the paper by Trudgeon and Prihoda[39] supports this suggestion. Numerous attempts have been made to overcome these problems, many of them on lines similar to that used to overcome the problem of crystallising in aromatic solvents where it was found that partial chlorination of the phthalocyanine inhibited crystal growth in such solvents.

The use of conventional surface-active agents has been conspicuously unsuccessful in curing the flocculation problem associated with phthalocyanine blues and even in cases where additives such as benzyl cellulose have been used to improve the flocculation resistance in nitrocellulose lacquers,[49] quantities in excess of 100% by weight on pigment were used. Moser[50] claimed that the use of aluminium tertiary benzoates prepared *in situ* gave useful non-flocculating properties but here again usage is high. The use of sulphonated phthalocyanine derivatives and mixtures of these with more conventional agents also gave non-flocculating pigments but the quantities of agent employed always lead to rather weak powders with often reduced tinctorial yield.

Numerous other substituted phthalocyanines have been used in admixture with the pigment usually at the pigment manufacturing stage but no explanation of the surface chemistry involved in rendering these mixtures non-flocculating, except presumably the idea of altering the polarity of the molecule was offered. As examples of this kind of treatment we can cite the claims of Siegel[51] on substituted sulphonamides and Hoelzle[52] on hydroxy-methyl-phthalocyanine derivatives.

The use of salts of orthocarboxybenzamidomethyl-substituted copper phthalocyanine has been claimed by Lacey *et al.*[53] and, in a very interesting paper,[54] they offer as an explanation of the deflocculating effect of these compounds a combination of a charge effect, insufficient to stabilise the particles, and a hydrogen bonding of the solvent to the additive. Recently several patent specifications[55] have described the use of basic pigment derivatives as deflocculating agents for pigments in both aqueous and non-aqueous media. The mechanism of the deflocculating action of these basic pigment derivatives has been examined by Black *et al.*[56] who have shown that the soluble compounds are well adsorbed on a variety of pigments from both aqueous acetic acid and white spirit solutions. Measurements of adsorption behaviour, deflocculation tests, infrared and cryoscopic studies show that the deflocculation of the soluble and insoluble basic pigment derivatives on finely divided pigments is due to hydrogen bonding between the basic groups and the acidic hydrogen atoms of the film forming binders in the paint.

In addition to flocculation, especially in reduced paints, two other paint defects associated with colloid chemical behaviour are *flooding* and *floating*. These two aspects have been described by Patton[57] in his excellent book as being due to the hydrodynamic behaviour of the pigment particles during evaporation of the paint solvent. Factors affecting hydrodynamic behaviour are particle size, density of the pigments and state of flocculation. These factors are interrelated and, as a result, more than one solution to such problems can be found. Crowl,[58] in a study of phthalocyanine blue and titanium dioxide pigmented paints, concludes that the finer the pigment the more prone it is to floating except where coflocculation occurs between the blue and the white pigments. Haselmeyer and Wahn,[59] on the other hand, suggest that floating is due to flocculation of one of the pigments in a paint. Addition of agents of long-chain amine salts of a polycarboxylic acid as suggested by Haselmeyer and Wahn may give controlled flocculation of the titanium dioxide or perhaps coflocculation between the blue and white pigments. Silicones, which have very low surface tensions and very high spreading pressures, would also be expected to level out the variations in hydrodynamic behaviour during solvent evaporation in paint films. Only small quantities of these agents are required to correct flooding and use of excess is to be avoided as this can lead to other defects.

Many papers have been published on the effect of conventional surface-active agents on pigments in alkyd and other paint media. The effect of surface-active agents on the flow properties of paint systems has been examined by Singer,[60] who finds that selectivity is very high between pigment-vehicle and agent and the improved rheological properties of his paints as measured by flow point could be due to improved wetting caused by specific interactions at the surface of the pigment or perhaps to removal by emulsification of the adsorbed water at the pigment surface. The effects of small quantities of water on the flocculation of pigments in non-aqueous media have been discussed and explained by Zettlemoyer.[61] The flocculating effects of water could be a possible explanation for the improved behaviour of surface-active agent treated cadmium pigments reported by Cooper[62] as he found that the amount of water adsorbed on the finely divided treated pigment was reduced from more than 1 to about 0·05% by weight.

The various papers reported here are sufficient to indicate that surface-active agents probably do not act by amphipathic adsorption in non-aqueous systems, and that for any satisfactory explanation of the effect of such agents in non-aqueous systems it is necessary to examine the possibilities of specific interactions between pigment and agent and even agent and medium. Organic metal compounds which can have possible interactions with the pigment surface and/or certain components in the solvent have been used for a considerable time in the paint industry. Resin treatment of pigments with abietic acid and metallic rosin salts formed *in situ* have been used for many years and the metal stearates also find uses in controlling rheological properties in the long-term storage of paints. It seems likely that the formation of multivalent salts of stearic acid can give rise to possible interactions especially as many of these salts are not clearly defined tri-stearates but may contain di-stearates and mono-stearates. Recently organometallic salts of other metals such as titanium and zirconium have been examined. Sidlow[63] states that butyl titanate has a useful effect in decreasing the time of milling of certain pigments in a long oil alkyd but that its use is limited to 0·5% by its antioxidant effect. This is another case exhibiting considerable selectivity since the dispersion of Prussian Blue and copper phthalocyanine in both the long oil alkyd and a linseed stand oil is retarded by the presence of this additive. More complicated titanates such as di-*n*-octylene-glycol-stearyl-butyl titanate are claimed by Russell[64] to be good dispersing agents for free or colloidal carbon in organic solvents. Metal amino alcohol derivatives of zirconium such as diethyl-di(triethanolamine)-zirconate (N,N-distearate) and metal polyhydric alcohol derivatives of the same metal which could possibly be polymeric in character are also said by Koehler and Lamprey[65] to decrease the grinding time of pigments in alkyd media.

The major part of the preceding section has dealt with alkyd paints and the associated low dielectric white spirit solvent. Other paint systems based on melamine-formaldehyde and urea-formaldehyde have solvents of greater polarity and faster evaporation rate required for stoving finishes. The greater polarity of the solvent can lead to marked alteration in the behaviour of adsorbed surface-active agents, and constituents in the solvent or solvent mixture can also exhibit competitive adsorption on the adsorption sites on the pigment surface. It is not surprising, therefore, to find that some agents which perform adequately in a alkyd/white-spirit paint do not appear to have the same effect or, even when compared to other agents, the same order of efficiency in a particular stoving medium. The different film-forming binders and the dispersing effect of constituents in these binders also complicates the picture and hence makes selection of a surface-active agent to achieve a particular technical effect more difficult.

In the more polar solvents and media surface-active agents of a more polar nature such as salts of organic bases with phosphoric acid have been claimed to give stable pigment dispersions. Dimethylcyclohexylamine mono- or dibasic phosphates are said by Dreher and Rack[66] to be good dispersing agents in nitrocellulose and spirit lacquers at very low concentrations. Substituted N-methyl pyrrolidones[67] are also claimed to be useful in nitrocellulose lacquers and also, somewhat surprisingly, in alkyd paints and in lithographic inks.

It has been mentioned previously that in non-aqueous media amphipathic adsorption is not usually encountered and that certain difficulties can arise due to competitive adsorption processes. Waite,[68] in a recent paper, has given an extremely interesting account of the development of graft copolymers which are amphipathic in an aliphatic hydrocarbon medium. By careful selection of the copolymers and control of the degree of grafting exercised by the chemistry of the disproportionation and combination reactions amphipathic polymers have been obtained. Such products have been claimed to have been successfully used in dispersion polymerisations, in the dispersion of pigments and preformed polymer particles and for the stabilisation of emulsions in non-aqueous media. Oil soluble polymeric dispersing agents for pigments have been claimed by Gardon *et al.*[69] from copolymers of long-chain methacrylates, methylacrylate, maleic anhydride and/or itaconic acid. The problem of adsorption of polymeric dispersing agents at pigment surfaces has been overcome by several authors either by heating pigments having negative surface charges with fatty acid polysalts of polyethylene imine which renders the particles hydrophobic,[70] by graft polymerisation of monomers on fresh pigment surfaces[71] or by cationic polymerisations on appropriately surface treated pigments.[72]

The dispersion of pigments in printing inks is important for several reasons but the effect of dispersion on rheological behaviour is perhaps the major criterion. Because of the application methods flow properties are all important in inks and this is certainly the first hurdle which a printing ink must satisfy in order to be considered for potential use.

There are a wide variety of printing ink media is use in modern printing processes from lithographic inks based on bodied linseed oil or synthetic resins dispersed in drying oils of high viscosity to gravure inks which are fairly thin inks. Pigments dispersed in a highly fluid vehicle which contains little or no drying oil give inks which dry mainly by adsorption and evaporation. The technical requirements for such inks vary considerably and many surface chemical effects are important in arriving at a satisfactory solution to the printing requirements. Many books have been written on the subject of printing inks and hundreds of articles published which have shed considerable light on the many complex processes which occur in the printing process and it is not proposed to enter into great detail here. It is, however, appropriate to observe that there are many different types of printing inks and that in general the problem of specificity of action of surface-active agents which is present in non-aqueous paint systems can also be found in the printing ink field.

Variation in the viscosity of the ink vehicle and the methods of incorporating pigments into such vehicles have considerable effects on the shear which can be applied to pigment particles and agglomerates and therefore on the speed and fineness of the ultimate dispersion. The rheological properties are important in the application of inks and associated phenomena such as *tack* or stickiness of the ink, which is important in obtaining an even distribution on the press and proper transfer to the paper are also controlled to some extent by the dispersion properties of the material. As in other media dispersion properties are controlled by particle size distribution, particle shape, degree of flocculation, etc., and various wetting and deflocculating agents have been used in such systems. As in paint systems where it has been shown that certain constituents in the medium have a deflocculating action certain constituents in printing inks also show similar effects. It has been known for some time that with lithographic varnishes increasing the acid value to about 5 improved the wetting and the dispersion of pigments. The addition of fatty acids was then tried and resulted in improved wetting and dispersion but the use of these agents is offset by attendant disadvantages such as livering and reduced drying properties. Less reactive surface-active agents such as metallic soaps of naphthenic acid, sulphonated castor oil, sulphonated phenyl derivatives of fatty acids, and many other polar compounds have been suggested in addition to natural products such as lecithin and rosin and its modifications. Cationic or amphoteric surface-active agents are said to offer considerable advantages in improving pigment dispersion and pigment soft-texturing. This latter property refers to the improved texture and dispersibility in non-aqueous media which arises when the pigment is made water-repellent in aqueous media by specific adsorption with reversed orientation by some surface-active agent in the aqueous phase just before drying the pigment. If the pigment is dried when it is water-wetted there is a high work of adhesion of water to the surface and as evaporation proceeds the particles tend to be drawn together to give irreversible aggregation and a poor texture. The use of surface-active agents which can give adsorption with reversed orientation, *i.e.* lipophilic part exposed to the water, improves the texture firstly by giving a flocculated structure in the water and hence reducing the shrinkage occurring on drying, and secondly by reducing the work of adhesion to the water.

A recent paper by Carr[73] indicates that texture can be rated numerically by measurement of the surface area of the pigment in metres^2/gram divided by the oil absorption in grams per 100 grams of pigment. The paper gives a theoretical basis for this evaluation and a large number of results obtained for a variety of pigments in linseed oil. In a recent article de Vries[74] states that stearyl-*n*-propylenediamine-dioleate reduces the grinding time of pigments in printing inks and gives improved opacity and a superior gloss. He also says that a water soluble alkyl amine acetate is useful as a soft-texturing aid in pigment production and that the amphoteric analogue of this agent, an N-alkylaminobutyric acid, has had considerable success in the soft-texturing treatment of Prussian blue.

Despite the successes which have been achieved with cationic agents these still depend on specific adsorption for their effects and with many pigments this is absent. The rheological properties of some benzidine yellow pigments in printing inks have been consistently poor but the advantages of such pigments are such that they are used, albeit with some difficulty, in trichromatic printing inks. Carr[45] has described some experiments designed to examine the flow behaviour of benzidine yellow in printing ink. The effect of triethanolamine oleate and a non-ionic surface-active agent added to the ink and present at the pigment coupling stage was investigated using a Ferranti–Shirley cone-plate viscometer. From the various flow-curves Carr concludes that the rheological properties are more improved when the surface-active agent is present at the pigment coupling stage but with the most difficult benzidine-yellow pigment no improvement in flow properties could be obtained using a wide variety of surface-active agents.

The use of vinyl pyrrolidone as a dispersing agent for carbon black is said by Honak[75] to be attributed to the establishment of a sphere of polar molecules round the particles. The use of titanium compounds as coating compounds for pigments to improve dispersion in lithographic varnish is the subject of a patent by Bernstein.[76] He claims that when tetrabutyl titanate is added to a pigment in hexane and subsequently hydrolysed at the pigment–hexane interface the hydrolysis product is $Ti(OH)_4$ and not the partially hydrolysed products obtained by alternative methods. The other hydrolysis product butanol is removed from the surface by solution in the hexane and the coated pigment can be filtered off and dried. Improved fluidity is claimed when such a treated pigment is incorporated into a printing ink and improved dispersion is also said to result in rubber.

The coloration of plastics and the problems associated with the difficulties of obtaining heavy depths of shade by dyeing man-made fibres has created an interest in the mass coloration of these products by pigments. In most cases it was found that the easiest way of obtaining suitable products for mass coloration was by preparing master-batches of high pigment concentration perhaps up to 70% pigment, by heating a proportion of the plastic material and mixing in the pigment in a heavy duty mixer so that the pigment was plastic milled into the product. Alternatively, master-batching the pigment with the plasticiser which could then be incorporated into the material at the appropriate loading has also been used. Variations on these techniques still find application and the use of milling with fine grinding media in water has been exemplified recently[77] using a pigment and a polyethylene wax to give a master-batch suitable for use in various plastics such as polypropylene and polyvinylchloride. The use of surface-active compounds is considerably more complicated in this field and is unlikely to be as successful as in paint and printing ink media due to the high temperatures which are reached in the melt-spinning or extrusion of many plastic materials such as nylon and Terylene. However, some surface-active compounds are claimed to be useful aids in the mass pigmentation of nylon. Geiger and Geiger[78] claim that if a mixture of a pigment paste with caprolactam and an oleyl alcohol/ethylene oxide condensate is prepared, then added to excess caprolactam which is then polymerised, a fine dispersion of the pigment is obtained in the final polymerised material. In another patent[79] it is claimed that basic derivatives of napthalene-β-sulphonic acid condensed with formaldehyde, made by reacting the derived sulphonyl chloride with diamines such as dimethylaminopropylamine are useful dispersing agents for carbon black when milled as their acetate salts in water and, moreover, such milled pastes when incorporated into nylon by autoclaving the paste with nylon 6·6 salt give finely divided pigment in the final melt-spun fibres.

Monsanto[80] state that master-batching of polyacrylonitrile degrades the polymer owing to the high temperatures which are required in order to plastic mill the pigment. Addition of pigment or dyestuffs to the monomer inhibits polymerisation, and although many finely divided aqueous pastes are known, these are unsatisfactory in that if more than 3% water is incorporated in the spinning solution gelling occurs and this causes blockage of the spinnerettes and consequently gives poor results. Monsanto claim to have overcome these difficulties for the mass-pigmentation of polyacrylonitrile and its co-polymers by milling an aqueous pigment batch paste with polyethylene glycols of molecular weight between 1000 and 2000 then drying off the water at 50°C. A finely divided pigment suspension containing up to 70% pigment is obtained and this suspension is said to give good results in the mass-coloration of polyacrylonitrile.

Other dispersants which have been used for pigments in non-aqueous media which, although interesting, do not fall conveniently into any of the previous categories have been exemplified by Maxey and Castle. Maxey[81] claims that the salts of N-cyclohexyl-N-palmitoyltaurine and the like are useful dispersing agents for pigments in the manufacture of wax crayons, candles and other waxes. The very large aliphatic portion of these molecules is presumably an important factor in the increased efficiency of these products over more conventional surface-active agents. Castle[82] claims that very fine dispersions of pigments can be obtained in various fluoro-alcohols and that such products can be used in the coloration of plastics, leather, etc. The perfluoro-hydrophobic group, where the hydrogen atoms in an aliphatic hydrocarbon have been replaced by fluorine atoms, has a very low surface free energy and compounds containing such radicals have been examined by several workers[83] but the present cost of such compounds would appear to limit their practical use as surface-active compounds to very specialised applications.

Easily dispersed pigments

The use of cationic or amphoteric surface-active agents to improve the soft-texturing of pigments dried down from aqueous media has been mentioned previously and, while this technique has proved to be successful in reducing the work required to overcome the deleterious effect of hydrophilic aggregation, the possibility of still further reducing the energy requirements for pigment dispersion in non-aqueous media has been a target for pigment manufacturers for many years. Recently considerable advances have been made in the production of easily dispersed pigments,[84] and several pigment manufacturers now offer ranges of these products which can be tailored for particular end uses such as paints, pigment inks and plastics.[85]

It is well known that long-chain alkyl amines will deflocculate some pigments in non-aqueous media due to specific adsorption of the amino groups at the pigment surface. Recently it has been found that certain long-chain amines can be made to chemically react with groups on the surface of fine pigment particles and that such a treated pigment is spontaneously dispersible in aromatic hydrocarbons without milling. Moilliet and Plant[86] have described one such reaction and ascribe the deflocculating ability to its good adsorption and to the presence of the long alkyl chain. In the same paper they describe another method of forming easily dispersible pigments whereby hydrophilic aggregation is almost totally prevented by the formation of a stable, open structure of pigment particles and an oil-soluble resin or polymer. It is essential that a stable structure of sufficient strength be formed in order to withstand the hydrophilic forces arising due to evaporation of water and several methods of doing this are indicated in the paper and the references therein. The structure-forming 'cement' would appear to require to be soluble in the medium into which it is desired to disperse the pigment in order to regenerate the fine particles but evidence is produced that the structure-forming resins do not act as deflocculating agents in the hydrocarbon solvents nor is their action due to improvements in oil wettability. Their sole function appears to lie in the formation of an open matrix which adheres to the primary pigment particles and does not allow these to be drawn together by the adhesion of water or water-soluble surface-active materials to their surface. Thus, while the pigment primaries are easily regenerated in a solvent which dissolves the matrix 'cement', other oil soluble deflocculating agents are required to give good suspension stability in the solvent medium into which the pigment has been easily dispersed. Most paint systems and many printing inks have already dissolved in them suitable deflocculators for pigments and this has enabled easily dispersible pigments to gain a considerable foothold in the marketing of dry pigment preparations.

UNIVERSAL TINTERS

In large scale manufacture of paint, either aqueous or solvent based, the coloration is carried out by means of preparing a tinter or stainer with a high pigment content which is then added to the bulk of the paint to give the finished coloured product. Recently, with the growth of the do-it-yourself shops, there has been introduced the idea of increasing the selection of colours by producing paint to a specific colour by tinting in the shop. This can be done for both emulsion paints and gloss paints by means of universal tinters, which as their name implies, can be used satisfactorily to tint both water and solvent based paints. Cole[87] in a review article has examined the basis of these products and concludes that the required compatibility with both water-based and solvent-based paints is generally achieved by the use of non-ionic surface-active agents of appropriate HLB number and a solvent having certain solubility characteristics. The use of a co-solvent which is compatible with both water and drying oils such as polyols or their ethers and salts of etherified alkylated ethylene oxide condensates of hydroxy acetic or propionic acid has been claimed by Ritter *et al.*[88] and the use of a water miscible non-ionic surface-active agent with a dehydrated castor-oil/soya-lecithin mixture has been claimed as a useful medium for preparing pigments as universal tinters by Secker.[89] Kocian[90] uses another approach by plastic milling the aqueous pigment with oleic acid, separating off the water (flushing), adding morpholine then a suitable co-solvent or solvent mixture such as ethylene glycol, diacetone alcohol and ethyl lactate.

All the products we have described so far are pastes of pigments in a liquid medium but Daubach *et al.*[91] have proposed that powdered universal tinters can be prepared by a suitable combination of an oil soluble dispersing agent of HLB between 4 and 12 and a water soluble dispersing agent. In an example quoted the oil soluble dispersing agent is a condensation product of a sulphonated phenol, urea, and formaldehyde which was then further condensed with phenol and formaldehyde and mixed with the triethanolamine salt of an alkyl benzene sulphonic acid. The aqueous pigment paste is plastic milled with the above components and can then be dried to obtain a powder containing the pigment in finely divided form.

As can be seen from the above examples there are various ways of obtaining pigment dispersions which are compatible in both aqueous and non-aqueous media. It is unlikely, however, that tinters prepared by several of the above methods will prove equally efficient as tinters in different alkyd paint media and it would appear that the particular tinter system could be optimised for perhaps only one emulsion paint formulation and one type of gloss paint although with the almost continuous variations which are possible in such systems this might be fairly difficult to detect.

PIGMENT FLUSHING

In many pigment manufacturing processes the production of the final pigmentary form, and by this term we mean particle shape, size and polymorphic form, is carried out by a chemical or physical reaction in an aqueous medium. In many cases this gives a flocculated suspension due to the presence of electrolytes and filtration and washing is required before a batch paste is obtained. If the batch paste is then dried a product of a poor texture is obtained and the aggregation of primary particles which occurs during the evaporation of the water has to be overcome by some mechanical milling process to obtain a satisfactory dispersion in a non-aqueous medium. We have previously mentioned the use of surface-active agents as soft texturing aids and the more recent surface-treated easily dispersed pigments in the reduction and almost elimination of hydrophilic aggregation. The very much older method of overcoming this aggregation is to *flush* the pigment from the aqueous phase into a non-aqueous water immiscible vehicle and thus avoid drying the pigment altogether. The easiest way to carry out this process is to make the pigment surface oleophilic by adsorption of a surface-active compound with reversed orientation. The manufacture of white-lead paints was carried out using this method more than a century ago[92] and it seems likely that the success of this depended on the formation of insoluble lead *soaps* on the surface of the pigment particles. Water soluble cationic surface-active agents which also form interfacial layers with reversed orientation on pigments in aqueous media are also advocated for promoting flushing and are probably the most widely used surface-active agents for this purpose. Bass[93] points out that cationic agents have the additional advantage that they can be applied under acid conditions and in the presence of heavy metal ions which are often present in pigment slurries. The flushing process is usually carried out in a heavy duty mixer of the Werner–Pfleiderer type and the separated aqueous phase is decanted. It is important that the surface-active agent does not act as an emulsifying agent otherwise the flushing process is impaired. The formation of an oleophilic surface layer round the pigment particle leads to flocculation of the pigment in the aqueous phase and this aids the flushing process, but as a result flushing tends not to be 100% efficient and some water is carried into the oil phase and this is usually removed by drying under vacuum. The fundamentals of the pigment flushing process have been examined by Gomm *et al.*[94] who, by considering the thermodynamics of the process together with the use of a technique devised to estimate the wettability of small particles, postulate that the equilibrium water wettability of the pigment is increased when precipitated anionic surface agents are used. These authors suggest that in spite of this increase in water wettability, the surface treatment promotes flushing due to the free energy changes arising at the pigment surface during the dissolution of the precipitated oil-soluble agent which overcomes the increase in water wettability and thus carries the pigment into the oil phase, albeit with some associated water.

The disadvantages of flushing as a method of preparing non-aqueous dispersions are that unless there is a close relation between the pigment manufacturer and the pigment user, the user has to be content with the vehicle that is incorporated in the flushed pigment or have such considerable business that he can afford to purchase batch paste and carry out his own flushing process into his preferred medium.

PIGMENT-RESIN PRINTING OF TEXTILES

The production of textile prints using finely divided pigments together with resins which when heated cross-link and bind the pigment particles to the fabric has been carried out for a number of years. Instead of the conventional water-soluble gums and modified starches used as thickeners in the dyestuff printing trade pigment printing of textiles uses either oil-in-water (O/W) or water-in-oil (W/O) emulsion thickenings. The thickening effect depending on the rheological properties of the emulsions, which being *dispersions* of liquid in liquid have flow properties somewhat similar to the flow properties of solids in liquids being influenced by concentration, particle size and flocculation of the disperse phase.

A useful summary of the general requirements in pigment printing of textiles has been given by Parikh[95] and, as the systems with their various advantages and disadvantages are fairly complicated, we will not discuss these in any great detail but select the various portions of the technology where surface-active compounds play an important role.

In the first place surface-active agents are required in order to obtain a stable emulsion whether it be O/W or W/O. In both these cases the first-mentioned constituent is the disperse phase. We have referred previously to the selection of emulsifying agents in the section on hydrophilic-lipophilic balance and there is nothing further we propose to say on this subject except to note that if the emulsifiers are not volatile or reactive the baking process used to polymerise the resin will not eliminate these products from the fabric and this can impair the fastness properties of the final print.

In order to obtain suitable colour yield, brightness and reproducible shades the pigment particles have to be fairly fine (around 0·5 μ) and well dispersed.[96] Flocculation of the pigment can give dull prints and other unsatisfactory finishes. The pigment can be dispersed in either the aqueous or non-aqueous phase but is dispersed in the continuous phase in order to achieve satisfactory contact with the fabric.

As might be expected in such processes the pigment must be well dispersed in the selected phase and must not exhibit any tendency to *flush* into the other phase as this decreases the colour value due to aggregation. Flushing has been used to obtain a fine dispersion in the non-aqueous medium by Interchemical[97] and here it would seem that the alkyd resin probably acts as the dispersing agent but the presence of a cross-linking phosphoramide might also play a part here. This patent claims W/O emulsions and ethyl cellulose is used as the emulsifier. The use of condensation products as binding agents cum dispersing agents is quite wide-spread and agents such as N(ethyl-2-hexoxymethyl) acryamide have been claimed by Badische,[98] and Bayer[99] report the use of a condensation product of stearylamine, triethanolamine and oleic acid. In the same patent it is reported that sorbitol trioleate proved particularly effective for dispersing carbon-black while cyclised rubber was used to stabilise copper phthalocyanine. In the O/W emulsions, for example, polymeric agents such as alkyl phenol ethylene oxide condensates reacted with an olefin/maleic-anhydride copolymer and used as their alkali salts, have also been used as stabilising agents together with anionic agents, such as sodium lauryl sulphate.[100]

PROBLEMS AND PROSPECTS

In the preceding sections we have attempted to cover a fairly wide range of problems connected with the use of surface-active agents in the dispersion of pigments in liquids. As can now be seen the field is indeed very wide and in our coverage we have interpreted the terms of reference of surface-active compounds, pigments and indeed liquids very freely indeed. The limited depth of coverage is perhaps regrettable in some ways but it is essential in that there have been literally thousands of references to pigment treating in the literature in the last few years. There seems little doubt that during this period considerable advances have been made in surface treatment and dispersion of pigments, many of them rather empirical, but some as a result of the considerable increase in our understanding of the reactions taking place at the pigment–medium interface. The greater sophistication of many of the pigment-using industries has proved to be a considerable incentive for the improved technology which has occurred within the last five or ten years, and it would appear likely that further improvements will be made in the ease of mixing and increased speed of dispersion required to speed up processing and improve productivity. Techniques of producing products such as the easily dispersible pigments which depend on a controlled surface treatment of the pigment at the manufacturing stage and the development of special coatings of chemisorbed surface-active materials together with 'tailored' polymeric deflocculants have shown considerable promise in overcoming the problems of dispersion in non-aqueous media. With increased knowledge of stabilisation mechanisms and the effect of differing solvent/surface-active agent interactions, it appears unlikely that one particular product can be made to satisfy the highest technological requirements in a variety of media. Products which at the present fulfil such requirements are arrived at by a judicial selection of properties to achieve a satisfactory compromise of the various conflicting requirements. This may be a satisfactory solution in the present structure of technological requirements and price situation, but from a purely scientific outlook the best results can only be obtained by not compromising too much.

The surface treatment of pigments has been carried out for hundreds of years but it is only within the last ten or fifteen years that the scientific understanding of the colloid chemistry involved has begun to help in the selection and synthesis of more effective products. If, as seems likely, the present rate of progress continues many of the problems which we have considered in this chapter which with the present stage of knowledge we are unable to understand, and hence solve, will probably be considered quite simple fifteen years from now!

BIBLIOGRAPHY

General

E. JUNGERMAN, *Cationic Surfactants*, Marcel Dekker Inc., New York, 1970.

J. L. MOILLIET, B. COLLIE and W. BLACK, *Surface Activity*, Spon, London (2nd edition), 1961.

M. SCHICK, *Non-Ionic Surfactants*, Edward Arnold, London, 1967.

N. SCHÖNFELDT, *Surface-Active Ethylene Oxide Adducts*, Pergamon, 1969.

Foams

J. J. BIKERMAN, *Foams, Theory and Industrial Applications*, Reinhold, New York, 1953.

E. MANEGOLD, *Schaum*, Strassenbau, Heidelberg, 1953.

REFERENCES

1. D. L. Anderson, *J. Am. Oil Chemists' Soc.*, **34** (1957) 188.
2a. H. Arai and S. Horin, *J. Colloid and Interface Science*, **30** (1969) 372.
2b. S. Saito, *Kolloid Zeits.*, **215** (1961) 16.
2c. S. Saito, *Kolloid Zeits.*, **216** (1968) 10.
3. B. V. Derjaguin and A. Titijevskaya, *Second Int. Congr. Surf. Activ.*, Butterworth, London, **1** (1957) 211.
4. W. Griffin, *J. Soc. Cosmetic Chemists*, **1** (1949) 311.
5. W. Griffin, *J. Soc. Cosmetic Chemists*, **5** (1954) 249.
6. J. T. Davies, *Second Int. Congr. Surf. Activ.*, Butterworth, London, **1** (1957) 426.
7. R. H. Pascal and F. L. Reig, *Off. Dig. Federation Soc. Paint Technol.*, **36** (1964) 839.
8. G. L. Weidner, *Off. Dig. Federation Soc. Paint Technol.*, **37** (1965) 1351.
9. J. Rapach, *Am. Paint J.*, **51** (1967) No. 47, 92.
10. G. G. Greth and J. E. Wilson, *J. Appl. Polymer Sci.*, **5** (1961) 135.
11. F. Testa and G. Vianello, *J. Poly. Sci.*, C **27** (1969) 69.
12. C. Bondy, *J. Oil Colour Chemists' Assoc.*, **49** (1966) 1045.
13. Farb. Hoechst, FP, 1,396,714.
14. J. H. Baxendale, M. G. Evans and J. H. Kilham, *Trans. Faraday Soc.*, **42** (1946) 668; J. H. Baxendale and M. G. Evans, *Trans. Faraday Soc.*, **43** (1947) 210.
15. W. D. Harkins, *J. Am. Chem. Soc.*, **69** (1947) 1428.
16. C. P. Roe, *Ind. Eng. Chem.*, **60** (1968) No. 9, 20.
17. C. E. McCoy, Jr., *Off. Dig. Federation Soc. Paint Technol.*, **35** (1963) 327.
17a. M. E. Woods, J. S. Dodge, I. M. Krieger and P. E. Pierce, *J. Paint Tech.*, **40** (1968) 541.
18. B. M. E. van der Hoff, in *Solvent Properties of Surfactant Solutions*, K. Shinoda, editor, Edward Arnold, London, 1967, page 285.

19. L. A. O'Neill, *J. Oil Colour Chemists' Assoc.*, **41** (1958) 780.
19a. Farb. Hoechst, Dutch Patent 67. 14978.
20. General Aniline and Film Corp. UKP, 835,637.
21. W. Bowyer, UKP, 885,604, to Imperial Chemical Industries Ltd.
22. W. Ritter, K. Hofer, K. U. Steiner and E. Hess, FP, 1,286,980.
23. M. Kronstein, *Paint Varnish Prod.*, **44** (1954) No. 5, 21.
24. C. R. Williams, USP, 3,235,526 to Monsanto.
25. T. Hunt, *J. Oil Colour Chemists' Assoc.*, **53** (1970) 380.
26. G. Cooper and J. A. Kitchener, *Quart. Rev.* (*London*), **13** (1959) 71.
27. L. Tasker and J. R. Taylor, *J. Oil Colour Chemists' Assoc.*, **48** (1965) 122.
28. E. S. J. Fry and E. B. Bunker, *J. Oil Colour Chemists' Assoc.*, **43** (1960) 640.
29. A. Strickland, *Paint Varnish Prod.*, **53** (1963) Nov., 61.
30. G. Landon and I. H. Ashton, *J. Oil Colour Chemists' Assoc.*, **49** (1966) 202.
31. L. Tasker and J. R. Taylor, *J. Oil Colour Chemists' Assoc.*, **49** (1966) 674.
32. F. D. Robinson and B. J. Tear, *J. Oil Colour Chemists Assoc.*, **53** (1970) 265.
33. C. T. Morley-Smith, *J. Oil Colour Chemists' Assoc.*, **41** (1958) 85.
34. L. H. Princen, *Off. Dig. Federation Soc. Paint Technol.*, **37** (1965) 766.
35. W. L. Kubie, J. L. O'Donnell, H. M. Tester and J. C. Cowan, *J. Am. Oil Chemists' Soc.*, **40** (1963) 105. See also USP 3,140,191.
36. L. R. Rogers and W. Todd, UKP 973,428 to Imperial Chemical Industries Ltd.
37. Farb. Bayer, UKP, 946,053.
38. J. J. Chessick, *Am. Ink Maker*, **40** (1962) Jan., 28.
39. L. Trudgeon and H. Prihoda, *Off. Dig. Federation Soc. Paint Technol.*, **35** (1963) 1211.
40. E. P. Lieberman, *Off. Dig. Federation Soc. Paint Technol.*, **34** (1961) 30.
41. Laporte Titanium Ltd, FP 1,437,065.
42. W. Carr, *J. Oil Colour Chemists' Assoc.*, **34** (1951) 400.
43. G. Florus and K. Hamann, *Farbe u. Lack.*, **62** (1956) 260 and 323.
44. G. Ozols, *Australian Paint J.*, **4** (1959) No. 7, 17.
45. W. Carr, *J. Oil Colour Chemists' Assoc.*, **45** (1962) 28.
46. M. K. Schwitzer, *Off. Dig. Federation Soc. Paint Technol.*, **33** (1961) 1111.
47. P. Kresse, *A. Paint J.*, **54** (1969) No. 5, 28.
48. M. van Loo and V. W. Bitter, USP 2,886,456 to Sherwin-Williams Co.
49. B. T. Stephens, USP 2,851,371, to Pittsburgh Plate Glass Co.
50. F. H. Moser, USP 2,965,662 to Standard Ultramarine and Colour Co.
51. A. Siegel, USP 2,861,005 to E. I. du Pont de Nemours.
52. K. Hoelzle, UKP 893,165 to Ciba.
53. W. H. McKellin, H. T. Lacey and V. A. Giambalvo, USP 2,855,403 to American Cyanamid Co.
54. H. T. Lacey, G. L. Roberts and V. A. Giambalvo, *Paint Varnish Prod.*, **48** (1958) Apr., 33.
55. A. Schoellig, R. Schroedel and H. J. Hasse, UKP 949,739 and 985,620 to Badische Anilin Soda Fabrik.
G. Barron, W. Black and A. Topham, UKP 972,805 to Imperial Chemical Industries Ltd.
56. W. Black, F. T. Hesselink and A. Topham, *Kolloid-Z.*, **213** (1966) 150.
57. T. C. Patton, *Paint Flow and Pigment Dispersion*, Interscience, New York (1964), pp. 443–8.
58. V. T. Crowl, *J. Oil Colour Chemists' Assoc.*, **50** (1967) 1023.
59. F. Heselmeyer and W. H. Wahn, *Paint Manuf.*, **38** (1968) No. 8, 16.
60. E. Singer, *Off. Dig. Federation Soc. Paint Technol.*, **32** (1960) 762.
61. A. C. Zettlemoyer, *Off Dig. Federation Soc. Paint Technol.*, **29** (1957) 1238.

62. E. C. Cooper, *Paint Manuf.*, **33** (1963) No. 2, 57.
63. R. Sidlow, *J. Oil Colour Chemists' Assoc.*, **41** (1958) 577.
64. C. A. Russell, USP 2,913,469 to National Lead Co.
65. J. D. Koehler and H. Lamprey, UKP 885,679 to Union Carbide Corp.
66. E. Dreher and F. Rack, UKP 884,147.
67. F. J. Prescott, *Paint Varnish Prod.*, **50** (1960) No. 12, 31.
68. F. A. Waite, *J. Oil Colour Chemists' Assoc.*, **54** (1971) 342.
69. J. L. Gardon, M. Kalandiak and LaVerne N. Bauer, USP 3,413,255 to Rohm and Haas.
70. P. F. Pascoe, *Farbe u. Lack*, **74** (1968) No. 3, 245.
71. A. B. Taubman, G. S. Blyskosh and L. P. Yanova, *Lakokras. Mat. Prim.*, **3** (1966) 10 from *Chemical Abstracts*, **65** (1966) 9167 g.
72. R. Kroker and K. Hamann, *Angew. Makromol. Chem.*, **13** (1970) 1.
73. W. Carr, *J. Oil Colour Chemists' Assoc.*, **49** (1966) 831.
74. R. J. de Vries, *Paint Manuf.*, **29** (1959) Feb., 59.
75. E. R. Honak, *Plaste Kautchuk*, **11** (1964) No. 6, 372 from Chemical Abstracts 63 (1965) 16611b.
76. I. M. Bernstein, USP 3,025,173.
77. Ciba, UKP 1,042,906.
78. G. Geiger and A. Geiger, FP 1,266,096 to Sandoz.
79. Imperial Chemical Industries Ltd., Belg. P. 675,964.
80. Monsanto, UKP 990,122.
81. W. J. Maxey, USP 2,919,993 to General Aniline and Film Corp.
82. J. E. Castle, USP 3,129,053 to E. I. du Pont de Nemours.
83. J. L. Moilliet, B. Collie and W. Black, *Surface Activity*, Spon, London, 2nd Edition, 1961, p. 442–3.
84. Pigment Dispersions Supplement, *Paint Tech.*, **34** (1970) No. 9, i–xx.
85. J. E. Todd, *Paint Oil Col. J.*, **159** (1971) 138.
86. J. L. Moilliet and D. A. Plant, *J. Oil Colour Chemists' Assoc.*, **52** (1968) 289.
87. R. J. Cole, *Rev. Curr. Lit. Res. Ass. Br. Colour Varn. Mfrs.* (1964) 635.
88. W. Ritter, K. Hafer, K. U. Steiner and E. Hess, FP 1,286,980 to Sandoz.
89. C. W. Secker, Jr., USP 2,996,397 to E. I. du Pont de Nemours.
90. L. Kocian, FP 1,329,489 to Sandoz.
91. E. Daubach, W. Fischer and L. Setzer, UKP 918,516 to Badische Anilin Soda Fabrik.
92. *See* T. A. Langstroth, Color Eng., **6** (1968) No. 4, 40.
93. D. Bass, *Paint Manuf.*, **27** (1957) 5.
94. A. S. Gomm, G. Hull and J. L. Moilliet, *J. Oil Colour Chemists' Assoc.*, **51** (1968) 143.
95. D. V. Parikh, *Am. Dyestuff Rep.*, **52** (1963) 590.
96. M. A. Maikowskie, *Melliand*, **51** (1970) 574.
97. Interchemical, UKP 877,865.
98. Badische Anilin Soda Fabrik., Belg. P. 674,791.
99. Farb. Bayer, UKP 877,369.
100. G. Leitner, FP 1,439,032 to J. R. Geigy.

CHAPTER 5

PRINCIPLES OF PRECIPITATION OF FINE PARTICLES

A. G. WALTON

INTRODUCTION

The precipitation of solids from liquids is of importance in many diverse areas of science, including oceanography, geology, metallurgy, physiology and chemistry. In industry, paint formulation, polymerisation and plastic manufacture, saline water conversion (*via* ice), photographic chemical manufacture and many others involve the principles of precipitation. In some cases precipitation is a process which is to be avoided (sedimentation and scaling); in others it is to be promoted (separation of trace elements). A colossal amount of data relating to one form or another of precipitation process has accrued in the literature. The majority of the observations reported have, however, been qualitative in nature and it is only within the last two decades or so that more quantitative information has become available.

Traditionally, precipitation phenomena have lain within the realm of activity of the analytical chemist who is concerned with separation of the various components which annoyingly get together in nature. Large commercial enterprises are founded upon the separation of materials by precipitation. This chapter is not, however, directed toward the analytical aspects of precipitation and separation, but is devoted to outlining the current status of knowledge in the area of the mechanism of precipitate formation. Even in this more limited area the literature is extensive and studies attempting to unravel the complexities of precipitate formation have been carried out for at least seventy years.

Perhaps one of the more important earlier observations was Ostwald's recognition that the normal solubility of a substance could be exceeded without precipitation occurring.[1] Other workers, of whom von Wiemarn[2] is most frequently recognised, have shown that the number, size and shape of precipitated particles are a function of the concentration of solute in excess of solubility. In other words, the relation between the reactant concentration and the solute solubility determines many of the physical (and chemical) characteristics of the precipitate. In the following paragraphs these parameters will be explored, predominantly for precipitation of inorganic and organic crystals from solution.

THE METASTABLE LIMIT

Ostwald's observation of stable solute concentrations in excess of the normal solubility led to numerous experiments in which an effort was made to determine the maximum stable concentration. This concentration was termed the metastable limit but is now more frequently called the critical supersaturation. Actually, the maximum stable supersaturation is a function of how long the solution has been prepared and consequently the critical supersaturation should be stated in terms of the period of stability. It will be assumed in the following examination of this metastable limit

TABLE 5.1
CRITICAL DEGREES OF WATER SUPERCOOLING IN THE PRESENCE OF POWDERED SUBSTRATES

Substrate	*Critical Supercooling (°C)*
Teflon	>16
Benzophenone	>16
Thallium Iodide	6·2
Lead Iodide	4·1
Silver Iodide	2·5
Silver Chloride	4·5
Mercuric Sulphide	5·6
Cadmium Sulphide	6·5

that such a stipulation is satisfied. The metastable limit might be established, for example, by direct mixing of reactants, by supercooling a saturated solution, or by any one of a number of chemical means.[3] The first question that arises is, then, what determines this metastable limit and how can it be related to the molecular chemistry of precipitation?

It is now known that the critical degree of supersaturation is established both by the nature of the precipitating phase and by the presence of impurities. Removal of more and more of the impurity particles by successive filtration raises the metastable limit, but whereas it is simple to achieve 3000% supersaturation for barium sulphate, it is impossible to exceed 300% for silver chloride. An ample demonstration of the catalysing action of impurities is shown in Table 5.1, where the supercooling required for solidification of water is shown as a function of the model impurities added. We must conceive then that the birth or nucleation of a precipitate involves the accretion of ions or molecules on an impurity surface; the consequent clustering eventually leads to the formation of the crystalline phase. However, since clustering occurs in undersaturated solutions, it is not immediately clear what causes the onset of nucleation. We will see later that from classical theory the energetics of forming a nucleus involves an energy barrier, which expresses itself physically as a critical supersaturation. Discussion of the nucleation mechanism in terms of atomistic processes, however, will be deferred until a more general thermodynamic approach has been explored.

FORMATION AND STOICHIOMETRY OF CLUSTERS AND METASTABLE PHASES

Let us suppose that there is some critical size of cluster at the critical supersaturation which, upon addition of an additional ion or molecule, grows irreversibly into a macroscopic crystal.[4] At concentrations below the critical level the cluster grows or dissociates reversibly

$$nA \rightleftharpoons A_n + A \rightarrow \tag{5.1}$$

This reaction can be expressed thermodynamically and is usually the basis for development of nucleation theory. If, however, we wish to explore the precipitation of a uni-univalent ionic material, eqn. (5.1) could be translated to

$$xA^+ + yB^- \rightleftharpoons AB^*_{(x+y)} + A^+ \text{ (or } B^-) \rightarrow \tag{5.2}$$

where $AB^*_{(x+y)}$ is the critical nucleus.

The standard Gibbs free energy for the critical nucleus is

$${}^*\Delta G^0_{AB} = -RT \ln [{}^*a_A{}^x][{}^*a_B{}^y] + RT \ln a_{AB} \tag{5.3}$$

where a is the thermodynamic activity. If we define ions in the cluster as being in the standard state then $\ln a_{AB} = 0$. It will be shown later that actually this definition incorporates an implicit surface term. From eqn. (5.3)

$$\ln {}^*a_A = -\frac{y}{x} \ln {}^*a_B - \frac{{}^*\Delta G^0_{AB}}{xRT}. \tag{5.4}$$

If we plot the critical concentration (activity) of the cation against that of the anion, a straight line of slope y/x results with an intercept of ${}^*\Delta G^0_{AB}/xRT$ providing that the nucleus is unaffected by the changes in solution stoichiometry, *i.e.* providing x is constant.

Normally it is to be expected that the critical nucleus will either be neutral or will contain sufficient ions that $y \sim x$ (though this is not always true). If $x = y$

$$^{*}\Delta G^{0}_{AB} = -xRT \ln [I.P._{crit}]. \tag{5.5}$$

If we take the solubility analogue of eqn. (5.5), namely

$$\Delta G^{0} = -RT \ln K_{sp} \tag{5.6}$$

then the critical supersaturation is given by

$$\ln \left[\frac{I.P._{crit}}{K_{sp}}\right] = -\frac{1}{RT}\left(\frac{^{*}\Delta G^{0}_{AB}}{x} - \Delta G^{0}\right). \tag{5.7}$$

Since the critical supersaturation is also constant for any fixed nucleus stoichiometry, the plot of eqn. (5.4) (with $y = x$) would be parallel to the solubility curve.

We can put these facts to good use in determining the stoichiometry of the nucleus as follows. If reaction (5.2) had been of unknown stoichiometry and $x = y$

$$x\alpha A^{+} + x\beta B^{-} \rightarrow (A_{\beta}B_{\alpha}{}^{*})_{x} \rightarrow. \tag{5.8}$$

Then eqn. (5.4) becomes

$$\ln {}^{*}a_{A} = -\frac{\beta}{\alpha} \ln {}^{*}a_{B} - \frac{^{*}\Delta G^{0}_{AB}}{xRT\alpha} \tag{5.9}$$

and the slope β/α gives the ratio of ions in the nucleus.

Many precipitates form first in a metastable state which reverts to the stable state on standing. Examples of this phenomenon are silicates, several hydrous oxides, *e.g.* metastannic acids, silicic acids and titania, and some forms of calcium phosphate. In some cases a knowledge of the stoichiometry of the cluster which initiates nucleation gives a direct insight into the nature and mechanism of precipitate initiation.

An application to the very important phosphate system is as follows. Let us suppose that the reaction forming a phosphate precipitate is

$$\alpha x Ca(OH)_{2}(soln) + \beta x H_{3}PO_{4} + x(\gamma - 2\alpha)H_{2}O \downarrow [Ca_{\alpha}H_{3\beta-2\alpha}(PO_{4})_{\beta}.\gamma H_{2}O]_{x}. \tag{5.10}$$

From eqn. (5.3)

$$^{*}\Delta G^{0}_{prod} = xRT \ln [Ca(OH)_{2}]^{\alpha} [H_{3}PO_{4}]^{\beta} \tag{5.11}$$

or

$$\ln (Ca^{2+}) (OH^{-})^{2} = -\frac{\beta}{\alpha} \ln (H^{+}) (H_{2}PO_{4}{}^{-}) + {}^{*}\Delta G^{0}_{prod}/xRT\alpha. \tag{5.12}$$

Hence we see that the ratio β/α gives the ratio of phosphate to calcium in the initial phosphate phase. Considerable controversy has existed over the nature of the phase which forms at pH 7·3 (physiological pH) and since this problem is closely related to an understanding of the mode of formation of bones and teeth, some attention has been paid to it. Hlabse and Walton[3] showed that at low pH (3–5) dicalcium phosphate, $CaHPO_4.2H_2O$, was the initial (as well as the stable) phase. Brown,[4] using this and other unpublished data by the same authors, concluded that octocalcium phosphate $Ca_4H(PO_4)_3$ was the metastable phase preceding

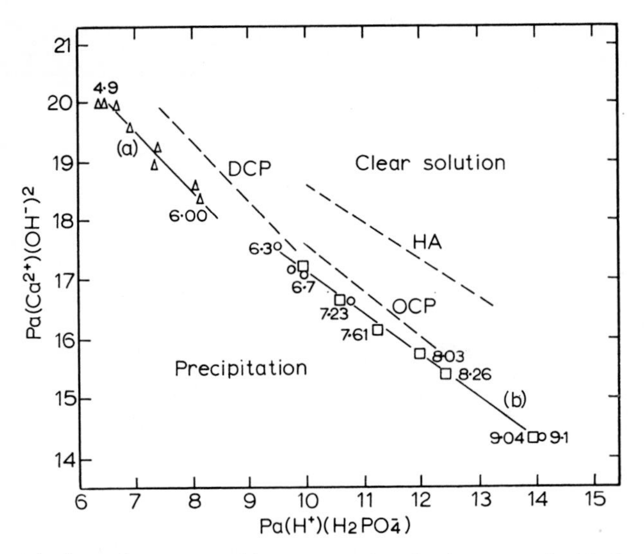

Fig. 5.1. *The logarithmic metastable concentration plot is compared with the solubility curves for dicalcium phosphate* (*DCP*), *octocalcium phosphate* (*OCP*) *and hydroxyapatite* (*HA*). *Curve* (a), *precipitation from homogeneous solution method, 25°C; Hlabse and Walton*[3]*: Curve* (b), 24 *hrs after mixing,* 37°*C; Walton* et al.[5] *At low* p*H* (*curve* (a)) *the slope is approximately that of DCP whereas at higher* p*H* (*curve* (b)) *a different phase is nucleated which appears to be intermediate in stoichiometry between HA and OCP.* (p*H values are indicated on the lines.*) (*It should be noted that the curves were calculated on the basis that $CaHPO_4$ and $CaH_2PO_4^+$ complexes were not present in solution to an appreciable extent. When association constants become available for these complexes a more precise determination of the stoichiometry of the initial phase will be possible.*

the development of hydroxyapatite ($Ca_5(OH)(PO_4)_3$, at physiological pH. More recent data by Walton, Bodin, Füredi and Schwartz[5] seem to indicate that the stoichiometry lies somewhere between octocal and hydroxyapatite. This method is fairly new and has not yet been applied to any extent (see Fig. 5.1). It seems likely that much more information upon the nature of metastable phases will eventually become available by use of this method.

Apart from determination of the gross stoichiometry of the cluster, it can be seen from eqn. (5.4) that a change in the total number of ions involved in the critical cluster may cause a change in intercept without causing a noticeable change in slope. An interesting example of a metastable limit showing such characteristics emerges from the work of Black, Insley and Parfitt[6] on silver chloride precipitation. In this case the supersaturation was achieved by hydrolysing allyl chloride in the presence of silver ions. Thus

$$CH_3CH{=}CHCl + H_2O \rightarrow CH_3CH{=}CHOH + Cl^- + H^+ \qquad (5.13)$$
$$Cl^- + Ag^+ \rightarrow AgCl \downarrow.$$

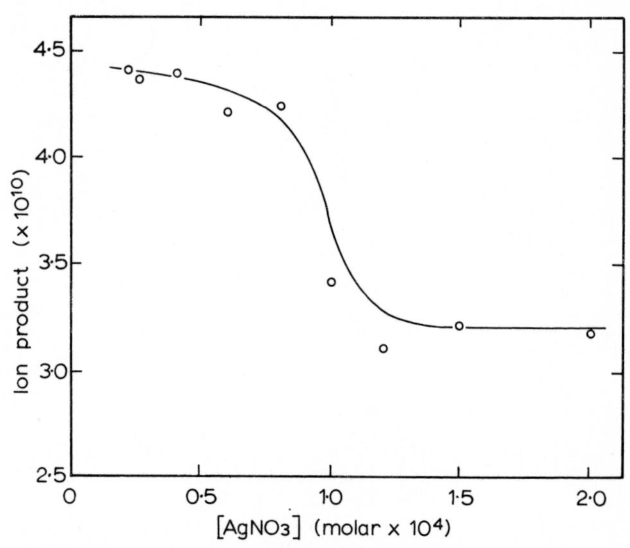

Fig. 5.2. The limiting critical supersaturation (I.P.$_{crit}$) *for silver chloride is shown as a function of the total silver ion concentration. The supersaturation is approximately constant except in the region of* $[Ag^+] \sim 10^{-4}$ *molar where a sudden change seems to indicate a change in the composition of the initiating cluster (after Black, Insley and Parfitt*[6]).

The system can be controlled by varying the concentration of silver ions present. In this manner critical ion products $[{}^*a_{Ag^+}{}^*a_{Cl^-}]$, over a range of Ag^+ and Cl^- concentrations, were obtained. Figure 5.2 shows the application of eqn. (5.4) to the silver chloride data. A significant kink is observed at the isoelectric point. Since it is unlikely that the structure of the nucleus changed by more than one molecule, it seems that the effect is indicative of rather few molecules in the cluster. It has been suggested that changes in cluster stoichiometry are brought about when one ion is in excess by change in excess electrical charge in the vicinity of the cluster surface. This point will not be explored in further detail here. We have seen then that the metastable limit is a function of the free energy of nucleus formation which is characteristic for any one solute and any one substrate.

NUCLEATION THEORY

Homogeneous nucleation

So far the onset of precipitation has been phrased in terms of thermodynamics relating to the metastable limit and no particular model for the nucleus has been assumed. It is evident that apart from the characteristics of the precipitating phase, the presence of impurity substrate must also be invoked. This type of initiation, known as heterogeneous nucleation, is the predominant process in many precipitation reactions, but at high levels of supersaturation it is conceivable that homogeneous nucleation may also occur. If the homogeneous nucleus is conceived as a spherical piece of solid phase then (provided the piece is large enough) the *solubility* of that nucleus, *i.e.* the activity of the solution in equilibrium with the nucleus, is given by the Gibbs–Kelvin relation

$$\frac{\rho}{M} RT \ln \left[\frac{I.P.}{K_{sp}} \right] = \frac{2\gamma_{S/L}}{r} = \frac{\Delta\mu}{\bar{v}} = \Delta G_v \qquad (5.14)$$

where r is the radius of the nucleus, M the molecular weight and ρ the density of the solid phase, and $\gamma_{S/L}$ the solid–liquid interfacial free energy. $\Delta\mu$ is the difference between the molecular chemical potential of solid and solution phases, $\bar{v}$ the molecular volume ($M/\rho N_o$), and ΔG_v the volume (Gibbs) free energy. N_o is the Avogadro number.

Thus it can be seen that the degree of stable supersaturation in a system can be related to the particulate size of the precipitate. In nucleation theory the creation of particle surface and its associated energy of formation relates to the energetics of cluster formation. The free energy $\Delta G^0_{AB}{}'$ for the formation of a critical metastable cluster is

$$\Delta G^0_{AB}{}' = \Delta G_v{}' + \Delta G_s \qquad (5.15)$$

where $\Delta G_v{}'$, the volume (Gibbs) free energy, is related to the heat of crystallisation and to the formation of *bonds* in the cluster. ΔG_s is the free energy associated with creation of the cluster surface. (Note that $\Delta G^0_{AB}{}'$ is the energy barrier to the formation of an AB cluster whereas ΔG^0_{AB} is the total free energy of the cluster). Equation (5.15) is only an approximation since it equates a standard free energy with the other free energy terms but a more rigorous statistical mechanical derivation yields a similar equation. From eqn. (5.14) the volume free energy is given by

$$\Delta G_v' = V\Delta G_v = -xRT \ln \left[\frac{I.P.}{K_{sp}}\right] \tag{5.16}$$

where V is the volume of the cluster.

The free energy associated with the formation of the surface of a cluster is difficult to define exactly because of the presence of edges, corners, and defects. For the present purpose it is (as usual) assumed that the contributions of these and of changes in surface energy with cluster size may be neglected and the free energy is given by

$$\Delta G_S = A\gamma_{S/L} \tag{5.17}$$

where A is the surface area of the cluster. For a faceted cluster a similar relation may be proposed

$$\Delta G_s = \sum A_i\gamma_i \tag{5.18}$$

where the area of the ith facet A_i is governed by the Gibbs–Wulff condition $\sum A_i\gamma_i$ = minimum for a given volume.

Insertion of eqns. (5.17) and (5.16) into (5.15) gives

$$\Delta G^0_{AB}{}' = -xRT \ln [I.P./K_{sp}] + A\gamma_{S/L}. \tag{5.19}$$

We can now see that the energy barrier to homogeneous nucleation $\Delta G^0_{AB}{}'$ is simply related to the excess of the cluster free energy over that of the same number of entities in the equilibrium solubility system, with the addition of a surface term.

The rate of homogeneous nucleation J is usually regarded as being controlled by the rate of addition of the post-critical ion to the critical nucleus and may be expressed by

$$J = A_k \exp (-\Delta G^0_{AB}{}'/kT). \tag{5.20}$$

A_k is a composite term including a diffusional energy barrier and the concentration of solution species, and k is the Boltzmann constant. The derivation of eqn. (5.20) involves the assumption that the embryo partition functions can be evaluated and that the formation of clusters does not significantly deplete the overall ionic concentration.

Maximisation of eqn. (5.19) with respect to x gives for a spherical nucleus

$$\Delta G^0_{AB}{}'(crit) = \frac{16\pi\gamma^3_{S/L}\bar{v}^2}{3k^2T^2 \ln^2 [I.P._{crit}/K_{sp}]}. \tag{5.21}$$

If experimentally the nucleation rate of one nucleus/sec per unit volume can be related to the critical ion product, then

$$\ln A_k = \frac{16\pi\gamma_{S/L}^3\bar{v}^2}{3k^3T^3\ln^2[I.P._{crit}/K_{sp}]}. \tag{5.22}$$

A_k is generally taken as $\sim 10^{25}$ so that all quantities are known except $\gamma_{S/L}$ and consequently the results of an experimental examination of homogeneous nucleation are usually expressed in terms of $\gamma_{S/L}$. This is a useful parameter for relating nucleation to solid–solvent interaction and would normally have values of 15–500 ergs/cm^2. The low values would indicate the interaction between a soft inorganic crystal and a non-interacting solvent or between a strongly interacting, high lattice energy crystal and a strongly polar solvent. Weak interaction between a high lattice energy crystal and solvent leads to high values for $\gamma_{S/L}$.

Heterogeneous nucleation

Since most precipitation reactions involve purities as heterogeneous nuclei it is desirable to pursue a similar theoretical approach for this situation. Various models have been assumed but the one which seems most appropriate for present purposes conceives heterogeneous nucleation as a sequence of ion or molecular diffusion, adsorption on the impurity, diffusion on the substrate surface and two-dimensional clustering at an active site.[7] If the location and configuration of ions or molecules in the critical cluster are determined by the spacing of atoms or ions in the substrate then nucleation is said to be coherent and

$$\ln[I.P._{crit}/K_{sp}] = 4\bar{d}^2\gamma_e^2/\beta kT\,(2Q_{ads} - Q_D + BkT) \tag{5.23}$$

where $\bar{d}$ is the average ion diameter, γ_e the cluster–solution edge free energy, Q_{ads} the adsorption energy, Q_D the energy barrier to surface diffusion, and B a kinetic factor. β is the ratio of lattice dimensions of solute crystal and substrate.

Presumably for precipitation processes many different impurities will be involved and $\beta \sim 1$ is an acceptable approximation. Experimental data in support of the above equation will be presented later but in terms of predicting the degree of supersaturation in precipitating systems two terms predominate namely γ_e and Q_{ads}. If the adsorption energy is high, *i.e.* the solute is strongly adsorbed by the impurity, then a relatively low supersaturation can be achieved. In general it might be expected that ionic crystalline fragments would be the best nuclei. We can also see that the level of critical supersaturation of ionic salts on an ionic substrate should be less than for organic crystals on similar substrates. This is in accord with the common laboratory experience that organic materials are not readily nucleated by foreign materials. Secondly, γ_e the edge energy, which is related to the lattice energy of a crystal, plays an important part in nucleation. Materials of high lattice energy would normally have a high γ_e and would consequently support higher supersaturation. Thus barium, lead, strontium and calcium sulphates and carbonates can be highly supersaturated whereas monovalent materials sodium chloride, silver chloride, etc. generally do not supersaturate to any extent. Finally, fairly soluble materials will normally have a high kinetic constant (B) and will thus support less supersaturation.

Both eqns. (5.22) and (5.23) are believed to apply equally well to organic crystals in which case the log of the critical supersaturation ratio $\ln [c_{crit}/c_0] = \ln S^*$ would replace $\ln [I.P._{crit}/K_{sp}]$; c_0 is the saturation concentration.

Criticisms of nucleation theory

Criticisms of nucleation theory are, and have been, many. Evidently nucleation rates are kinetic but the theory is based upon equilibrium thermodynamic reasoning. Fortunately the criteria of a critical concentration corresponding to a critical nucleus defined in terms of a generation rate of one nucleus per second, represent a situation sufficiently close to thermodynamic equilibrium that the above objection is not of major importance. This is not true, however, when large nucleation rates are involved. Probably the most serious of the criticisms brought to light so far has been that the Gibbs–Kelvin equation is not applicable to small clusters and that surface energetics lose their meaning for such small units. There is at present no known way of avoiding the use of the Gibbs–Kelvin equation since it is the only equation linking physical size and solubility of particles. With regard to the surface energetics various authors have considered changes of surface energy both for molecular clusters[8] and for ionic crystals.[9,10] Although there is some divergence of opinion this effect does not seem to be as important for three-dimensional nuclei as was originally supposed.

On the other hand the change of edge energy with cluster size is significant for ionic crystals[11] and probably plays some part in determining the morphology of crystalline precipitates.

NUCLEATION OF POLYMERS

One of the commercially important precipitation processes involves the precipitation of polymeric particles prior to the formulation of plastics. In the polymerisation process we may regard an n-mer as composed of n monomer units, thus

$$nA \rightarrow A_n \tag{5.24}$$

Although it is tempting to think of such a process as nucleation of an insoluble particle this is not so, at least not in terms of the classical concept of nucleation. Reaction (5.24) is not reversible and cannot, therefore, be treated by equilibrium thermodynamics. The step involved in the nucleation of polymers from solution is concerned with configurational changes in the molecule. Thus the nucleation of a crystalline polymer might be described in terms of chain folding, and the driving force for chain folding, nucleation and crystallisation being the supersaturation in terms of the molecules of specific length. Non-crystalline polymeric particles probably undergo similar but more random configurational changes on nucleation.

The nucleation of polymerising particles from solution has some unusual features. Since the rate of addition of monomer units to the *n*-mer is usually much more rapid than the diffusion of the *n*-mer to a suitable nucleating impurity, the system is able to undergo spontaneous homogeneous nucleation. The homogeneous nucleation process is typified by the large numbers of particles which can be produced in such systems (up to 10^{15}/ml and higher). Polymeric products from solution polymerisation processes are hence usually agglomerates and the control of agglomeration often controls the quality of plastics formulated from such polymers.

SECONDARY OR ANCILLARY NUCLEATION

It is known that if a seed crystal of ice is introduced into supercooled water, many ice particles are precipitated prior to complete solidification. Similarly, in the seeding of clouds, the introduction of relatively few silver iodide particles often leads to the precipitation of a vast number of rain drops. This proliferation of particles is brought about by secondary or ancillary nucleation. Examples more akin to the central theme of precipitation from solution are those of the seeding of supersaturated solutions with crystals of the solute phase. Generally, the introduction of one organic crystal induces the formation of many secondary crystals and recent experimental evidence indicates that the same is true for the seeding of inorganic precipitates.[12] It is believed that nuclei generated on the surface of the seed crystals are carried into the bulk solution by convection and then act as ideal heterogeneous nuclei. Jackson and co-workers have shown in a series of remarkable micrographs[13] that for ammonium chloride crystallisation, it is the formation of small dendrites, which fragment and fall into bulk solution, that causes ancillary nucleation. Actually, of course, this is not nucleation in the normal usage, but is rather ancillary crystal growth.

THE ROLE OF THE IMPURITY SUBSTRATE—EPITAXY

If the model for heterogeneous nucleation is correct, then the ions or molecules of a solute phase must take up some special configuration relative to the substrate in the most efficient energetic process. Macroscopically, such a process manifests itself as epitaxy or oriented crystallisation. Epitaxy of crystals on planar substrates has been known for many years and hundreds of examples are known. In Fig. 5.3 is shown the epitaxial arrangement of RbI crystals on a mica substrate. It can be seen that the rubidium iodide crystals which have the sodium chloride cubic lattice array, grow with their (111) planes adjacent to the (001) mica plane. Figure 5.4 shows a schematic diagram of the arrangement of the RbI ions

Fig. 5.3. Oriented overgrowth of rubidium iodide crystals on (*muscovite*) *mica. The crystals grow with their* (111) *planes adjacent to the* (001) *plane of mica.*

relative to those in mica. The most efficient substrates for epitaxy are those in which the lattice parameters match those in the depositing phase, as supposed in eqn. (5.23). Usually crystals do not grow epitaxially on substrates when there is a mismatch of $>15\%$, though isolated examples are known where there is no lattice matching criterion. One interesting example of this phenomenon is the epitaxy of polymers on ionic crystals. Willems and others[14,15] observed that polyethylene crystals deposited from solution onto the (001) face of sodium chloride crystals grew epitaxially along the (110) plane of the substrate (see Fig. 5.5). The initial suggestion was that the chain folds matched the lattice requirements of the substrate. However, more recent evidence[16] indicates that a range of polymers (polyethylene, polypropylene, polyoxymethylene, nylon, isotactic polystyrene) will all grow epitaxially on sodium chloride and that polyethylene will orient on LiF, NaBr and NaI (and probably many more) completely independent of the lattice matching requirements. The mechanism and energetics of polymer nucleation have recently been studied in the author's laboratory

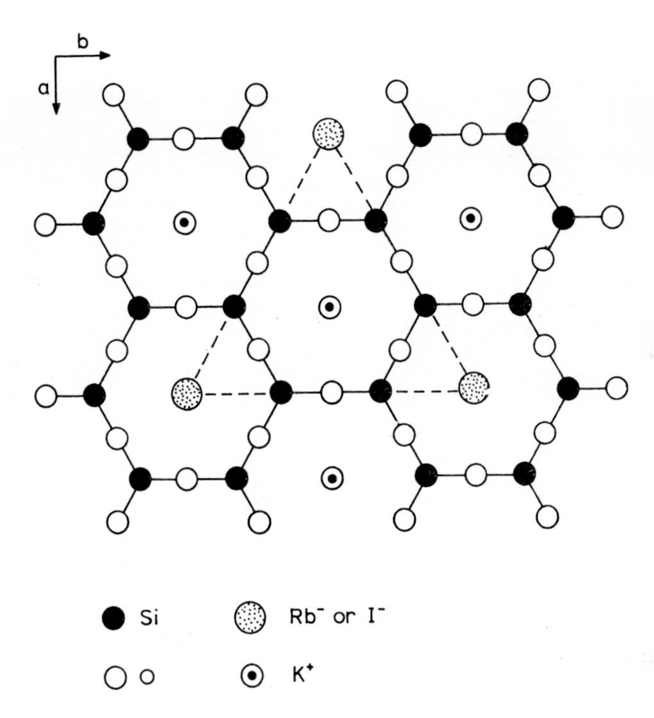

Fig. 5.4. The surface structure of mica is such that the heterogeneous nucleation of crystals of close lattice match is directly controlled by ions (or ion holes) in the mica surface. An arrangement of a few ions in the sub-critical cluster is shown. In practice surface nucleation probably occurs in the region of a surface defect (probably caused by the inclusion of one or more foreign ions which distort the surface structure).

with the conclusion that although the matching between the folded chain structure of polyethylene and the substrate plays a small part, it is predominantly the effect of induced dipoles which orients the molecules.[16-18]

The use of solid crystalline additives to control plastic crystallinity is in its early stages, but developments seem feasible in terms of the model experimental approach. Orientation of polymer molecules in physiology is also a likely area of development.

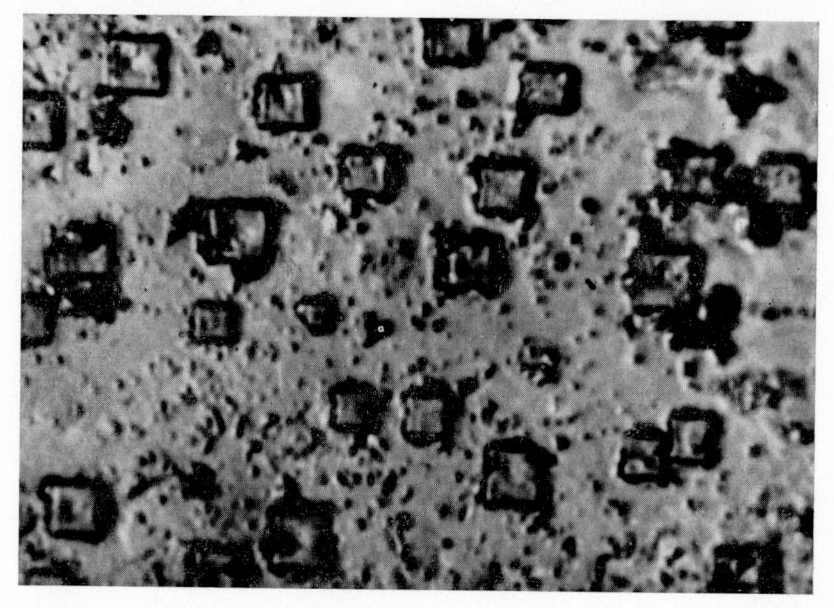

Fig. 5.5. Epitaxial growth of polypropylene crystals on sodium chloride apparently shows <100> *orientation on the* (001) *face. However the molecules of polypropylene lie diagonally across the square crystallites and are consequently ordered in the* <110> *direction.*

EXPERIMENTAL TESTS OF NUCLEATION THEORY

Heterogeneous nucleation

Probably the simplest way of determining the nucleation capability of additives is to supercool the melt, liquid, or solution, in the presence of the powdered additive and, by use of some differential thermal analyser, to detect the maximum subcooling and the subcooling or solidification profile. This latter information gives an insight into the spectrum of nucleating properties of the additive and impurities. Although this technique is suitable for some melts, particularly polymeric, it is not particularly suitable for the freezing of liquids (because of ageing effects limiting reproducibility) or for solutions.

In addition, since there would normally be a range of particle sizes and nucleating efficiencies in the additive, the data cannot be expected to provide an adequate test of heterogeneous nucleation theory. Consequently, in the relatively few attempts to devise definitive tests for heterogeneous nucleation theory, discrete and well defined substrates have been used. Most popular among substrates has been mica which can be easily prepared and is remarkably free from surface defects. Newkirk and Turnbull[19] examined the nucleation of ammonium iodide on to the surface of various micas and determined that, as expected, lattice matching between substrate and deposit plays a part in the energetics of heterogeneous (epitaxial) nucleation. The critical supersaturation increases with increase in lattice mismatch (25°C ambient temperature) as is shown in Table 5.2.

TABLE 5.2

CRITICAL SUPERSATURATION RATIO S^* FOR NUCLEATION OF NH_4I ON VARIOUS MICAS AS A FUNCTION OF LATTICE MATCH

(after Newkirk and Turnbull[19])

Mismatch (%)	*ln (Supersaturation ratio)*
1·13	0·61±0·01
1·53	0·62(5)±0·01
2·73	0·81±0·04
3·40	1·00±0·09
3·60	0·96±0·14

More recent work[7] has shown that the lattice matching requirement is born out for the nucleation of a series of alkali halides on mica, and furthermore the critical supersaturation decreases with increase in operating temperature, as required by eqn. (5.23). The theoretical predictions for two-dimensional heterogeneous nucleation cannot, however, be expected to apply exactly because we are dealing with a situation in which macroscopic thermodynamic considerations have been applied to clusters of relatively few ions. Particularly serious is the assumption that γ_e is independent of cluster size. Some current work being performed in the author's laboratory[11] has shown that not only does the edge energy of two-dimensional clusters change with size, but the gross changes in the energetics of cluster formation caused by modified edge energetics can cause major morphological effects.

One experimental method of avoiding the preceding difficulties would be to examine the nucleation of non-polar organic crystals on to a defined substrate. It is our experience, however, that because of the low interaction energy between organic deposits and substrates, the critical supersaturation is extremely large and cannot be handled conveniently by current methods. These conclusions are, of course, in line with common experience relating to the difficulty in nucleating many types of organic crystals from supersaturated solution.

One exception to the preceding observation is known to this author. Succinic acid can be nucleated on mica at reasonable (though not precisely reproducible) degrees of supersaturation. Some data for this system are given in Table 5.3 showing that again the critical supersaturation decreases with increase in temperature.

The alkali halides grow with their (111) planes adjacent to a mica (001) plane and thus form a rather limited class of epitaxial arrangement. Data are almost completely lacking in other systems.

TABLE 5.3

NUCLEATION OF SUCCINIC ACID ON MICA

Saturation Temperature	20°*C*	25°*C*	30°*C*	35°*C*
Critical Supersaturation Ratio (S^*)	1·628	1·60	1·57	1·55

Homogeneous nucleation

The test of homogeneous nucleation theory for crystal initiation from solution is particularly difficult. It is evident that, since impurities catalyse the nucleation of crystals from solution, one method of preventing heterogeneous nucleation would be by removal of the impurities. In practice it has proved impossible to achieve this end. Weiss, for example, has attempted by extensive purification, to remove impurities active in the nucleation of barium sulphate.[20] Normally, barium sulphate forms about 10^7 particles/ml when precipitated from aqueous solution and by scrupulous care Weiss was able to reduce this number to less than 10^2/ml. However, despite all the care taken, there remained these active sites or motes which, it was suggested, probably did not arise from solid contaminants, but rather from some chemical property of the solution.

One method of circumventing this problem is to break up the precipitating solution into more droplets than there are impurities. Hence, some droplets will contain no impurities and nucleation, if any, must be homogeneous. This technique has several problems associated with it. First, it is evident that the supersaturation within the drop must occur after drop formation since homogeneous nucleation is so rapid that it is not feasible to separate the drops before precipitation has occurred. Thus, it is necessary to use either the supercooling of saturated solutions or the homogeneous generation of ions to produce the required supersaturation. The latter of these methods does not seem to have been used in conjunction with the drop dispersal technique. Second, for accurate assessment of the critical supersaturation, it is necessary to know concentrations and temperature fairly precisely, which means that evaporation of the droplets must be avoided. It is usual then to disperse the droplets of a saturated salt in some inert medium (*e.g.* an oil). The medium itself will not catalyse nucleation if the two components (oil and aqueous solution) are completely immiscible and if the interfacial energy of the precipitate–oil interface is higher than that of the solution–precipitate interface (which it always is for ionic precipitates).

A third requirement for homogeneous nucleation studies by the drop dispersion method is that the size of the droplets is small enough to minimise the heterogeneous effect (droplets should be smaller than 5×10^{-3} cm in diameter) and yet contain sufficient solute that upon crystallisation the particles attain sufficient size to be detected. A suitable detection device for such a system might for example incorporate filming the microscopic field, illuminated by polarised light. Crystallisation then leads to the appearance of *twinkling* particles in the drops. The third of these criteria limits the use of the technique to fairly soluble salts since the particles in more sparingly soluble salts become too small to be detected. The solubility of the salt under consideration must indeed be greater than 10^{-2} M and must have a solubility which is strongly temperature dependent.

With the above limitations, very few systems are suitable for study by this method. White and Frost[21] were probably the first to attempt the use of this technique and applied it to the nucleation of ammonium nitrate. Their value for the interfacial energy of the solution–solid interface derived from a modification of eqn. (5.23) was 35 ergs/cm^2, which seems reasonable. Melia and Moffitt[22] have developed the technique used by White and Frost, their apparatus being shown in Fig. 5.6. They studied the nucleation kinetics of ammonium chloride and bromide and concluded that the kinetic constants are in the range 10^3–10^5, which seems rather low.

Apart from the difficulties with technique, the problems arising from sources of spurious nucleation in the droplet method are considerable. For example, although the medium itself may be inert, it may contain many impurities which could catalyse nucleation at the oil–solution interface. Furthermore, since this interface is likely to be diffuse at the molecular level, there exists the distinct possibility that local fluctuations in solute density could cause spurious nucleation in the interfacial region. If effects of the preceding type are present, calculated interfacial energies would be too low if homogeneous nucleation were assumed to be the predominant process.

Since the above procedures have been based upon the supercooling of aqueous solutions, it seems appropriate to examine the nucleation relations for such a situation.

For organic molecules

$$RT \ln S^* = \frac{\Delta H \Delta T_c}{T} = \frac{2\gamma_{S/L}\bar{V}}{r^*} \tag{5.25}$$

where ΔH is the heat of solution, ΔT_c the critical subcooling for the homogeneous nucleation of clusters of radius r^* and interfacial free energy

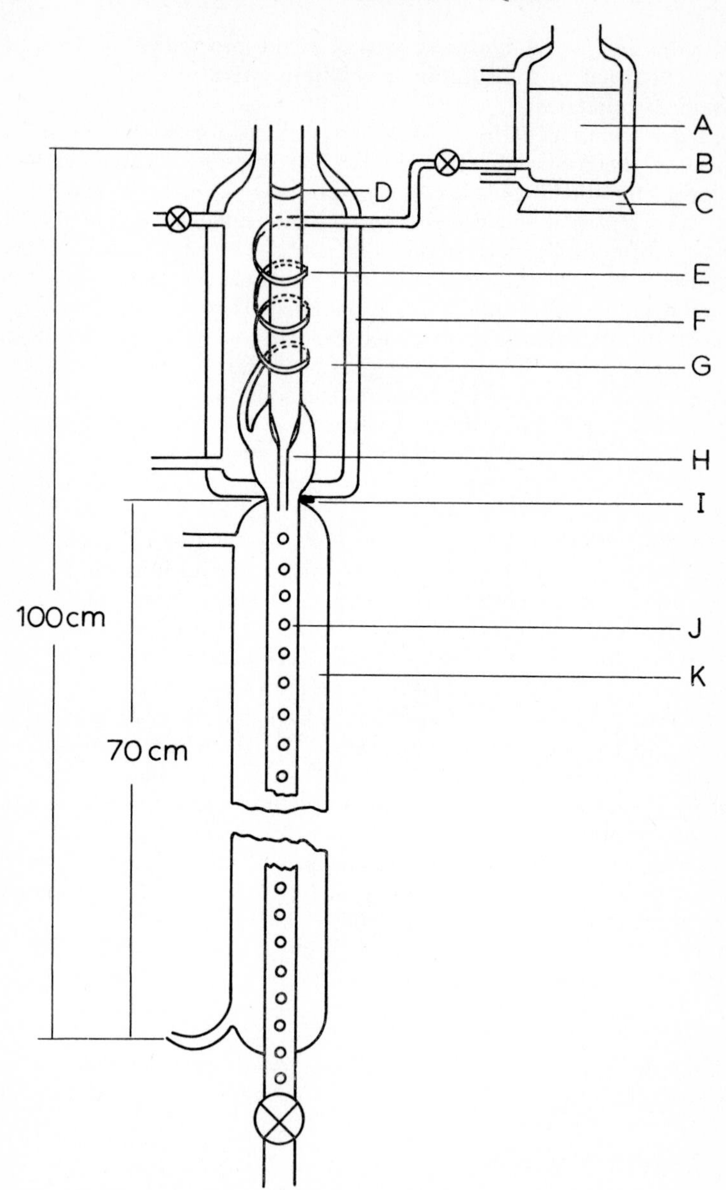

Fig. 5.6. Apparatus for studying the nucleation of crystals in aqueous droplets (*Vonnegut dispersion method*): A, *oil reservoir*; B, *hot water jacket*; C, *electric hot plate*; D, *liquid paraffin*; E, *solution under test*; F, *air jacket*; G, *hot water jacket*; H, *capillary jet*; I, *constriction*; J, *droplets of solution suspended in oil*; K, *water jacket* (*after Melia and Moffit*[22]).

$\gamma_{S/L}$. $\bar{V}$ is the molar volume and S^* the critical supersaturation ratio. For ionic systems, through modification of eqn. (5.14)

$$nRT \ln S^* = \frac{2\gamma_{S/L}\bar{V}}{r^*} = \frac{\Delta H \Delta T_c}{T} \tag{5.26}$$

where n is the number of ions in a neutral molecule. Thus eqn. (5.22) becomes

$$\ln A_k = \frac{16\pi\gamma_{S/L}^3 \bar{V}^2 T N_0^3}{3R\Delta T_c^2 \Delta H^2} \tag{5.27}$$

which is the equation generally used in homogeneous nucleation studies by the supercooled drop method.

Often information is required about the nucleation of precipitates other than in carefully controlled model systems. One might, for example, wish to know something about the nature of nucleation in a system such as that existing when components A and B are mixed together to form precipitate AB. Evidently the manner of mixing will play an important part in the rate of formation of the precipitate, but if the components are mixed and stirred rapidly, it might be assumed that a fairly uniform, instantaneous supersaturation can be achieved. If this supersaturation is relatively low, nucleation will be almost entirely heterogeneous, and since the number of precipitate particles is limited to the total number of effective nuclei, it is to be expected that the number of precipitate particles will be virtually independent of the supersaturation. This is, in fact, in agreement with observations made for many systems where, if the impurities originate from the solvent, the number of precipitate particles is either approximately constant or increases in short *jumps*.[23–26] This latter effect is due to the spectrum of nucleating efficiencies for the various impurities. In either case no sudden change of more than 10^2 particles/ml occurs.

In some systems, foreign nuclei originate mainly from the precipitating components. Such a situation arises when component A is dissolved in a solvent and precipitated by addition of a second miscible solvent in which it is insoluble. For this case different degrees of supersaturation can be achieved, at any one solvent combination, by changing the original amount of A dissolved. If the foreign nuclei originate from A (as they usually do), a plot of the number of precipitate particles as a function of the initial supersaturation is a straight line passing through the origin.

In each of the two preceding types of systems, a new type of phenomenon occurs at high degrees of supersaturation. This new phenomenon is that of homogeneous nucleation which becomes energetically favourable at high degrees of supersaturation. Homogeneous nucleation is characterised by a burst of particle generation which is detected on the macroscopic scale by a drop in particle size and an increase in particle numbers of several orders of magnitude. The critical supersaturation for the onset of homogeneous nucleation can easily be detected by examination of the particle size or by particle counts and thus the need for elaborate equipment may be avoided. Several systems have been studied by these means and data have been treated by means of eqn. (5.22). Critical supersaturation ratios, interfacial energies and nuclei sizes for several ionic precipitates are given in Table 5.4.

TABLE 5.4[a]

INTERFACIAL ENERGIES (AGAINST AQUEOUS SOLUTION) AND CRITICAL CLUSTER SIZES CALCULATED FROM HOMOGENEOUS NUCLEATION DATA (ASSUMING COMPACT SPHERICAL NUCLEI OF DIAMETER d^*)

Precipitate	*Critical Supersaturation Ratio* (S^*)	*Interfacial Free Energy* (*ergs/cm*^{2})	*Critical Size* d^* (Å)
$BaSO_4$	1000	116	11
$PbSO_4$	28	74	13
$SrSO_4$	39	81	12
$PbCO_3$	106	105	11
$SrCO_3$	30	86	12
CaF_2	80	140	9
MgF_2	30	129	9
AgCl	5·5	72	15
AgBr	3·7	56	15
Ag_2SO_4	19	62	14
$Ca(C_2O_4)$	31	67	13
$CH_2(NH_2)COOH$	2·1	40	30
Cholesterol	13	17	28

[a] Taken from *Formation and Properties of Precipitates* by A. G. Walton. John Wiley and Sons, New York, 1967, p. 30.

SOLUTION PHASE NUCLEATORS

We have already seen that careful precautions to remove impurities from solution do not reduce the heterogeneous nucleation rate to zero at moderate levels of supersaturation. This evidence alone might suggest the possibility of active agents other than solid particles. The possibility of chemical nucleators is also suggested by the many physiological crystallisation processes which occur in association with proteinous matrices in the human body. The protein molecules cannot be logically treated in the same terms as a rigid crystalline substrate. Recently Schiffmann and co-workers[27,28] have discovered that small quantities of EDTA (ethylenediaminetetracetic acid) cause supersaturated solutions of calcium phosphate to precipitate from normally *stable* supersaturated solutions. This effect has also been noted for a number of other chelating ligands containing nitrogen, acetate and or hydroxyl groups.[29] The mechanism for this phenomenon is rather obscure, but if we regard the nucleation step for inorganic crystals as being the collapse of the solvation structure to yield a solid in a cavity the heat change is given by

$$\Delta H = nH_{\mathrm{solv}} - Zn^{2/3}H_{\mathrm{imm}} \tag{5.28}$$

where Z is a geometric factor, n is the number of entities in the cluster and H_{solv} and H_{imm} are the heat of solvation per ion and heat of immersion per unit area. Hence, any agent which decreases the solvation energy of any ion or ions in the cluster will reduce the energy barrier to nucleation. It seems probable that the solution phase nucleators are effective because of a combination of desolvation and specific configurational binding.

EFFECT OF NUCLEATION MECHANISM ON PRECIPITATE CHARACTERISTICS

The mechanism of nucleation is directly related to the size and number of precipitate particles. If we put on one side for later discussion the morphology of individual particles and possible fragmentary processes, we can infer the character of the precipitate from the previous discussion. For example, if heterogeneous nucleation is the predominant mode of precipitate initiation, the total number of precipitate particles cannot exceed that of the number of initiating impurities. Now if all the impurities display equal catalysing ability, the number of particles will be independent of the initial concentration of reactants (c_0), and the particle size increases with increase in c_0. At sufficiently high degrees of initial supersaturation, homogeneous nucleation occurs and with the vast increase in particle numbers, the average particle size decreases. Hence, for the preceding type of system we have a particle size maximum at the critical supersaturation for the onset of homogeneous nucleation. This critical supersaturation is of course a function of the interfacial energy of the material in question; the particle size is shown schematically as a function of initial supersaturation ratio of the reacting components in Fig. 5.7. The three curves A, B and C correspond to materials of low, medium and high interfacial energy respectively (*e.g.* 50, 100 and 200 ergs/cm^2). Actually, many precipitates are known to give size distribution curves corresponding to those in Fig. 5.7. Von Weimarn,[2] for example, reported many systems showing the size maximum early in this century. It also seems likely that the *crystallisation maximum* referred to extensively by Tezak and co-workers and deduced from turbidity and light scattering measurements also refer to the same phenomenon. Evidently some control over the particulate size can be maintained by manipulation of appropriate degrees of supersaturation.

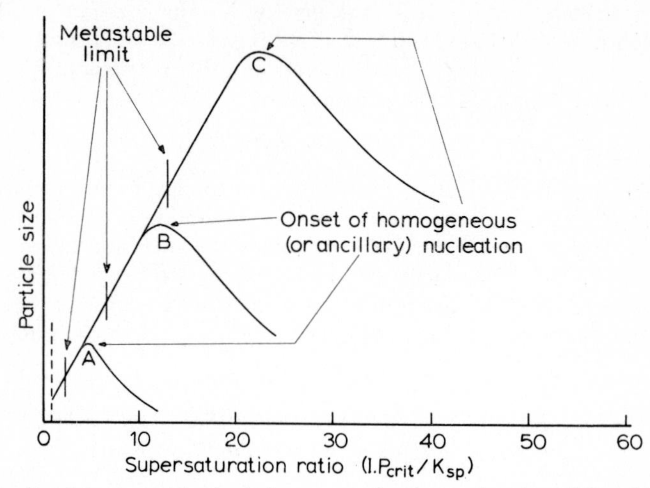

Fig. 5.7. *Precipitate particle size (after say one hour) as a function of the initial supersaturation ratio. Curve* A *is for low;* B, *for intermediate; and* C, *for high interfacial energy materials. In the area of heterogeneous nucleation, the number of initiating impurity nuclei is assumed constant.*

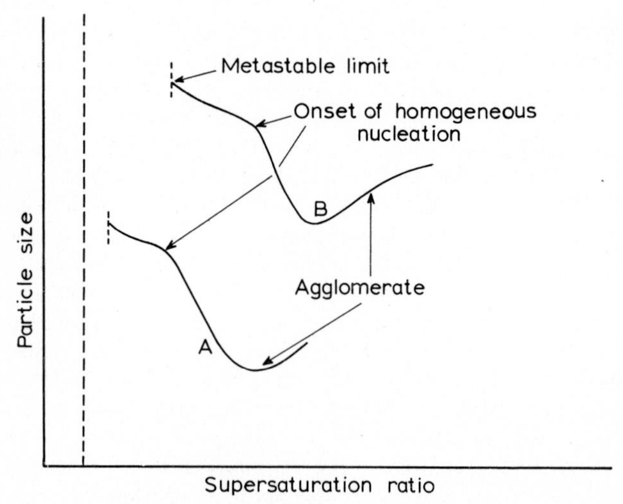

Fig. 5.8. *This figure shows the effect of a wide spectrum of initiating impurity nucleators on the final average precipitate particle size. Curve* A *is for low, and* B, *for high interfacial energy materials.*

Maximum particle size can be achieved by increasing solubility and working at the crystallisation maximum concentration.

Not all precipitates show distinct size maxima and the reasons are fairly apparent. In practice the activity of the various nucleation catalysing impurities is not uniform and a spectrum of efficiency is displayed. A simple example of this situation might occur in a system where at a supersaturation ratio of 2, 10 impurity particles caused nucleation; at ratio 3, 100 particles; 4, 1000 particles and so on. In this case the final precipitate particles will always decrease in size but will still usually exhibit detectable inflections at the onset of homogeneous nucleation (Fig. 5.8). One further complicating feature in many systems is that in the high concentration areas where homogeneous nucleation has occurred, agglomeration also occurs giving an illusory increase in particulate (agglomerate) size. This effect is also shown in Fig. 5.8.

Control of precipitate morphology is then closely related to control of supersaturation. Specific effects will be considered later.

PRECIPITATION KINETICS

The many fundamental measurements which may be made in nucleation systems give a direct indication of the molecular mechanism of crystal and precipitate birth. Unfortunately, the same cannot be said of precipitation kinetic studies. Here the situation is extremely complex and it is impossible to reach any really valid conclusions about the mechanism of precipitation from gross measurements on the formation rate. Nevertheless, precipitation rates are of considerable importance in determining the feasibility of commercial operations and many attempts to derive empirical relations have been made in order to systematise the appropriate approach.

A short historical digression seems appropriate at this point since it gives a direct indication of the categories into which growing precipitates may be placed. In the early theories by Balarev[30] and others,[31] it was supposed that the nucleation of precipitates was a continuous process supplying considerable numbers of primary particles which underwent coalescence and ripening, leading to highly irregular clumps of coalesced crystallites with a large *inner surface*. Occlusion of solution and impurities was a natural consequence of such a theory. Simple microscopic observation seems to confirm these suggestions.

However, if the nucleation processes described in the previous section are correct, heterogeneous nucleation is the predominant formation mechanism and it is unlikely that sufficient crystallites are generated to undergo extensive agglomeration. Hence, the morphological development of precipitates at low degrees of supersaturation must be predominantly determined by the mechanism and rate of precipitation rather than by agglomeration. It is common experience that the most perfect crystalline precipitates are produced by careful control of the environment and under these circumstances, there are systems in which the kinetics of growth can be conveniently related to the development of crystal surface. The obvious limit of the ideal system is that of the single crystal growing from a slightly supercooled solution.

The theories of single crystal growth are outside the scope of this chapter, but it is by now well established that growth by the Frank screw dislocation mechanism occurs commonly in the formation of crystals from very low supersaturation. At higher degrees of supersaturation, surface nucleation becomes feasible. It is to be expected then that precipitate crystallites would also grow by similar processes.

TABLE 5.5*

FORMATION OF COARSE CRYSTALLINE PRECIPITATES BY PRECIPITATION FROM HOMOGENEOUS SOLUTION

Precipitate	*Method*	*Reagent*
Sulphates (Ba, Pb, Sr, Ca, etc.)	Generation of anion	Dimethyl sulphate hydrolysis, sulphamic acid hydrolysis, thiosulphate-persulphate reaction
Sulphides (Sb, Bi, Mg, Cu, As, Cd, Pb, Sn, Hg)	Generation of anion	Thioacetamide hydrolysis
Phosphates (Zr, Hf)	Generation of anion	Hydrolysis of triethyl phosphate, tetraethyl pyrophosphate, metaphosphoric acid
Chlorides (Ag, Tl)	Generation of anion	Hydrolysis of allyl chloride
Phosphates (Ca)	Change of pH	Hydrolysis of urea or cyanate
Oxalates (Mg, Ca, Zn, La)	Generation of anion	Hydrolysis of dimethyl or diethyl oxalate
Oxalates (Mg, Ca, Zn)	Change of pH	Urea + bioxalate
Calcium Salts (fluoride, sulphate)	Generation of cation	Hydrolysis of ethylene chlorohydrin in presence of Ca–EDTA complex
Carbonates (rare earths, Ca, etc.)	Generation of CO_2	Hydrolysis of trichloro-acetates, cyanate, etc.
Chromates (Pb, Ag)	Change of pH	Hydrolysis of urea or cyanate

* From *Formation and Properties of Precipitates* by A. G. Walton. John Wiley and Sons, New York, 1967, p. 152.

The control of precipitate morphology by causing growth from low degrees of supersaturation has been an active area of research. Of the possible methods of performing such a manipulation (evaporation, supercooling and precipitation from homogeneous solution), precipitation from homogeneous solution (PFHS) has met with the most consistent success. PFHS methods involve the chemical generation of either anion or cation (but usually anion) in the presence of the excess of the other ion. For example, generation of sulphate ions into strontium nitrate solution causes the precipitation of well formed strontium sulphate crystals. A summary of some of the precipitates which are formed as coarse crystalline materials by PFHS methods is given in Table 5.5.

Efforts to determine the kinetics of precipitate formation under the preceding circumstances have been attempted by several workers. We may imagine the sequence of steps involved in PFHS to be the build-up of supersaturation, a burst of heterogeneous nucleation and then collapse of supersaturation as the precipitate particles grow. This sequence is shown in Fig. 5.9.

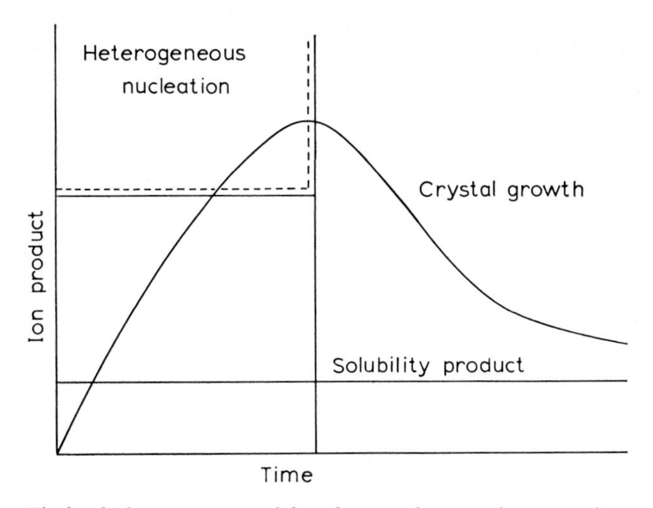

Fig. 5.9. The level of concentration of the solute in solution is shown as a function of time for a slow generating reaction (PFHS). The reaction might, for example, be the hydrolysis of allyl chloride into excess silver ions, the ion product [Ag^+] [Cl^-] would then be represented on the ordinate axis.

Since the crystals produced by PFHS methods are very regular in shape, we may assume that agglomeration is usually absent. Simple models for the growth rate might be based upon the limiting features of diffusion of solute through solution, diffusion across the boundary layer, or diffusion across the solid surface, in which case precipitation may be said to be diffusion controlled. Alternatively, growth might be controlled by surface reaction or nucleation in which case precipitation is said to be interface controlled.

Various subdivisions of these limiting processes may be described and approximate growth laws evolved. For example, the growth rate should be first order in solution supersaturation of solute, if simple diffusion is rate controlling, but between first and second order if growth is by screw dislocation. With simple diffusion to active sites, the growth rate should be a function of the solid surface area; with spiral growth the rate may be independent of the surface area.

For interface controlled growth[32] the rate should be of order (in terms of supersaturation) equal to the number of ions in a neutral molecule (*e.g.* 2 for AgCl, 3 for Ag_2CrO_4, etc.), or one, for organic precipitates if surface reaction is rate limiting, and should be proportional to the surface area; for surface nucleation the rate is very sensitive to concentration and high orders can be expected. (N.B. The rate law described for surface reaction controlled growth is applicable only at low degrees of supersaturation. At higher supersaturations, the order increases). Thus, most examinations of precipitation kinetics have been channelled towards rate laws of the form

$$R = f(A, S) \tag{5.29}$$

where R represents the rate, A is surface area and S the supersaturation.

Using the PFHS method, La Mer and Dinegar[33] found that the growth rate of sulphur precipitates was diffusion controlled. For inorganic systems Collins and Leineweber[34] found that precipitation of $BaSO_4$ (generation of sulphate by $S_2O_6{}^{2-}$, $S_2O_3{}^{2-}$ reaction) was diffusion controlled and a similar conclusion was reached by Jaycock and Parfitt[35] for the precipitation of AgI from ethanol (generation of iodide by C_2H_5I hydrolysis). However, in both these studies, complicating side reactions make analysis difficult.

The above analyses were based upon a rate law of the form

$$\frac{dm}{dt} = k'A^{\frac{1}{3}}S \tag{5.30}$$

where m is the mass of precipitate and k' the rate constant. An alternative form of eqn. (5.30) which has found common use contains the parameter α which is the fraction of the total amount of precipitate formed at time t

$$\frac{d\alpha}{dt} = k''\alpha^{1/3}(1-\alpha). \tag{5.31}$$

However, we notice that for PFHS processes the amount of solute remaining in solution is controlled not only by the precipitation reaction itself, but is counterbalanced by the generation of new material. Thus, if the rate of precipitation is followed by a technique which measures the amount of solute remaining in solution (*e.g.* conductivity), the rate of disappearance is slower than it would be in the absence of a generating reaction, *i.e.*

$$R = f(A, S) - f'(R'). \tag{5.32}$$

Conversely, if the rate of precipitation is followed by a technique which detects the appearance and size of the precipitate (*e.g.* light scattering[35]), the rate is faster than it would be in the absence of a generating reaction. Under these circumstances, the resolution of eqn. (5.32) in its integral form is extremely difficult.

Walton and Klein[36] have examined the kinetics of AgCl formation by PFHS (hydrolysis of allyl chloride) in the region for which the generating reaction just balances the precipitation rate. Under these circumstances, S is constant and the rate was found to be proportional to A, which is normally typical of an interface controlled reaction. The reaction order could not be determined under such circumstances.

Perhaps a logical manoeuvre to circumnavigate the preceding difficulty would be to find a PFHS system in which the generating reaction could be *turned off* after precipitation had been initiated. Such a method is possible. Klein and co-workers[37] have, for example, used the electrodecomposition of thiocyanate ions to produce sulphate, and although this type of reaction does not appear to have been used for kinetic studies, useful co-precipitation data have been obtained by the method.[38]

Another method whereby the precipitation of solute from solution on to a well defined crystalline substrate can be attained is by the use of seed crystals. Although the seeding of supersaturated solutions is not strictly analogous to precipitation *ab initio*, the advantages of growth studies on such systems are many. Firstly, known amounts of seed crystals may be used in successively reproducible runs. The rate of decrease in supersaturation can be conveniently followed by conductivity measurements (for sparingly soluble salts in solutions which have well-defined ionic species) and the order of the reaction is readily established. This situation is, in a sense, the opposite of the PFHS case where the solute concentration was maintained constant and the volume of precipitate changed. With seed crystals the percentage volume change is small, but the solution concentration changes.

Seed crystal growth has been studied for over half a century, but most quantitative developments have occurred in the past twenty years. Davies, Nancollas and co-workers[39–42] have performed extensive analyses in terms of eqns. (5.29)–(5.31) with the conclusion that the rate of growth is given by

$$\frac{dm}{dt} = k'n_s\,([A^+] - [A_0{}^+])^a\,([B^-] - [B_0{}^-])^b \tag{5.33}$$

where n_s is the number of seed crystals and A_aB_b is the neutral solute. For equivalent concentrations of A and B

$$\frac{dm}{dt} = k'n_s(c - c_0)^x \tag{5.34}$$

where x is the number of ions in the neutral molecule.

Systems examined and found to conform to the above relation have been $AgCl(x = 2)$, $Ag_2CrO_4(x = 3)$, $MgC_2O_4(x = 2)$, $BaSO_4(x = 2)$. Just as PFHS methods involving constant supersaturation cannot yield information indicative of the *reaction order*, seed crystal experiments do not divulge the rôle of the seed surface area. For example, it is not possible to tell whether each individual seed grows as a function of its total surface area or whether its growth rate is independent of its size. However, the evidence, from seed crystal growth data, points to the fact that the growth process is dominated by an interfacial mechanism.

The applicability of a generalised growth relation for seed crystals has led many workers to search for a similar relation which might apply to precipitation by direct mixing processes. Here much more serious difficulties are encountered. Apart from supersaturation inhomogeneities occurring during, and shortly after the mixing process, growth is generally very irregular and often dendritic.

There is thus no evidence that crystallites will maintain their shape through the precipitation and in all probability, the growth rate for any one crystal varies from point to point on the surface because of surface irregularities. It is shown later that fragmentation of crystals can also occur under rapid growth conditions and even for discrete surfaces there is no evidence that growth will continue at a uniform rate in the presence of impurities that are likely to be in solution. The situation would thus appear to be impossibly complex.

In spite of these difficulties precipitation reactions do appear to behave remarkably uniformly. For example, Turnbull[43] was able to show that the fraction of $BaSO_4$ precipitated with time showed a sigmoidal relation which was dependent upon the initial concentration, and mode of reactant mixing (Fig. 5.10). Evidently the curve is related to the appearance of solid surface and is autocatalytic in nature. Empirically the kinetic relation is similar in form to eqns. (5.31) and (5.34), and might be written

$$\frac{d\alpha}{dt} = k''\alpha^{2/3}(1-\alpha)^p. \tag{5.35}$$

Turnbull found that the precipitation curves could be superimposed by varying k'', thus indicating the uniformity of the process. Subsequently, similar relations have been found applicable to a number of different precipitation systems, though barium sulphate has been more intensively studied than any other material. Contention has centred around the value and meaning of p. Doremus,[44] from an analysis of Turnbull's data, reports that $p = 3$ or 4, and is indicative of surface reaction controlled precipitation. Nielsen[45] also reports $p = 3$ or 4 from his own data. Other workers[46–48] found $p = 2$, 1 or 0 with diffusion control also being reported.

This confusing situation may in part be due to the detection techniques used. Conductivity methods used to establish $p = 3$ or 4 have been shown to suffer from the fact that platinum electrodes catalyse the precipitation process. It is also evident that methods which measure the rate of disappearance of solute from solution do not identify the nature and extent of the precipitate. Light scattering techniques, which measure the volume of suspended solid, are susceptible to the problematical interpretation of data for non-uniform and non-spherical particles. Thus, not only is the phenomenon of kinetic precipitation complex in itself, but techniques devised to follow the process have not been entirely adequate. Possibly

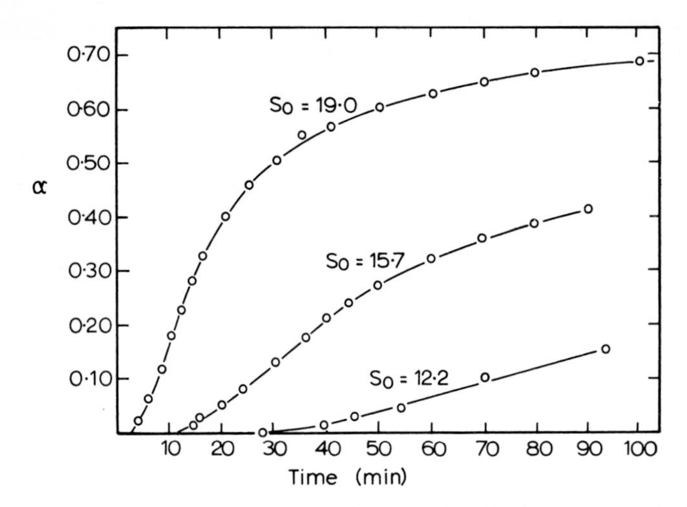

Fig. 5.10. Precipitation curves for barium sulphate produced by direct mixing. The fraction precipitated α is shown as a function of reaction time. The precipitation rate is seen to be strongly affected by the initial supersaturation ratio S_0 *(after Turnbull*[43]*).*

some form of dilatometric technique may eventually be feasible. It is this author's opinion that an equation of the form of eqn. (5.35) is a suitable relation to be used in conjunction with precipitation kinetics if the solid phase is discrete and well formed. It is unlikely, however, that the parameter p is independent of the supersaturation, though it should reduce to x of eqn. (5.34) at low degrees of supersaturation.

PRECIPITATE MORPHOLOGY

The morphological characteristics of a precipitate have been seen to be dependent on the nature of both nucleation and growth processes. The dominant parameter in nucleation and growth is the degree of supersaturation. At very low degrees of supersaturation, small crystals may be expected to grow with a shape close to that which corresponds to the minimum interfacial free energy, *i.e.* the Gibbs–Wulff criterion $\Sigma A_i\gamma_i$ = minimum, should be satisfied.

Confining our attention initially to the equilibrium form of crystal, it might be assumed that the lowest energy configuration would correspond to flat planar surfaces, thus minimising the effective surface area. Actually, according to the Gibbs concept, the interface is *diffuse*, and Mutaftschiev[49] has calculated that for the crystal–liquid interface, the diffuse layer can be represented by defects which lie at a depth of one molecule or ion either side of the surface of tension. Thus, at normal temperatures we may conceive that the interface undergoes thermal fluctuations which render it *rough* in terms of molecular dimensions. There is also some evidence that thermal instability causes surface steps to form. This point would seem to be important in relation to the mechanism of crystals growth since it is often assumed that surface steps are necessary for growth propagation.

Although the equilibrium form of crystal surfaces involves roughening at the molecular level, the gross structure of pure crystals involves macroscopically flat surfaces. The equilibrium shape thus will conform to a configuration involving low surface energy surfaces. The surface energy may be calculated in some cases for the idealised crystal–vacuum interface, and from such calculations, Stranski and others[50–52] have deduced the equilibrium shape of some simple crystals. For example, the (100) planes of alkali halide crystals are of lowest surface energy and the equilibrium shape is thus cubic.

However, it does not necessarily follow that the equilibrium shape of crystals grown from *solution* is the same as that calculated for crystal–vapour equilibrium. If the relative energetics of the crystal surface–environment are unchanged, as they would be if the solvent was inert, the crystals would indeed have the same equilibrium form as that calculated for the vapour environment. It is rare that solvents are inert and unreactive towards inorganic crystals, because the solute would not be soluble in such a solvent. In organic systems hydrocarbon solvents might be regarded as inert and unreactive towards non-polar materials, but here again we may expect that in general solvent–solid interactions will modify surface energetics and will hence modify the shape of crystals produced from a specific solvent.

The energetics of a crystal–liquid interface may be considered to be composed of two major terms, the solid–vapour surface free energy ($\gamma_{S/V}$) and a solid–liquid interaction term ψ_{12}. These parameters may be related by

$$\gamma_{S/L} = \gamma_{S/V} + \gamma_{L/V} + \psi_{12}/A \tag{5.36}$$

where $\gamma_{L/V}$ is the solvent–vapour interfacial free energy (the surface tension) and A the surface area of solid in contact with the liquid. The interfacial energy of a particular crystal facet is then governed by the specific surface free energy and solvent interaction for the facet. For sodium chloride (100) planes the appropriate relation would be

$$\gamma_{S/L}(100) = \gamma_{S/V}(100) + \gamma_{L/V} + \psi_{12}(100)/A. \qquad (5.37)$$

It can be seen that the larger the solid–liquid interaction, the lower is the solid–liquid interfacial free energy (ψ_{12} is negative by convention) and hence that face is more energetically favourable and more predominant. It turns out that since the (100), (010) and (001) planes of the crystal sodium chloride structure have the same surface configuration of ions, they will also have the same (solid–vapour) surface free energy and solid–liquid interaction. The balance of these forces allows the equilibrium shape of NaCl-type crystals, grown from solution, to be of cubic habit unless some other planes are energetically more favourable.

If we consider the (111) planes of a sodium chloride crystal, we find that the planes are occupied by ions of one type only, and hence, the surfaces composed of such planes will carry a net electrostatic charge. If this surface is exposed to a polar solvent, strong interaction will occur, and hence the ψ_{12} term of eqn. (5.37) will be large, and it might be expected that (111) planes will be stabilised by polar solvents. Undoubtedly (111) planes are stabilised by polar solvents, but in all known cases $\gamma_{S/V}(111) \gg \gamma_{S/V}(100)$ and this effect overcomes the solvent–solid interaction. It is possible, therefore, to justify the fact that the equilibrium form of NaCl-type crystals in solution is usually cubic.

The situation for non-cubic crystals is obviously much more complicated. In this case the arrangement of ions in the crystal faces is not the same. If the solvent does not interact discriminately with any one face the crystal, in solution, should take on the theoretical low surface free energy ($\gamma_{S/V}$) form. The low surface energy form often reflects the underlying crystal structure and precipitate crystals can often be identified by relation of shape to crystal structure (cubic, hexagonal, tetragonal, etc.). In some cases specific solid–liquid interactions are likely to cause certain planes to become exaggerated and in crystals of the same crystal structure, grown from the same solvent, differences in habit may be observable due to the polarisation of the solvent by the smaller ions. It might be deduced, therefore, that differences in habit are caused by the solvent–solid interactions. Typical of the preceding considerations are the differences between crystals of pentaerythritol, $C(CH_2OH)_4$ grown from hydroxylic and non-hydroxylic solvents. This material has a tetragonal structure with $-CH_2$ groups lying predominantly in the (100) planes and $-OH$'s in the (110) planes. In aqueous solution the (100) surface is absent and a (110) bipyramid is formed.

It is not possible to calculate an interfacial free energy directly, but its cousin, the interfacial energy ^{S}U may be calculated in some cases (and $^{S}U = \gamma$ when $T = 0°K$). Most calculations have been limited to cubic systems[53,54] though recently some developments have occurred which allow more complicated systems to be examined.[10] For organic crystals it is straightforward to calculate ^{S}U since it may be related to the sublimation energy E_0 (and with appropriate volume corrections to heats of sublimation)

$$^{S}U = \frac{E_0}{2n} N_x \tag{5.38}$$

where n_1 is the number of atoms or molecules per unit area of surface and N_x is a geometrical factor accounting for the appropriate packing of molecules in a surface.

Although a general value of ^{S}U may be obtained for any specific solid, it is not easy to obtain the surface energy of specific faces. Estimates might perhaps be made through wettability measurements with all the attendant thermodynamic problems, but there is as yet no published material relating surface energetic measurements to crystal shape.

The principles of organic crystal habit are, nevertheless, fairly clear. For example, organic molecules which possess both polar and non-polar segments will have a habit in which polar planes are predominant if grown from a polar solvent and non-polar (or less polar) planes predominant if grown from a non-polar solvent. A schematic diagram is shown in Fig. 5.11. Wells[55] has been instrumental in pointing out the habit modifying influence of several organic solvents. Examples taken from Wells' work are: Anthranilic acid crystals are bipyramidal when grown from ethanol, and prismatic with pyramidal faces when grown from acetic acid. Iodoform forms bipyramidal crystals when grown from aniline and prismatic crystals when grown from cyclohexane.

For a two-dimensional crystal of equilibrium shape, it is easy to show that

$$\gamma_{S/L}(1)l_1 = \gamma_{S/L}(2)l_2 \tag{5.39}$$

where $\gamma_{S/L}(1)$ is the interfacial free energy of side length l_1, etc. By analogy we might expect that

$$\gamma_{S/L}(1)A(1) = \gamma_{S/L}(2)A(2) = \gamma_{S/L}(3)A(3) \ldots, \tag{5.40}$$

i.e. the product of the interfacial free energy and the area of that surface is a constant.

It might be argued that since each plane in a crystal has a discrete surface, and hence interfacial free energy, it should be represented in the equilibrium form of a crystal. Such a conclusion might indeed have some validity if true equilibrium could be attained. However, we already know that crystals do not have an infinite number of facets, and we must suppose therefore that crystal habit is defined both by surface energetics and the mode of growth. It now remains to be shown that the growth rate does indeed have a profound effect upon precipitate and crystal morphology.

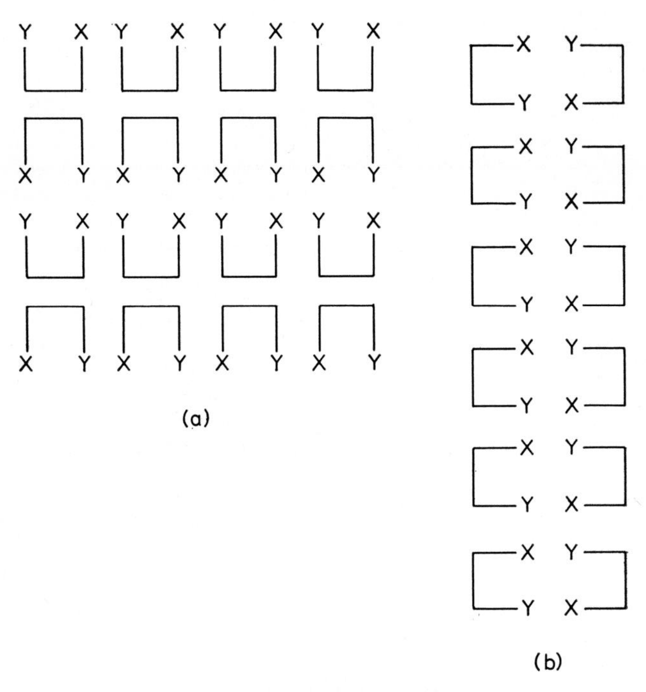

Fig. 5.11. *Schematic, two-dimensional crystal grown* (a) *from a polar solvent and* (b) *from a non-polar solvent.*

It is already evident, qualitatively, that the most coarse precipitates are formed by PFHS methods which control growth rates and solution concentration. Since the growth rate of a crystal is dependent upon the availability of solute, it is apparent that if surface adsorption and diffusion did not occur, crystal growth would be equally rapid in all directions. In fact, the supply of solute to the active growth site, whether it be screw dislocation or surface step or nucleus, is controlled by the amount of material adsorbed on that particular surface. Since the adsorbed concentration is proportional to the binding energy at that surface, which is in turn a function of the surface energy, it is to be expected that the adsorbed concentration and rate of growth will be greatest on the high energy surfaces. Consequently, growth is most rapid in the direction perpendicular to the high energy surface.

One result of this sequence of events is the fact that high energy planes grow out of existence and do not contribute in the manner described by eqn. (5.40). A two-dimensional representation of the sequence is shown in Fig. 5.12. For very slow growth rates well formed crystals are produced with low energy faces dominant; this habit does not, though, necessarily resemble the equilibrium form. At higher degrees of supersaturation and at correspondingly higher growth rates, the crystallites sometimes change habit, showing normally unfavourable faces and eventually become dendritic.

The reasons for the development of unusual planes is not clear, but if the effect of adsorbed impurities is negligible, it is possible that the energetics of nucleating unusual facets becomes feasible at intermediate degrees

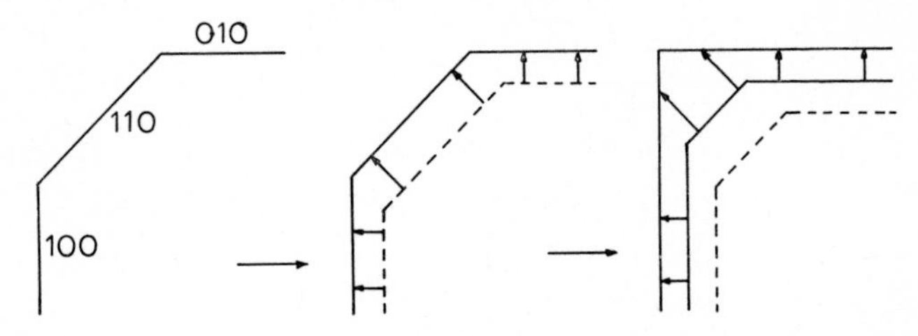

Fig. 4.12. *Schematic diagram of the elimination of a high energy* (110) *crystal face. Growth occurs most rapidly on high energy faces such that eventually they become eliminated* (*for slow growth*).

of supersaturation. The effect of impurities is readily placed within the preceding framework of surface energetics and interactions. Surface active agents generally adsorb on the (111) planes of NaCl-type crystals, supposedly because of the high electrostatic forces. The situation is then the opposite of the normal growth sequence since growth now virtually ceases perpendicular to the (111) surface and the crystal grows with octahedral habit, the (100) planes having grown out of existence.

Many modifications of the preceding type are known, the modification of sodium chloride by urea being a commonly quoted example. The electrostatic explanation of (111) plane activity is not entirely satisfactory and recent advances have suggested that preferential adsorption of surface active agents is also a function of matching configuration with the substrate. Presumably there are at least two types of effects induced by surfactants. Firstly, the specific type in which growth on definite planes is inhibited, and secondly, the non-specific type in which growth on all planes is inhibited equally. This latter effect may, for example, be noted upon addition of citrate ions to precipitating barium sulphate, whereupon the precipitate crystals become spherical. Other similar effects have been noted for lead and lanthanum iodates.

The addition of surface active agents causes, then, not only the changes in habit, but also the underlying growth rate, and slower growth rates usually lead to increased perfection of the precipitated crystallites. Consequently, it is fairly common industrial practice to add traces of surfactant to the solvent in crystal growing plant. It might also be argued that in PFHS methods, one of the reactants may adsorb and react at the surface, thus slowing the growth rate and producing more compact crystallites. It would be interesting to study the growth rate of seed crystals in the presence and absence of the PFHS generating reagent to determine whether significant surface adsorption occurs.

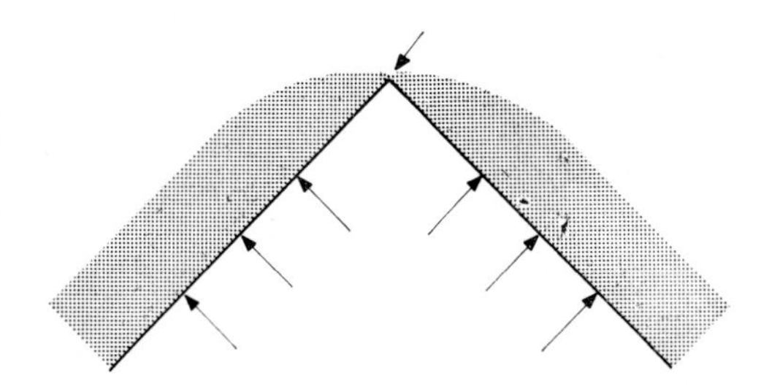

Fig. 5.13. Schematic representation of solvent molecules released from the forming solid hindering growth on plane surfaces.

Growth of ionic precipitates from solution should be regarded as a two-way process; solvated ions diffuse to the surface and subsequently, at some stage, desolvate. The released solvent molecules diffuse away from the surface causing a counter current. At high degrees of supersaturation the steady state concentration of solvent molecules in the interfacial region reaches a maximum, and the concentration contour resembles that in Fig. 5.13.

The limitation on the growth rate imposed by this boundary layer probably does not extend to crystal corners which are not blocked and hence growth at the corners becomes predominant as shown in Figs. 5.14(a) and (b). Thus, the development of dendritic crystals ensues with the dendritic arms extending in well defined crystallographic orientation to the parent crystal. Continued growth of dendrites often leads to solution (and impurity) entrapment as shown in Fig. 5.15. It is also often found that for fast growing crystals, addition of surfactants promotes dendrite formation, probably by the *traffic jam* mechanism.

We have seen, therefore, that on the one hand, addition of surfactants to solutions of relatively slowly growing crystals often improves their crystal form whilst on the other hand at higher degrees of supersaturation and faster growth rates, the same surfactant may cause dendritic growth with the same solute. It is not surprising that control of precipitate and crystal morphology by additives has become something of an industrial *art*.

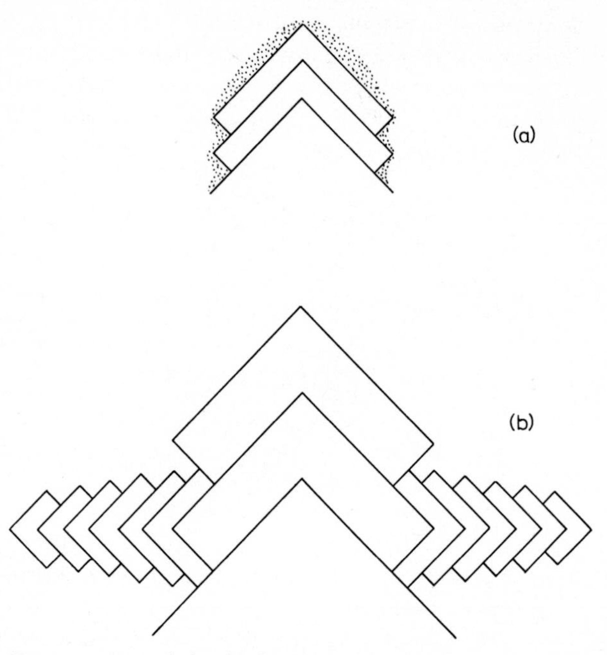

Fig. 5.14. *Representation of the development of a crystal dendrite. The corners not blocked by solvent diffusing from the surface grow preferentially.*

A further feature of dendritic growth which is of some importance in an understanding of precipitation is that of dendrite fragmentation. Jackson and co-workers[13] have shown that under rapid growth conditions, dendrite arms separate from the parent crystal and thus generate secondary nuclei. This is not a result of mechanical fracture, but is rather a crystallisation dissolution phenomenon.

Jackson's observation seems to tie together studies of so-called secondary or ancillary nucleation processes in widely diverse fields. For example, the seeding of clouds with silver iodide or some other suitable particles would clearly be of no commercial value if it required one particle for each raindrop which was produced. In fact, up to 10 000 drops are often produced for each seed particle. The explanation would now seem to be that dendritic ice crystals are nucleated on the seed particles and subsequent fragmentation causes a chain nucleation process. In the liquid phase, seeding supercooled water with ice also produces many secondary crystals.

Relatively few experimental characterisations have been made of ancillary nucleation, but it is common experience that the seeding of supersaturated organic solutions leads to many more precipitate particles than

Fig. 5.15. Development of dendrites leads to entrapment of foreign materials, solvent, etc., at re-entrant angle traps.

seeds originally introduced. Melia and Moffitt[12] have described the effect of stirring and supercooling upon the generation of secondary inorganic crystals, the number of precipitate particles being strongly dependent upon both stirring and supercooling. Obviously, mechanical fragmentation can be quite important in morphological modifications.

Ageing and recrystallisation

An important factor in the development of precipitate morphology is the process of ageing. We will put into this category the features which are secondary to the nucleation and growth processes. In this classification appear such phenomena as flocculation, coalescence, ripening and surface remodelling.

It has already been shown that careful control of supersaturation by such methods as PFHS leads to well formed precipitate crystallites. Under these circumstances flocculation can be assumed to be completely absent, thus refuting the old concept that the normal mode of precipitate formation is *via* flocculation of primary crystallites.

It is also unlikely that precipitation, by any method, from solutions where heterogeneous nucleation is the predominant mode of initiation, involves extensive flocculation. The reasoning behind such a statement is as follows: From the von Smoluchowski coagulation relations (see Chapter 1) it may be readily deduced that at least 10^7 particles/ml are required before agglomeration becomes appreciable in reasonable periods of time. Since there are generally fewer than 10^7 initiating impurities/ml, and since growth is predominantly on these impurities, agglomeration must be minimal.

Many of the forms which appear under the microscope to be agglomerates probably result from concurrent outgrowth of several crystallites from the same impurity, dendritic growth and, or, artifacts products by the preparation procedure. If dendritic fragmentation and ancillary nucleation have occurred, genuine agglomeration may be observable. On the other hand, homogeneous nucleation almost always leads to heavy agglomeration of very finely dispersed particles. Agglomerated precipitates are also commonly observed in polymerisation systems where the monomer → polymer reaction generates huge numbers of primary particles.

In some plastic formulation processes it is imperative to maintain the precipitate in its finely dispersed form, and this is usually achievable (up to 10^{15}–10^{16} particles/ml) by addition of surfactants to the reaction material. In other formulations the product is required as large, discrete particles and here it is desirable to promote flocculation if possible.

Evidently the promotion of agglomerates involves (a) the frequency of collision between particles, and (b) the effective sticking together of particles. With a given situation nothing can be done to promote (a) other than mechanical methods such as stirring and heating, both of which are unfavourable economically. If the dispersion is not stabilised by some form of surface active agent, the sticking probability is a function of surface energetics. Hence, inorganic crystals, which generally have a high interfacial energy at the solvent–solid interface, generally stick together rather well because the interfacial energy, and thus the energy of the system as a whole, is reduced substantially. The addition of inert solvents of low dielectric constant tends to increase the solid–liquid interfacial energy and thus promotes agglomeration.

However, the sticking together of precipitate crystallites is much more complicated than is indicated by a simple surface energetic argument. There is reason to believe that the surfaces of freshly formed crystalline precipitates are rather loosely packed and partially solvated. As the surface ages, consolidation occurs and the more compact surface is much less reactive.

Such deductions may be made from the extensive studies of radio tracer uptake made by Kolthoff, Mirnik and co-workers.[56,57] It is the freshly formed *loose* surfaces which seem to be *stickiest*. A good agent for promoting coalescence will thus be one which wets or *loosens* the surface and this often amounts to finding a solubilising solvent. Thus, the achievement of large, compact, polymeric agglomerates for plastic formulation may sometimes be achieved by adding traces of a suitable solubilising solvent to the reaction medium.

The flocculation of inorganic precipitates is, of course, a colloidal phenomenon described in detail in other parts of this book. *In situ* studies of the conditions leading to the flocculation and stabilisation of inorganic solids have been made by Tezak and co-workers. Their extensive work in this area will not be reviewed in detail here. However, it is germane to point out that their data which is qualitatively in accord with the usual concepts of the flocculating action of electrolytes and of reduced stability in solvents of low dielectric constant shows that the intimate nature of the interaction at the solid–liquid boundary is extremely important in characterising the stability of the system.[58] Thus, it seems fair to state that the solvent effects upon stability do not originate entirely from an electrostatic origin and cannot therefore be adequately predicted from models which only incorporate this form of interaction.

Another area relating to the formation of precipitates from solution which has been examined extensively by Tezak and co-workers is that of complex formation. Precipitation diagrams which relate the amount and composition of the precipitate to the reactant conditions are constructed and such data can be applied to the extraction of various materials of commercial importance. Many of the uranides have for example been characterised by these methods. For further details the reader is best referred to a recent review on this subject.[59]

Flocculation and coalescence is then one form of precipitate ageing which is often modified by the nature of the solution phase. Other forms of ageing phenomena include recrystallisation and ripening. Both these processes are favoured by increased solubility since they involve transport of material through solution. The equilibrium between solid precipitate and surrounding solution is a dynamic situation in which material is continually dissolving from positions of high energy (edges, corners) and re-depositing at surface positions of lower energy (steps, dislocations). Hence the crystallite tends slowly towards its equilibrium shape of lowest interfacial energy.

The ripening procedure may be regarded as a more drastic form of recrystallisation in which major parts or all of particle crystallites dissolve with subsequent deposition upon other larger and more well formed crystals. The simplest form of this ripening may be demonstrated by reference to the Gibbs–Kelvin equation, eqn. (5.14). It can be seen that spherical particles of small radius are more soluble than are larger particles. (Such an effect has indeed been well characterised).[60]

Thus, two particles of different sizes cannot be in equilibrium with the same solution. In a situation where the solution is close to saturation, the small particle dissolves and the large particle gets larger. Eventually, of course, after infinite time, any suspended precipitate would reform into a single crystal of equilibrium shape. Under more practical circumstances, it is the parts of the precipitate crystallite with a small radius of curvature which dissolve preferentially, dendrite arms being among the first features to disappear.

Ageing then causes a reduction in total surface area of a precipitate and various ions accelerate, and some surfactants inhibit, such a sequence. Ageing, in collaboration with coalescence has one other important effect, that of locking adsorbed impurities into the internal region of a consolidated crystal. The methods, parameters and problems of co-precipitation are outside the scope of this article, but in principle any foreign ions present in solution are adsorbable to a greater or lesser degree upon the precipitate surface. The greater the surface area, the more impurity is adsorbed and the more is eventually precipitated. As particles coalesce the adsorbed material is trapped on the inner surface of the precipitate crystal. Thus, processes which will require a highly pure product must avoid the large precipitate surface areas produced by homogeneous nucleation, and/or dendritic growth.

Morphology and supersaturation

The structure, composition and morphology of precipitates are strongly dependent upon the mode of formation and development. The most important single parameter in controlling all aspects of precipitation is the supersaturation. Supersaturation is a measure of the driving force to both nucleation and growth, and the nature and rate of these processes determine the morphological characteristics of the precipitate. A few examples are perhaps appropriate as a demonstration of the effects which nucleation and growth exert.

Let us suppose that the onset of homogeneous nucleation for inorganic precipitates occurs at a critical supersaturation ratio of 100 (most materials have values in the 10–100 range) and ask what effect solubility has upon particle size for three materials of solubility 10^{-7}, 10^{-4} and 10^{-1} M. Since the maximum particle size occurs at 10^{-5}, 10^{-2} and 10 M respectively, and the normal number of nucleating impurities is of the order

10^6/ml, we may conclude that the maximum particle sizes that can be obtained by direct mixing experiments are 1 μ, 10 μ and 0·1 mm respectively. Although this calculation is obviously very crude, common experience bears out the general agreement with the order of magnitude.

If instead of asking what the maximum particle size would be, we ask what would be the approximate size if 1 molar solution were used, the difference in particle size becomes much more pronounced.

TABLE 5.6*

DEPENDENCE OF PRECIPITATE MORPHOLOGY UPON THE SUPERSATURATION AND INTERFACIAL ENERGY

Initial Supersaturation Ratio	*Interfacial Energy*	*Nucleation*	*Growth*	*Morphology*
1–2	High	None	None	—
	Low	Heterogeneous	Slow–predom. screw disloc.	Discrete, well-formed crystals, no agglomeration
2–5	High	Heterogeneous	Slow–predom. surface nucl.	Discrete, well-formed crystals, no agglomeration
	Low	Heterogeneous	Dendritic	Poorly formed or dendritic crystals. No agglomeration
10–50	High	Heterogeneous	Dendritic	Poorly formed or dendritic crystals. No agglomeration
	Low	Homogeneous or ancillary	—	Stability dependent. Agglomeration evident
>1000	High	Homogeneous	—	Stability dependent. Agglomeration evident
	Low	Homogeneous	—	Colloidal

* From *Formation and Properties of Precipitates* by A. G. Walton. John Wiley and Sons, New York, 1967, p. 184.

The first material of solubility 10^{-7} M will have undergone homogenous nucleation and will probably be almost gel like with agglomerates consisting of primary particles of size order 10–20 Å. The second material will also have undergone homogeneous nucleation with particular agglomerates predominating, primary particle size 100–1000 Å. Finally, the third material will contain discrete crystals of about 30 μ size.

A schematic relation between precipitate morphology and the interfacial energetics is given in Table 5.6. Attempts to avoid the morphological implications of the nucleation and growth sequence have generally been directed at delivering more material to the precipitate for any given supersaturation. Suitable methods include chemical means (PFHS), slow cooling or evaporation, or increase of solute solubility (change of solvent, increase in pressure). For high degrees of supersaturation, large particular agglomerates may be obtained under suitable flocculating conditions (stirring, heating, addition of flocculating ions), and the reverse process may be achieved by stabilization with surfactants which produces very finely dispersed material.

Nucleation and precipitation in electric fields

Control over the nucleation and subsequent crystallisation processes is important in many diverse areas of industry. Although this chapter deals only with precipitation from solution, nucleation and solidification of melts are the backbone of the metallurgical and plastics industries and in many respects the nucleation processes may be phrased in similar language.[61] The number of nuclei affect the eventual bulk mechanical properties of metals and plastics and hence many methods of controlling nucleation have been sought. Such methods include addition of seed nuclei, working in the homogeneous or heterogeneous nucleation range or minimisation of nucleation by addition of surfactants. An unusual method of controlling nucleation in melts has been the application of electric or magnetic fields. Most of the published work in this area has originated in Russia where studies have shown that the application of pulsed electric or magnetic fields causes showers of nuclei to form in undercooled melts.[62–64] In model systems such as betol, piperine, etc., it is generally found that the application of fields in the range 103–104 V/cm and 10^2–10^3 cycles/sec produces the most effective nucleation conditions, *i.e.* produces most nuclei at a given degree of undercooling.

Similar effects have been reported in supersaturated (undercooled) solutions of ammonium chloride and bromide,[65] in this case the rate of nucleation was found to be proportional to the square root of the applied field.

In recent years, the application of asymmetric fields to suspensions of particles and to undersaturated solutions has been the focus of some interest.

The application of a strong asymmetric field such as that demonstrated in Fig. 5.16 has the effect of inducing a dipole in any particles suspended between the plates, causing them to migrate towards the position of maximum field intensity. This is particularly true of particles of high dielectric constant in a liquid of low dielectric constant, the field being concentrated in the vicinity of the particles. The theory of this phenomenon has been fairly well worked out by Pohl and co-workers.[66]

Possibly the most intriguing aspect of the application of high asymmetric fields to aqueous solutions of salts is concerned with the competing polarisation of the solvent by the ions themselves and external field. It has been suggested that the use of electric and magnetic fields might be useful in

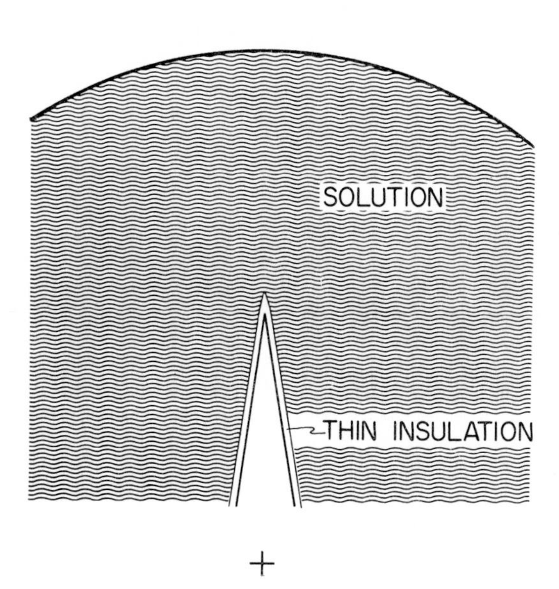

Fig. 5.16. Schematic diagram of an asymmetric electric field emanating from the tip of an insulated point electrode. Even for a low field gradient the field and polarising effect are high in the region of the anode.

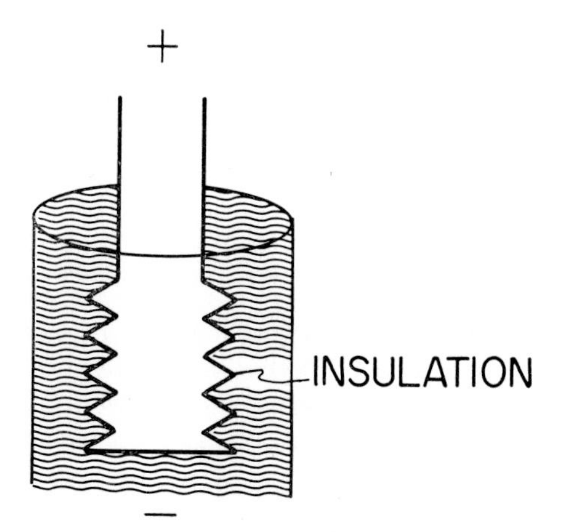

Fig. 5.17. Electrode configuration used in commercial (Worthington Corp.) electrical polariser.

separating ions and in commercial application for scale removal.[67] These possibilities are highly controversial[68] but using a device similar to that described in reference 67 which applies 4000 volts from a positive insulated central anode to a grounded outer cathode (Fig. 5.17), this author and co-workers find that salts may be nucleated from undersaturated solutions. This has been found true for all inorganic salts but does not occur for organics. The mechanism of the process is open to question. Some energy is expended electrically (approximately 1–2 μ amps of current is expended between electrodes) and it seems possible that the heat of solution is counteracted by polarisation. Quantities of salts precipitated from aqueous solutions in twenty hours are shown in Table 5.7. It appears that faradaic

TABLE 5.7

PRECIPITATION OF SALTS IN ELECTRIC FIELDS (WORTHINGTON CORP. ELECTROSTATIC SCALE CONTROL UNIT) IN TWENTY HOURS (STEEL CATHODE) FROM UNDERSATURATED SOLUTIONS

Material	*Initial Conc.* (*g*/100 *ml*)	*Amount of Ppte.* (*g*/100 *ml*)
Calcium phosphate (pH4)	0·26	0·014
Mercurous chloride	7·05	0·152
	0·74	0·021
Calcium sulphate	0·241	0·110
	0·071	0·005
	0·0014	0·000 12

conductance and/or electrolysis cannot explain this phenomenon but on the other hand contaminant from the cathode (iron or copper) is usually found in the precipitate. The amount of such contaminant seems to be dependent on the salt being precipitated. Virtually no contaminant is found with sparingly soluble salts such as calcium carbonate/sulphate whereas a major component is found with the corrosive soluble alkali halide salts.

This method of causing nucleation and precipitation of trace salts from solution (followed by rapid removal) appears to have some novel possibilities in the pollution control field.

It is also noteworthy that many physiological precipitation processes may originate in the transient bioelectric field induced by conformational changes in the underlying proteinaceous substrates.

SUMMARY

The fundamental mechanism of nucleation is by now fairly well understood, though the application of bulk thermodynamic reasoning is still subject to criticism. Perhaps the least well understood area of precipitation is that relating to precipitation kinetics, and here only qualitative agreement between experiment and theory have so far proved feasible. Other poorly understood phenomena include the ageing of precipitate surfaces, and electrically induced precipitation.

Although the qualitative description of precipitation phenomena has been known for at least fifty years, the contributions of the last two decades have added a much greater insight into the quantitative aspects of the mechanism of precipitate formation; some of these aspects have been outlined in this chapter.

REFERENCES

1. W. Ostwald, *Z. phys. Chem.*, **22** (1897) 284; **23** (1897) 365; **34** (1900) 444.
2. P. P. von Weimarn, *Chem. Revs.*, **2** (1925) 217.
3. T. Hlabse and A. G. Walton, *Anal. Chim. Acta*, **33** (1965) 373.
4. W. E. Brown, *Biology of Hard Tissue*, Ed. A. Budy, New York Academy of Science, 1967, p. 310.
5. A. G. Walton, W. J. Bodin, H. Füredi and A. Schwartz, *Canadian J. Chem.*, **45** (1967) 2695.
6. J. J. Black, M. J. Insley and G. D. Parfitt, *J. Photogr. Sci.*, **12** (1964) 86.
7. M. C. Upreti and A. G. Walton, *J. Chem. Phys.*, **44** (1966) 1936.
8. G. C. Benson and R. Shuttleworth, *J. Chem. Phys.*, **18** (1951) 130.
9. A. G. Walton, *J. Chem. Phys.*, **39** (1963) 3162.
10. A. G. Walton and D. R. Whitman, *J. Chem. Phys.*, **40** (1964) 2722.
11. E. Hauser, L. Sogor and A. G. Walton, *J. Chem. Phys.*, **45** (1966) 1071.
12. T. P. Melia and W. P. Moffitt, *Ind. Eng. Chem. (Fund.)*, **3** (1964) 317.
13. K. A. Jackson and J. D. Hunt, *Acta Metallurgica*, **13** (1965) 1212.
14. J. Willems, *Disc. Faraday Soc.*, **25** (1954) 111.
15. E. W. Fisher, *Disc. Faraday Soc.*, **25** (1954) 204.
16. J. A. Koutsky, A. G. Walton and E. Baer, *J. Polymer Sci.*, **4** (1966) 611.
17. J. A. Koutsky, Ph.D. Thesis, Case Institute of Technology (1966).
18. J. A. Koutsky, A. G. Walton and E. Baer, *Polymer Letters*, **5** (1967) 177.
19. J. B. Newkirk and D. Turnbull, *J. Appl. Phys.*, **26** (1955) 579.
20. R. Weiss, Ph.D. Thesis, Columbia University (1962).
21. M. L. White and A. A. Frost, *J. Colloid Sci.*, **14** (1959) 247.
22. T. P. Melia and W. P. Moffitt, *Nature*, **201** (1964) 1024.
23. R. A. Johnson and J. D. O'Rourke, *Anal. Chem.*, **27** (1953) 1699.
24. E. Suito and K. Takiyama, *Bull. Chem. Soc. Japan*, **27** (1954) 121.
25. A. G. Walton and T. Hlabse, *Talanta*, **10** (1963) 601.
26. D. Mealor and A. Townshend, *Chem. Communications*, **10** (1966) 9.
27. E. Schiffmann, B. A. Corcoran and E. R. Martin, *Arch. Biochem. et Biophys.*, **115** (1966) 87.
28. E. Schiffmann, *Biology of Hard Tissue*, Ed. A. Budy, New York Academy of Science, 1967, p. 128.
29. A. G. Walton, B. Friedman and A. Schwartz, *J. Biomed. Maths. Res.*, **1** (1967) 337.
30. D. Balarev, *Der Disperse Baue der festen Systems*, Dresden (1939).
31. J. Traube and W. von Behren, *Z. physik. Chem.*, **138** (1928) 85.

32. A. G. Walton, *J. Phys. Chem.*, **67** (1964) 1920.
33. V. K. La Mer and R. H. Dinegar, *J. Amer. Chem. Soc.*, **72** (1950) 4847.
34. F. C. Collins and J. P. Leineweber, *J. Phys. Chem.*, **60** (1956) 389.
35. M. J. Jaycock and G. D. Parfitt, *Trans. Faraday Soc.*, **57** (1961) 791.
36. A. G. Walton and D. H. Klein, *Kolloid Z.*, **189** (1963) 141.
37. D. H. Klein and B. Fontal, *Talanta*, **10** (1963) 808.
38. D. H. Klein and B. Fontal, *Talanta*, **12** (1965) 35.
39. C. W. Davies and A. L. Jones, *Trans. Faraday Soc.*, **51** (1955) 812.
40. J. R. Howard and G. H. Nancollas, *Trans. Faraday Soc.*, **53** (1957) 1449.
41. G. H. Nancollas and N. Purdie, *Trans. Faraday Soc.*, **59** (1963) 735.
42. G. H. Nancollas and N. Purdie, *Quart. Revs. London*, **18** (1964) 1.
43. D. Turnbull, *Acta Metallurgica*, **1** (1954) 684.
44. R. H. Doremus, *J. Phys. Chem.*, **62** (1958) 1068.
45. A. E. Nielsen, *Acta Chem. Scand.*, **13** (1959) 1680.
46. K. H. Lieser and A. Fabrikanos, *Z. physik Chem.*, **22** (1959) 406.
47. P. J. Lucchesi, *J. Colloid Sci.*, **11** (1956) 113.
48. A. G. Walton and T. Hlabse, *Anal. Chim. Acta.*, **29** (1963) 249.
49. M. B. Mutaftschiev, *Compt. Rend.*, **259** (1964) 572.
50. I. N. Stranski, *Z. Krist.*, **105** (1943) 287.
51. I. N. Stranski, *Disc. Faraday Soc.*, **5** (1949) 13.
52. K. Moliere, W. Rathje and I. N. Stranski, *Disc. Faraday Soc.*, **5** (1949) 21.
53. F. van Zeggeren and G. C. Benson, *J. Chem. Phys.*, **26** (1957) 1077.
54. G. C. Benson and T. A. Claxton, *Can. J. Phys.*, **41** (1963) 1287.
55. A. F. Wells, *Disc. Faraday Soc.*, **5** (1949) 197.
56. The extensive work of Kolthoff and co-workers has been reviewed by H. A. Laitenen, *Chemical Analysis*, McGraw-Hill, New York (1960).
57. M. Mirnik, *Kolloid Z.*, **163** (1959) 25.
58. W. Ostwald, H. Kokkoros and K. Hoffmann, *Kolloid Z.*, **81** (1937) 48.
59. H. Füredi, Chapter VI in *Formation and Properties of Precipitates*, by A. G. Walton, John Wiley, New York, 1967.
60. B. V. Enüstün and J. Turkevich, *J. Amer. Chem. Soc.*, **82** (1960) 4502.
61. A. G. Walton, Chapter 5 in *Nucleation*, Ed. A. Zettlemoyer, Dekker, New York, (1970).
62. F. K. Gorskii and L. T. Prishchepa, *Fiz. Tverd. Tela, Akad. Nauk. Belorussk. S.S.R.*, **386** (1962).
63. F. K. Gorskii, *J. Exptl. Theoret. Phys. (USSR)*, **4** (1934) 522.
64. L. I. Chesnokov, see *Chem. Abs.*, **56,** 11006a.
65. M. I. Koslovskii, *Fiz. Tverd. Tela, Akad. Nauk. Belorussk. S.S.R.* **404** (1962); see *Chem. Abs.*, **58,** 8468c.
66. See, for example, H. A. Pohl, *J. Appl. Phys.*, **29** (1958) 1187 and **30** (1959) 72.
67. L. Spector in *Plant Engineering*, March (1971).
68. B. Q. Welder and E. P. Partridge, *Ind. Eng. Chem.*, **46** (1954) 954.

CHAPTER 6

TECHNICAL ASPECTS OF DISPERSION AND DISPERSION EQUIPMENT

I. R. SHEPPARD

INTRODUCTION

When considering an industrial process in a logical manner it is important to know the classification into which it falls from an academic or theoretical standpoint, irrespective of the industry or type of process involved, if we

TABLE 6.1

Machine	*Class*	*Industries**
Pug Mixer	Low Shear	1, 2, 3, 4, 5, 6, 7, 8, 10
Planetary		2, 4, 5, 6, 8, 9, 10
Z Blade		1, 3, 4, 5, 6, 8, 9, 10
Cavitation	High Shear	1, 2, 4, 7, 8, 9
Stator Rotor		1, 2, 3, 4, 5, 6, 7, 8, 9, 10
Hynetic or Kady		2, 3, 4, 5, 7, 9
Colloid		1, 4, 6, 7, 8, 9
Horizontal	Ball Mills	2, 3, 4, 5, 7, 8, 9, 10
Vibratory		2, 3, 4, 7, 8, 9, 10
Planetary		2, 4, 7, 8, 9, 10
Attritor		3, 4, 6, 7, 8, 9. 10
Sand and Perl		4, 7, 9
Microflow		4, 8, 9
Single	Roll Mills	4, 7, 8, 9
Double		8, 9, 10
Treble		1, 2, 4, 7, 8, 9, 10
Multiple		4, 7, 8, 9, 10

* The numbers in this column refer to the particular industries indicated by the first two columns of Table 6.2.

are to produce the best possible product as economically as possible with maximum reliability and control. This is particularly true when dispersion operations are under consideration as this knowledge will not only affect the type of machine used and the cost of production in terms of power, labour, floor space, overheads, etc., but also may reduce the cost of the product in terms of actual materials.

This chapter is aimed at interpreting the theoretical considerations expounded in earlier chapters in terms of practical industrial application.

If we review the industrial applications and proprietary machines in which the basic concepts of dispersion are applied, the newcomer to the field will probably at first become completely at loss and unable to extract a set of suitable conditions. Reference to Tables 6.1 and 6.2 in which the

TABLE 6.2

Ref. No.	*Industry*	*Type of Product*				
		Surface coatings		*Solid phase*	*Liquid phase*	*Intermediate process*
		On site	*Off site*			
1	Adhesives	×			×	
2	Ceramics	×		×		
3	Chemicals			×	×	×
4	Paint		×		×	
5	Paper	×			×	
6	Pharmaceuticals	×		×	×	
7	Pigments/Dyestuffs				×	×
8	Plastics	×		×	×	
9	Printing Inks		×	×	×	
10	Rubber			×	×	

various machines are associated with the various industries and the industries themselves further classified into type or condition of product, shows the problem to be extremely complex.

Fortunately the basic principles of dispersion technology know no boundaries as far as machine or industry are concerned and as will be shown it is possible to apply these principles almost universally.

DEFINITIONS AND OBJECTIVES

In previous chapters the fundamental principles of dispersion have been considered. By utilising this theoretical information it should be possible to assess the degree to which each of the interacting phenomena is applicable in any given case. Unfortunately the majority of dispersion processes are carried out under conditions in which instrumentation and technology make complete identification of the fundamental conditions impractical. This is frequently due to the fact that the producer has a vast range of systems under his control and due to frequent changes in the materials and product, the exact fundamental condition cannot be established satisfactorily. It is, therefore, necessary to establish a straightforward technique whereby identification of the basic problems can be established to enable suitable equipment to be selected on which to perform the dispension operation.

Important data are therefore collected relating to the raw materials and also to the final product required, by answering the following questions.

(*i*) Is it necessary to reduce the particle size of the solid material during the process?

(*ii*) Is the dry material in the form of an agglomerated powder when presented to the process or is it in a more acceptable semi-prepared, possibly surface treated, form?

(*iii*) Does the suspending liquid possess either a natural or designed affinity for the solid surface?

(*iv*) What is the viscosity of the final product and can a dispersion of lower viscosity and higher dispersible solids content be extracted by removal of soluble solids, to provide more efficient conditions at the dispersion stage?

(*v*) What technique is to be used to assess the product and to what degree must dispersion continue?

(*vi*) What is the relative cost by volume of the suspended solid and the suspending liquid? Does any portion of the formula have a higher cost by volume than another? If so can this be extracted as a separate process to receive separate attention and greater degree of dispersion to increase its usefulness and reduce the cost of the overall process?

The answers to these questions will decide the type of machine used and the technique to be adopted. In the author's experience questions (*i*) and (*ii*) can best be answered by using techniques involving rheology,[1] optical microscopy, etc.[2]

The rheological technique of examination is condensed in Fig. 6.1 where it is shown that for two pigment systems the rheological change taking place during a dispersion operation at high non-soluble solids content using a treated pigment is the reverse of that taking place in a similar medium during a particle size reduction operation. When a dispersion operation is being performed on a solid material in which either natural cemented aggregates (Prussian Blue ICI 16266) or a pre-densified material (Peerless Carbon Black Beads) are used, a rheological effect is obtained which is a composite of the two basic reactions obtained from particle size reduction and dispersion.

Questions *(iii)* and *(iv)* can best be answered using the Daniel Flow Point Technique in which solutions of the soluble solid in solvent are titrated with the insoluble solid phase at different concentrations. (A constant weight of the insoluble powder is weighed into a number of beakers and the different solutions titrated into it until a given end point or flow characteristic is obtained. This is usually accepted as being the point at which the dispersion obtained after stirring with a glass rod *snaps back* after one drop has fallen from the end of the rod.) When the volume of solution

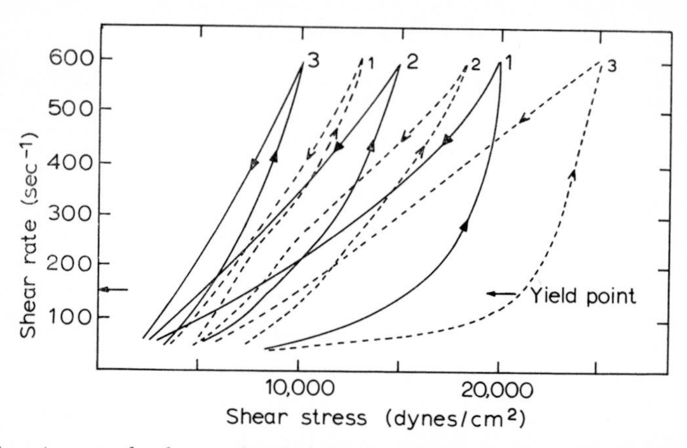

Fig. 6.1. An example of a set of typical Ferranti-Shirley rheographs showing the change in rheology with dispersion of a titanium dioxide dispersion (solid lines), and during a particle size reduction operation on microdol (dotted lines). This shows that the tendency is for dispersions to become less shear-thinning or more Newtonian with increased dispersion whereas a particle size reduction exercise has the reverse effect. The yield point of the microdol dispersion is indicated.

employed is plotted against concentration, as in Fig. 6.2, a response curve is obtained. If a sharp or high amplitude type of curve results the solution can normally be accepted as having a high affinity for the solid whereas if a flat response is shown the solution is considered to have little affinity.

If it is possible to extract a formula based on the Daniel Flow Point from the final formula, particularly if this is at the maximum non-soluble solids content, this will mean that a dispersion base formula can be used in production of the finished formula. This normally means that more efficient dispersion conditions in lower powered equipment can be utilised thus resulting in considerable saving of time, power and labour.

Unfortunately the Daniel Flow Point technique is not reliable for fine colloidal dyestuffs and other fine materials, and it is in these cases that fine dispersion is most often rewarding and various other techniques have been developed. One example of these alternative formulative techniques[2] will be described later in the chapter.

Questions (*v*) and (*vi*) can best be appreciated from Fig. 6.3 which shows the varying response to increase in dispersion which can be detected using a range of different testing techniques. There are, of course, many other bench methods of assessment but in the majority of cases the final criterion of acceptability is by application. It is not necessary, however, that

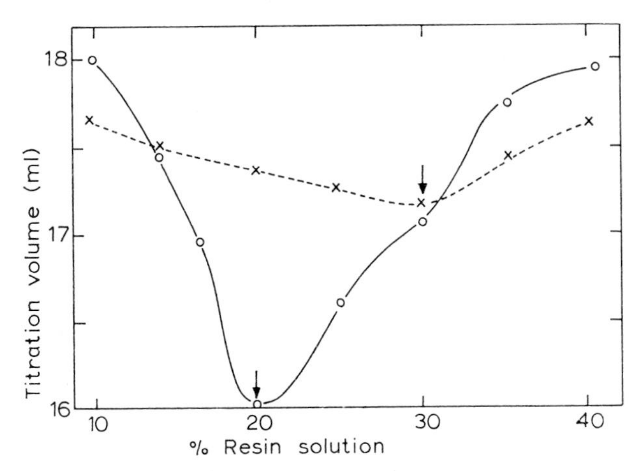

Fig. 6.2. A typical set of Daniel Flow point results. The system having the greater affinity for the pigment (shown by solid line) has a more pronounced optimum point (arrow) than that with little affinity (dotted line).

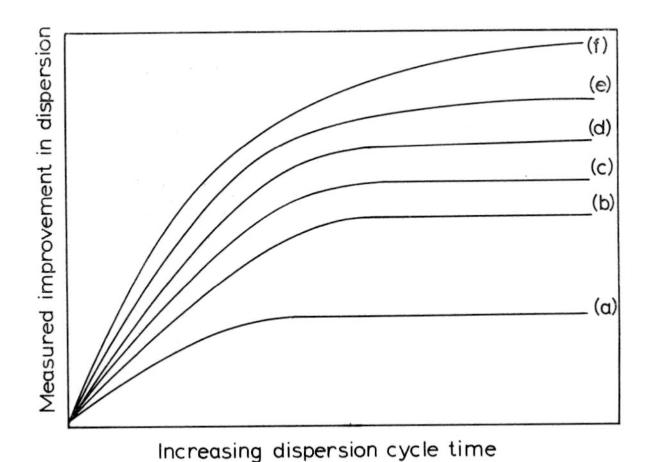

Fig. 6.3. Comparison of Testing Methods: (a) *Hegman Seals*: (b) *photometric sedimentation*; (c) *optical microscopy*; (d) *centrifugal sediment*; (e) *colour of finished product*; (f) *rheology*.

acceptability alone is sufficient; the economics of absolute dispersion must also be considered, as discussed below.

It is, therefore, necessary to establish in simple terms the main objectives of the dispersion operation, to characterise the function of the operation by rheology or other means and to study the economics of the operation extremely closely before selecting the type of equipment to be used in the process.

CLASSIFICATION OF TYPE OF SYSTEM INVOLVED

Dispersion of solids in liquids under diverse conditions, ranging from the pigmentation of adhesives and high viscosity ink systems to the dispersion of clays in water for use in ceramics or paper treatment, require very different treatments in terms of equipment. The three main classifications are

(*i*) aqueous phase dispersion,
(*ii*) solvent and other resinous phase dispersion and
(*iii*) emulsification and emulsion phase dispersion.

In each case the system can be further classified in terms of rheology (here it is important to know the materials) by means of

(*iv*) shear thinning characteristics (*see* Fig. 6.4),
(*v*) change in rheology with increased dispersion,
(*vi*) apparent viscosity,
(*vii*) true viscosity,
(*viii*) adhesive and cohesive characteristics and
(*ix*) degree of stability.

Example 1: (*i*), (*ii*), (*iii*)

An aqueous dyestuff filter-press cake has very different rheological characteristics, exhibits completely different shear-thinning characteristics, cohesion, and adhesion, before and after the introduction of a surface active agent or wetting agent, and the type of machine utilised for maximum efficiency is therefore different according to the degree to which the additive satisfies the surface of the material.

For systems which are to be stored in a liquid state complete stability results in a very hard sediment. In the author's experience the term *stability* must be sub-divided into two conditions; machine-dependent or dynamic stability and partial or complete static stability.

Machine dependent or dynamic stability is essential if maximum dispersion efficiency is to be obtained, but depending on the machine used and the shear rate applied this condition can be varied as the action of the machine appears to produce an effect on the surface of the solid similar to that produced by having a higher proportion of the surface active component in the formula. This is possibly due to the production of similar static charges on the surface of the particles under shear, the opposing charges being discharged through the suspending fluid to earth. These charges then aid in the dispersion of the material but rapidly discharge themselves when

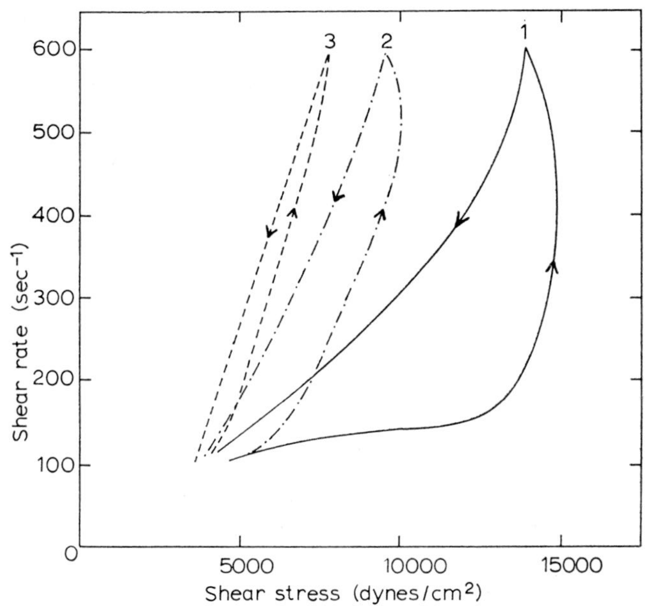

Fig. 6.4. The effect of shear-thinning with a set of results obtained using the Ferranti-Shirley Viscometer. The curves were plotted from three separate sets of readings taken on the same sample, in the order shown. The changes in rheology are due entirely to the additional shear applied for each determination.

the system comes to rest. It is therefore necessary to *stabilise* the system before discharge from the dispersion equipment by the addition of further surface active material. This is then adsorbed onto the surface to replace the charge and produce a system having static stability.

Partial or complete static stability is considered in more detail in Chapters 1 and 7.

Example 2 (*ii*), (*iii*)

Aqueous emulsion based paste inks have a completely different processing requirement to solvent based gravure inks for use in a very similar printing process. This is obviously due to the difference in flow characteristics at approximately the same pigment solids concentration produced by the different suspending liquids. In the one case an aqueous emulsion in which the emulsifying agent and other additives have an anionic-cationic effect on the pigment surface, and in the other where the more polar portions of the resin's polymeric structure orientate themselves at the pigment surface to insulate surfaces charges on one particle from those on the other. In dispersing the pigment in an emulsified medium care must be taken to avoid two effects, namely; excess adsorption of the emulsifying agent by the pigment thereby reducing emulsion stability, and reducing this stability of the emulsion by application of high shear rates.

Example 3 (*i*), (*ii*), (*iii*)

When producing a dispersion of calcium carbonate for use in an emulsion paint or paper coating base, the dispersion exhibits shear thinning characteristics which are reversible on the removal of shear forces. However, when dispersing titanium dioxide in a similar medium containing a wetting agent, the viscosity decreases as the dispersion continues and less shear thinning properties are exhibited. This is not reversible on removal of shear forces. As the surface area of the material in the dispersed phase increases so does *dispersion thinning*, as opposed to shear thinning, become more apparent (Fig. 6.3).

Example 4 (*vi*), (*vii*)

Examination of some materials under static or low shear rate conditions would suggest use of heavy duty dispersion equipment involving high horsepower requirements over relatively long periods, whereas it can be shown in many instances that the rheology of the system is such that more efficient and economical processing conditions can be achieved under high shear rate conditions above the *yield point* of the material (Fig. 6.2). This phenomenon sometimes produces a product of lower viscosity due to increased degree of dispersion, which enables the producer to use pumps for subsequent discharge and transfer instead of screw conveyors or extruders. An extremely good example of this is in the treatment of dyestuff intermediates before spray drying. Filter press cake is fed to a high shear rate unit and is discharged as a free flowing slurry, as compared with the paste-like product which would be produced after some dilution from a low shear rate trough type mixer.

RANGE OF EQUIPMENT AVAILABLE

The machinery available for use in dispersion equipment has been classified in the following sub-divisions, which are summarised in Table 6.1. It is not intended in this chapter to enter into the full theory of each individual unit discussed, since this would only duplicate previously published work in the majority of cases, but to describe the unit under discussion and explain briefly the theory of operation and also its application and limitations. Only those pieces of equipment within the author's own experience are discussed.

Developments in dispersion equipment appear to pass through cycles of activity, large numbers of new dispersion tools are developed as demand is realised for new techniques. These are frequently brought about due to finite changes in raw material dispersibility following improvements in pigment production techniques. A further cause of machine development is the need to constantly reduce labour by increased automation. The need for a change in machine design being first established by a particular industry, approaches are made to equipment manufacturers and a basic design concept is established and proved by pilot plant trials. If successfully developed and accepted for more general application the new machine type is then put into use by industry on a wider scale. Almost invariably additional refinements are then developed either by the original manufacturer or by competitors, and the new equipment either becomes consolidated as a standard production tool or has its application more closely defined within a given range of applications.

Within the last 15–20 years a number of new design techniques have been developed but within the last ten years little change has taken place. During this period the new techniques have been proved, modified and established. The basic types of equipment listed below represent the four basic concepts of dispersion equipment design, and development in the machinery field principally occurs within these four groups.

(*i*) *Low shear rate equipment*. All equipment employing slow speed mixing at high viscosity or high pigment concentration, these being listed in order of power requirements. The basic principle of operation is to formulate the dispersion base to give the maximum resistance to the movement of the blades within the capabilities of the unit. The motor power is transmitted to the batch through blades which present a large working surface to the batch, and the energy so transmitted is utilised at high shear stress in the normally fairly cohesive mass at extremely low shear rates. In all cases the blades pass close to the walls of the container to ensure complete intermixing and impart additional shear at this point.

(*ii*) *High shear rate equipment*. This type of equipment is typified by the use of impellors moving at high peripheral speeds, which generate high shear rates and high rate of interparticle shear by virtue of careful control of dispersion base rheology, or by the use of a variety of different shear rate control baffle arrangements.

(*iii*) *Ball mills*. This classification covers all types of equipment in which the energies transmitted are applied to the dispersion by use of free moving members irrespective of size, shape or means of activation.

(*iv*) *Roll mills*. Equipment which uses one or more rollers designed to carry the dispersion base between either two rollers or a roller and a blade at a closely controlled gap under pressure.

Before studying each individual machine with a view to selection of the most suitable unit for any given industrial application, we should review the operation in terms of selection and availability, of continuous or batch processing, and batch size. Normally when reviewing a given process it is possible to determine the degree to which versatility and flexibility is required as compared with throughput. The amount of control testing and number of different materials involved quite frequently are the decisive factors when large throughput is required in terms of large batch or continuous equipment.

Versatility and flexibility are normally associated with relatively small unit batch equipment except where an acceptable product can be achieved with easily cleaned low capital cost continuous plant. Thus, if on reviewing a problem an output of a single product is required at the rate of 10 000 gallons per week and the finished material consists of only three components, then a continuous system utilising automatic metering and continuous premixing would be the obvious choice unless the cost of testing the product and the frequency of testing necessary was greater than the cost of controlling and operating a plant working on a 1000 gallon batch basis. It is normally accepted that the cost of control equipment for batch operation on an automatic basis is less than for continuous metering. If, however, the simple three component system was replaced by a more complex 6 or 10 component system, the cost and sensitivity of the metering/continuous system would probably outweigh the cost of manually operating the plant on a large batch basis. If the same output of 10 000 gallons per week was required with a varied range of products in differing batch sizes from 100–1000 gallons a different range of batch operated equipment at a higher capital cost would probably prove most suitable. The economics of machine type size and capital cost selection is considered later in the chapter.

Low shear rate equipment

Pug mills and trough mixers

The Pug mill is characterised by a long trough with a U-shaped cross section. A driven shaft passes through glands at each end of the trough and is fitted with blades of varying design from simple radial bars to the more complex ribbon and interrupted ribbon form. It is a multipurpose machine used for a wide range of industrial applications, and the power rating is usually fairly low. It is used in the ceramic, plastics, paper, pharmaceutical and paint industries.

The applications, almost invariably, involve the mixing and partial dispersion of a solid of comparatively large particle size to a fairly heavy paste-like consistency. As blade area and viscosity increases so does the power rating of the unit. The majority of the work done is by the clearance between the tank walls and the blades but as the viscosity increases more and more is done by the application of shear stress at the lower shear rates within the moving mass. Design of the blades is extremely critical if the batch is to be maintained in circulation throughout the process. In some industries this type of equipment is used at elevated temperatures using oil, steam and electrical jackets and they are also frequently used under vacuum in order to remove entrained air from the product.

Planetary mixers

These machines have slightly less viscosity restriction than the Pug mill in that the design of the unit is such that only a relatively small proportion of the batch is being worked at any one time (Fig. 6.5(a)). Due to the use of a scraper blade and the contra-rotation of the mixing head the point at which the mix ceases to flow occurs at a higher viscosity.

The incorporation of a removable tank in the design is also a considerable asset in efficient transfer of materials to the next process as almost all materials processed in this type of machine possess very poor flow characteristics. This type of machine is frequently used as a pre-mixer and partial disperser for roll milling. Modern engineering techniques have enabled machines to be made with a stationary tank which can be readily heated or cooled and to which vacuum attachments can be made. These machines are used in the paint, ink, plastics, and pharmaceutical industries.

Z Blade mixers

This type of machine is typified by the use of two shafts on which blades of different shapes are fitted according to the type of process to be conducted. The origin of the machine was in the bakery trade where it has been used for many years for the mixing of bread doughs. Usually the shafts are designed to rotate at different speeds and in different directions, usually inwards, as shown in Fig. 6.5(b), but the drive is arranged to enable the direction to be reversed to assist discharge by tipping. Figure 6.6 shows a Baker Perkins Model heavy duty mixer. The blades vary in design from those of the traditional Sigma type, which are designed to provide rapid homogenisation under high shear stress conditions, to the smooth outline of dispersion blades and three wing blades which are capable of more efficient dispersion or work input and provide less homogenisation.

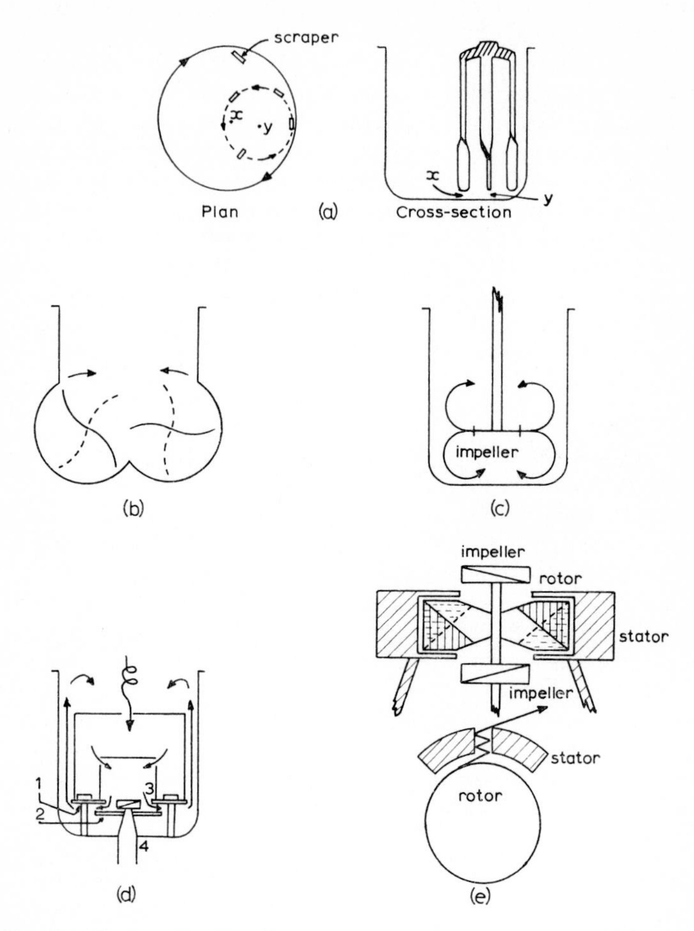

Fig. 6.5. (a) *The position of the blades within the tank of a planetary mixer and their contra-rotation in relation to the tanks own movement.* (b) *The trough of a typical sigma-bladed mixer and the rotation of the blades. The waisted-trough design is used to increase shear at the wall and to prevent build-up above the area of maximum energy.* (c) *The theoretical flow pattern obtained in a disc cavitation mixer.* (d) *The use of baffles to increase efficiency of dispersion and circulation in stator rotor mixers*; (1) *adjustable stator,* (2) *rotor,* (3) *variable shear rate gap,* (4) *mechanical seal.* (e) *The Hynetic or Kady principle showing a cross section through the dispersion head and the theory of right angular impingement on radial stator slots.*

In all cases the dispersion is produced using forces transmitted to the batch by large skilfully designed and orientated blade areas. Due to the low rotational speed and high power input and the extraordinarily high stress values set up under these conditions, the design of these mixers is very robust. The equipment is, therefore, very costly to manufacture particularly in stainless steel. Hence, it is necessary to make a careful study of optimisation of conditions to avoid selecting a machine of an excessive power for a given duty. The batch is normally formulated to pass through a heavy *plastic* state in which it is so cohesive that the walls of the trough are left clean during the maximum efficiency stage. Due to the

Fig. 6.6. *A Baker Perkins size* 15 *Type VI Class BB heavy duty dispersion mixer.*

high level of work input much heat is generated and water cooling becomes necessary in many cases to avoid degeneration of organic constituents. The viscosities at which the process is conducted are normally so high as to prevent adequate definition in terms of rheology except by use of special equipment which is designed to simulate operating conditions, such as the Brabender Plastograph. The sigma-bladed type of machine is widely used for lighter duty applications where premixing and homogenisation is required under high viscosity conditions. These conditions are frequently controlled by the final formula. The dispersion bladed system is used where homogenisation is of lesser significance and higher dispersion efficiencies are necessary. This is normally associated with the dispersion of fine pigments and dyestuffs in systems possessing little natural affinity for the pigment surface. This type of unit will frequently produce a dispersion at an extremely high pigment solids, which cannot be achieved by any other means.

The three-wing bladed machine is normally referred to as a Masticator and has an amazingly high power input ratio. This is almost entirely used in the rubber and plastics industries where the heat produced during the process is aiding some other function such as vulcanisation or plasticiser absorption. It can, however, be used when necessary for dispersion of pigment systems but in these cases it is normally found necessary to perform a pre-mixing operation to produce an homogeneous feed material. This gives a more consistent product and maintains a high output from high capital cost equipment.

High shear rate mixers

Disc cavitation mixers

These mixers are characterised by a flat disc of approximately $\frac{1}{3}$–$\frac{1}{2}$ the tank diameter revolving at peripheral speeds of 1000–4000 feet per min. The blades are fitted with small deflectors or subsidiary blades which direct the flow from the periphery to provide adequate homogenisation during the premixing stage from dry powder and separate liquid, and also provide for a regular circulation of the dispersion base through the high shear zone around the impellor, during the later stages of dispersion. This type of dispersion tool was introduced some years ago and is now manufactured under a wide range of proprietary names. It is typified by Fig. 6.7 showing the Steele and Cowlishaw Hydiscolver. This machine, like its many counterparts, is designed to operate within a wide range of peripheral speeds and being overdriven can be placed into any container of appropriate size. The need for a variable speed drive arrangement is due to changes which take place during the sequence of premixing and dispersion in the flow characteristics of the batch (Fig. 6.4), and also provides for modification in shear rate to adjust for slight variations in batch formulation away from optimum conditions.

The theory of the machine depends upon transmitting power from the drive to the batch *via* the spinning disc. Some work is performed at the impellor tip but the majority of the work is released as dispersion energy due to the differential rate of movement of the streams of material within the batch (Fig. 6.5(c)). A further small proportion of the work is performed at the tank walls. In order to utilise this type of equipment most efficiently it is necessary to control the dispersion formula quite accurately, so that the maximum possible amount of inter-particle shear takes place. This principle of dispersion is best used under circumstances in which the surface has been pretreated but it can also be used to disperse any system in which the feed material is ground below the maximum size required.

It is also increasingly being used as an economical flexible and versatile tool for pre-mixing for other equipment such as ball mills, Perl mills and sand mills as it provides a simple and effective means of increasing the output of this type of plant. In order to increase the range of application of the principle, various manufacturers have produced composite machines based on this principle but with additional equipment for the homogenisation of heavy systems fitted to the same tank. These range from fixed tank equipment such as the Master Mix and the Premier Colloid contra-rotating mixer with dispersator, to the portable tank Molteni mixer disperser which incorporates the use of a semi-planetary system in

Fig. 6.7. A Steele & Cowlishaw 25 h.p. Hydiscolver with variable speed drive and lifting ram driven by hydraulic system.

which the container is free to move about its axis driven by the forces transmitted through the batch due to the blade being inserted in an off-centre position. Inclusion of scraping blade and little or no manual attention is required even at high solids content. On most recent models the tank can be locked in position during the latter stages of dispersion while a vacuum is applied to remove entrained air. This is an extremely useful facility with many applications.

Dry disc dispersers

A special application of cavitation type equipment of the 'underdriven' type has been developed. This is for use with free flowing pigments and other powders. As this can be regarded as a form of 'dry dispersion' it becomes of some interest within the scope of this chapter. In the heavy duty form this type of equipment is used as a pigment and plasticiser predisperser to produce a premix of PVC granules before extrusion at elevated temperatures when the product becomes semi-liquid. This is of interest in that it replaces techniques used previously which involved heavy duty dispersion-bladed trough mixers and masticators.

Stator rotor mixers

This type is probably best typified by the Steele and Cowlishaw Hydisperser (Fig. 6.5(d)). This type of machine is capable of performing similar dispersion functions to the normal disc cavitation mixer, but is made more versatile and less controlled by formulation than its predecessor due to the fitting of a shear plate or baffle. This means that more work is done between the rotor and the baffle and less precise formulating is necessary. The design of the baffle also controls the flow in the tank and this means that the impellor can be designed to promote flow, leaving direction of flow to the baffle system. Due to its underdrive and excellent mechanical seal this particular machine can be designed for very large scale batch operation up to 5000 gallons. It is therefore also in demand for less versatile applications where large throughput forms the major requirement such as in paper coating preparation and emulsion paint manufacture.

Hynetic or Kady mills

A further extension of the high shear principle is the inclusion of the collision theory in baffle design. The dispersion head of the Hynetic or Kady mill is designed to incorporate the features associated with the stator rotor mixer but also provides for high speed impingement of streams of dispersion base on right angled faces in the stator (Fig. 6.5(e)). This extends the use of this type of machine beyond the range normally associated with high shear equipment into systems where more tightly cemented aggregates are to be dispersed, and also into the range of systems containing untreated pigments and to systems in which the suspending liquid exhibits less affinity for the pigment surface. Unfortunately like the other machines in this section the viscosity range of the unit is quite critical for maximum efficiency and in some cases with this unit the optimum condition for efficient operation is very delicately balanced and must be closely controlled. The machine is quite widely used for production of paint, printing ink and paper coatings apart from other applications on such materials as dyestuffs and detergents.

Colloid mills

This class of equipment is typified by the Premier Colloid mill. The action consists of using centrifugal force to feed a fine film of material between an abrasive rotor and stator working in close proximity. The shear imparted to the film at speeds up to 8000 feet per min produces the dispersion. Clearances can be reduced to as little as 0·0005″ or 12 μ and adjustments can be made simply by hand and are controlled at a given setting automatically. This equipment benefits from use of an optimised formula but is used for a wide range of dispersion and emulsification operations without modifications from final formula. It differs from the remaining high shear mixers in that being continuous it requires a pre-mixer and also a post-mixer when a dilution is required but offers the advantage that it will produce some particle size reduction and will handle many aggregated systems of both organic and inorganic materials.

Ball mills

Horizontal ball mills

This machine (Fig. 6.8) is so well established in the dispersion industries for a vast range of applications that it would seem to require little comment. In recent years, however, many manufacturers have found that this type of machine can be used to such good effect that they prefer it to more advanced equipment on the basis of its inherent flexibility, reliability, large batch size and its ability to be used with minimum labour cost. The main basis of operation is, of course, that a large number of free moving solid bodies are placed in a rotating drum and the processed material is loaded in so that it more than fills the interstices. If the formula of the mill base and the rotational speed are correct then the cascading action of the balls produces both a high shear rate and also attrition, and impact resulting not only in dispersion but also in particle size reduction. The shear rate developed and the rate of attrition and impact increase with mill diameter, so that the large mills will disperse much more quickly and to a finer ultimate degree of dispersion than the smaller ones. This could be said to be due to the increase in the hydrostatic head of material increasing the point pressure between any two balls and to the increased distance available for acceleration in free fall. It is, however, well known that a mill having a total capacity of 1000 gallons of 7′ 0″ diameter will produce a given dispersion in less than a third of the time required in a mill of 2 gallons capacity, and half the time required in a 60 gallon capacity mill of 2′ 6″ diameter. This has resulted in many experiments with small capacity equipment to discover means by which small production exercises can be speeded up. The use of balls or grinding media of smaller size or greater density have a measurable effect but steel balls having three times the density, while producing the same product in about one third the time cycle, frequently contaminate the product. Smaller sized balls, while

Fig. 6.8. *A Steele & Cowlishaw S.G.M. type ball mill of* 3′ 0″ *length* × 3′ 0″ *diameter and capacity* 80 *gallons.*

increasing efficiency by increasing shear area and the numbers of point contacts, also provide for an increased retention area which extends discharge and cleaning times. Many other ways of increasing efficiency of this well tried method have, therefore, been attempted.

Vibratory ball mills

The use of vibratory milling has been adopted to increase the impact and attrition rates during processing. The material to be processed is fed into a container filled with grinding media until it just fills the interstices. A vibration source is then connected to the sprung container and the frequency and amplitude adjusted to give optimum conditions. This produces an incalculable number of tiny collisions between the grinding members which result in particle size or agglomerate size reduction. The rate of acceleration is obviously much greater than can be achieved under free fall conditions in a rotating cylinder and dispersion and particle size reduction take place at an increased rate. Unfortunately, large scale machines have not been developed and so for large production the horizontal mill is still in use, but the vibratory unit is extremely useful for high speed small and intermediate production and can be used in some circumstances on a continuous basis.

Planetary ball mills

Another approach to increased ball milling efficiency is to increase the rate of acceleration of the balls by rotating the mill on a planetary gear. By mounting four pots in this manner in their Mk II, 24 h.p. high speed planetary ball mill Steele and Cowlishaw Ltd have simulated a force

Fig. 6.9. A Steele & Cowlishaw high speed planetary ball mill Mark II 24 h.p. fitted with 4 × 10 gallon porcelain pots.

equivalent to ten times that affecting the normal horizontal mill and this enables balls of smaller diameter to cascade through dispersion bases of much higher viscosity without resistance (Fig. 6.9). Due to the increased energies transmitted in this way dispersion is effected in anything from one tenth to one three hundredth of the time required in a mill of similar diameter under normal horizontal working conditions, and the ultimate degree of dispersion achieved is also much improved. Unfortunately problems of engineering design prevent large scale equipment being manufactured and the largest machine available of this type is one fitted to hold 4–10 gallon capacity pots. This size is ideal for handling fairly large quantities of high cost materials in cases where separate treatment of staining pigments to maximum colour development is economic.

Attritor mills

This type of equipment presents yet a further use of the technique. The container is again filled with free moving balls or pebbles but instead of rotating the container the grinding media is moved by stirring arms designed to agitate the whole mass. As the size of ball or pebble which can be caused to move under these conditions with free kinetic energy is smaller than under normal horizontal ball milling conditions, thus increasing the area of working surface and number of point contacts, and because the whole ball mass is kept in constant movement, a much increased rate of dispersion is obtained.

A further advantage is that flow conditions during processing can be observed and corrections to formulation made without stopping the machine. The design of the drive also permits speed variations to off-set the need for formulation variation. The ability to make additions during processing without interrupting the transmission of energy to the batch also avoids the risk of flocculation on dilution of dynamically stable processing formulae into statically stable final products. Unfortunately the design of really large machines of this type is impracticable because the efficiency ratio, as compared with a horizontal ball mill of similar working volume falls from a 10:1 or 30:1 ratio with small laboratory equipment to a 5:1 or 3:1 with the largest scale production equipment. This is due to the fact that as the volume of the mass increases the area of the blades presented to the mass reduces per unit volume unless extremely complex design changes are made to the blades. If an attempt were made to maintain the volume/blade surface relationship constant the power rating and heat development rate of the larger units would be impracticable. Nevertheless, this type of equipment is extremely useful for heat sensitive systems where small to intermediate sized batches are to be produced, providing that sufficient of the final formula remains to wash down the grinding media after each batch or at each colour change. These units are quite widely used for the dispersion of inks, plastics, and sulphur. The mill can, of course, be used on a continuous basis in some circumstances or by special design.

Sand or Perl mills

These two types of units depend on the extension of basic principles of the attritor to higher rates of rotation and the use of still smaller grinding media. If we consider the packing of spheres into a cylindrical container we find that a vessel of 1 gallon total capacity holds 808 $\frac{3}{4}''$, 6856 $\frac{3}{8}''$, 22 232 $\frac{1}{4}''$ or 173 824 $\frac{1}{8}''$ diameter balls. By using higher speeds of impellor rotation greater kinetic energies can be transmitted through the much larger number of *point contacts* between the smaller spheres. In a batch machine the problem of using such small grinding media is threefold.

(*i*) The large surface area of the grinding mass wetted by the dispersion which makes cleaning and discharge prohibitively difficult.
(*ii*) The cost and availability of small sized grinding media makes their use in large scale batch equipment impracticable.
(*iii*) The wear rate tends to increase with surface area and a greater rate of contamination is therefore experienced both for fine material worn away from the surface and also from grinding media of reduced size passing through refining screens.

Both the sand mill and the Perl mill (Fig. 6.10) are continuous machines and are designed to separate the sand in the one case and the small beads in the other, from the dispersion base. Due to the high efficiency of the system the size of container and, therefore, quantity of grinding media required for a given output is quite small in relationship to a batch unit of the same output potential. This disposes of problems (*i*) and (*ii*). Careful formulation and design of impellor and container reduces the contamination rate. In the case of the sand mill and to a lesser extent in the case of the Perl mill, which has on average a larger size of grinding media, formulation and degree of premixing employed is critical in obtaining best results. It is normal to find this type of equipment in use only where fairly large throughputs of suitable products are required as it is only in these cases that an adequate study of the formulation requirements can really be made. Large throughputs also reduce the disadvantage of having to clean not only the continuous dispersion unit itself but also the premixing unit and the post mixer or dilution tank as well.

Microflow mills

This unit was introduced by Messrs. Torrance Ltd some years ago. It operates on the principle that a feed of material is passed at a carefully controlled rate to a stationary cylinder containing free moving much smaller solid cylinders made of steel or ceramics. A carefully machined and specially designed cage is rotated inside the large horizontal cylinder

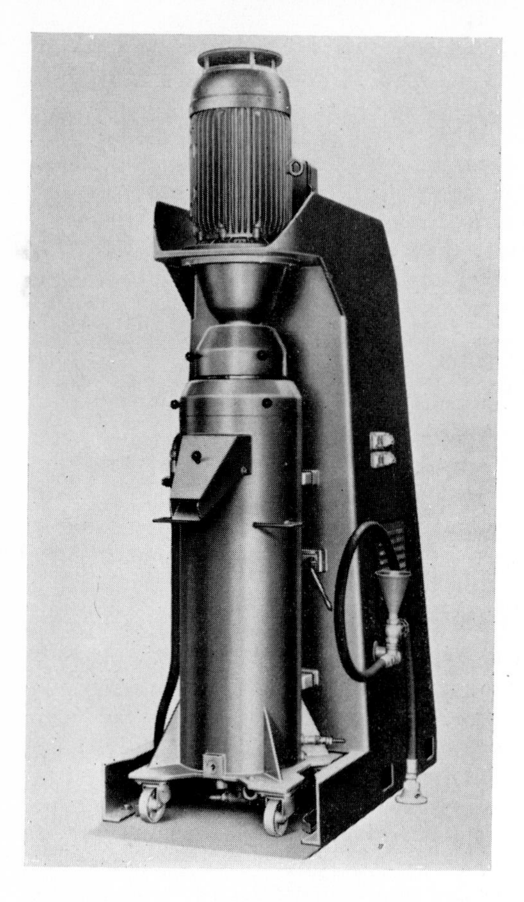

Fig. 6.10. *The Draiswerke G.m.b.H. Perl mill PM* 125 *for continuous processing.*

and this sets up a uniform movement of the free moving cylinders around the walls. Due to the design of the unit the fast moving cage and the stationary container wall exert rapidly changing accelerations and speeds of rotation on the free cylinders, which are held against the outer wall by centrifugal forces. The dispersion base is also held against the wall and is subject to shear forces from the outer side of the cage together with forces applied by the free moving cylinders themselves. As the design gives continuous operation and a low retention factor even when operating on fairly viscous systems it is becoming established as a small production unit for high cost materials.

Roll mills

These machines are generally only used in cases where no other alternative is available. The reason for this is that as a class of dispersion unit they have an extremely high labour component in that a fairly skilled operative can at most only operate two machines and that cost of cleaning and maintaining the unit per unit of output is generally high as compared with a ball mill or high speed disperser.

In some industries these machines are used for applications in which quite small throughputs of a wide range of different products are required. One of the main reasons behind their selection in these cases is the lack of sensitivity they exhibit to formulation changes and the lack of any real need for optimisation of dispersion conditions. This is due to the basic theory of the units which is given below.

In the majority of cases these units are used on systems in which a solid material is to be dispersed into a viscous medium from which no satisfactory formula can be extracted for use in more efficient equipment. Despite the fact that this type of unit provides a continuous process it is quite a low output process and, therefore, the premixing requirements and dilution processing do not normally lend themselves to automatic or even semi-automatic control.

The basic theory of operation common to all roll mills is that the design of the unit allows for a feed of premixed dispersion base into a roller which is carefully machined to a smooth surface, and the thin film is passed between either two rollers or a roller and a bar at an extremely high hydraulic or mechanically applied pressure. The gap allowed between the two surfaces is, therefore, extremely small and hence high shear rates are developed. Large agglomerates are crushed between the working surface and it is common practice to pass the more resistant dispersion bases through the unit at progressively higher pressures until an adequate degree of dispersion is achieved. In all machines of this type some effect is also obtained by the complex eddy currents set up immediately behind the *nip* and this is promoted in many of the multi roll units by driving rolls at different speeds or by applying a brake to free running undriven rolls.

Single-roll mills

These machines are widely used. Probably the best known unit is the Keenock Machine now made by Vickers Armstrong. This consists of a hopper feeding a roll and bar system. Depending on the basic design the adjustment to the gap is made by the use of either hydraulic or mechanical jack settings. It is normal to find that only the centre section of the discharged paste is dispersed to an acceptable degree and deflectors are provided to channel the material on either side of the main stream into separate containers for reprocessing. If abrasive materials are used frequent re-setting of the gap is required and the rate of wear on the blade and roll is often quite high and these, therefore, need regrinding and sometimes resurfacing or replacement. The main application is for the production of special products in the paint industry but even with the use of water cooled rolls solvent loss can be high and identification of the solvent lost is difficult.

Two-roll mills

These mills come in various forms, and the simplest is a pair of driven rolls which are designed to intermesh. They can be driven at similar or dissimilar speeds depending on the application and the pressure applied can also be varied considerably. The basic system of operation is to feed a supply of pre-dispersion in a cohesive state onto the rolls and allow it to coat both rolls completely while leaving a suitable volume in a reservoir between the rolls. This reservoir is then constantly changed until the required degree of dispersion is obtained when the operator cuts or scrapes the product from one or both of the rolls and the sheet produced is then discharged. Alternatively the material can be passed through the *nip* in one pass only and scraped off continuously. The main application of this type is in the rubber and plastics field.

Triple and multiple-rolls

In these machines three or more rolls are mounted in a heavy framework. With the most modern machines the three roll system has been adopted (Fig. 6.11). The centre roll and the front roll are geared together and are driven through a variable speed fluid coupling to give a wide variation in

Fig. 6.11. *The Torrance* 24 *in by* 12 *in streamline triple roll mill with hydraulic pressure control.*

operating speed. The back roll is again of variable speed but independently driven through a separate variable speed unit. A scraper tray is mounted to remove the dispersion from the front roll and a hopper is mounted to feed the pre-mixed paste between the back and centre rolls. The pressure on the back and front rolls can be adjusted independently using an hydraulic system. The dispersion base passes from the hopper between the back and centre rolls and the speeds and pressures are adjusted so that the centre roll carries the material onto the front roll from which it is scraped by the scraper tray. As the rolls are made of high grade abrasive resistant material and are accurately machined and are driven by a heavy duty drive, a considerable amount of work is done on the material as it passes through the two *nips*. All three rolls and the hopper are water cooled and control of the process obtained by observing hydraulic pressure and temperature gauges mounted on the unit. It is frequently necessary to pass the dispersion base through this type of unit more than once in order to attain an acceptable dispersion. The main application of this type of unit is for the dispersion of pigments into oleo resinous systems for the manufacture of high viscosity inks and PVC pastes.

RANGE OF INDUSTRIES AND TYPES OF PRODUCT

From experience in a wide range of industries using dispersion processes it is possible to further classify the process involved in terms of the type of dispersion under consideration; these fall into the following classification irrespective of the industry.

(*i*) Surface coatings to be applied on site.
(*ii*) Surface coatings to be applied elsewhere.
(*iii*) Solid phase dispersion.
(*iv*) Liquid phase dispersion.
(*v*) Processes in which dispersion is an intermediate operation.

In Table 6.2 we list the industries to be considered here and indicate the type of product obtained. The industries are listed in alphabetical order and for each a brief example will be selected to show the type of operation involved.

Adhesives

In most cases a twofold operation is being performed at varying levels of viscosity. The objective is to create a partial solution or colloidal dispersion of a polymerised organic material in a solvent base or an emulsion in an aqueous base while at the same time dispersing a pigment or extender in order to modify the finished product rheology, reduce the cost of the finished product or to colour and opacify the product.

Due to the presence of polymer–solvent solutions and/or emulsions together with pigments and extender this type of system can obviously be regarded as similar to those processed in other industries such as paint and ink.

Ceramics and refractories

Production of a ceramic clay body

In this case the objective is to produce an even dispersion of non-plastic fusible materials in a clay dispersion. The non-plastic materials are of considerably larger particle size than the clay and the clay itself is normally densely agglomerated. Any surface active agent used must be either volatile at the drying temperature or non-reactive at firing temperature. The degree of dispersion of the plastic clays controls the strength of the freshly formed wet article and also its unfired strength whereas the particle size and degree and uniformity of dispersion of the fusible non-plastic elements controls the strength, shrinkage and flexibility of the finished article. Traditionally the process is carried out by slow speed wet mixing, filter pressing and pug milling, but an increasing number of manufacturers now consider the use of more modern high speed units to increase the degree of dispersion and thereby improve the final properties both before and after passage through the kiln.

Production of glazes

The process involves the dispersion of fusible ceramic materials with small quantities of plastic clays and soluble silicates. It is traditionally performed in horizontal ball mills capable of reducing the particle size of the non-plastic elements. In recent years there has been a tendency to adopt a technique in which the non-plastic materials are individually prepared to the required particle size before dispersion in high speed dispersion units. This overcomes difficulties in identifying oversize materials when all the materials are ground and dispersed as a single operation.

Chemicals

It is almost impossible to select a truly typical example of dispersion from such a wide field. The most frequent requirement is chemical reaction and/or extraction. The dispersion may be one of a solid phase into a reactive liquid, or of two liquids which react to produce a solid. In all such cases the design and speed of the machine used are frequently the only variables available, as consideration of dilution, *p*H, reaction rate and temperature control define the actual formula during the dispersion operation. The machines used must be capable of continually exposing fresh reactive surface while at the same time maintaining a high degree of homogenisation in terms of concentration, *p*H and temperature and without aeration of the product. In many cases where a solid phase product is obtained it has been shown that control of shear rate will effect the particle or crystalline size of the product and much work is being done on variable shear rate systems such as the Premier Colloid mill and Hydisperser in this field.

Paint

The paint industry is probably the oldest established industry involved in dispersion. Therefore it has a very wide range of applications and also equipment to consider. These range from sophisticated industrial finishes to oil-bound distempers and water paint, and from edge runner mills to modern sand and Perl mills and Hynetic or Kady mills.

In such an old established and competitive industry the machine selected for a new factory for a given application will frequently differ from that selected in a fully equipped established one, as it is usually considered that purchase of new capital equipment should be avoided where possible as many of the new continuous production techniques would not only require new equipment but new buildings to house them.

Due to the wide variety of products we will consider a selection of applications.

(*i*) The dispersion of pigments into short oil length alkyds and other modified resins in which there is little affinity for the pigment surface. In the majority of cases the pigments are available in readily dispersed forms but relatively high energy equipment such as the ball mill, sand or Perl mill, or Hynetic or Kady mill are used because it would seem that it is necessary to compensate for the lack of natural affinity by the use of mechanical energy.

(*ii*) The dispersion of decorative quality enamels based on long oil length alkyds and similar resins can be performed on high speed disc cavitation and stator rotor equipment to an acceptable level of dispersion, but many manufacturers still prefer to use large capacity batch ball mills or sand mills for this type to reduce labour costs and increase pigment colour or opacity levels.

(*iii*) Emulsion paints, flat and satin finished solvent based paints, undercoats and distempers. Many manufacturers still use ball mills and other equipment for this type of production, but with the availability of micronised pigments and pre-ground extenders more and more are transferring to high speed equipment such as the disc cavitation mixer, but more particularly to large underdriven stator rotor mixers, as the batch size requirements of this type of product has increased at least four fold during the last ten years.

(*iv*) Stains and intense shades based on high cost organic pigments. When we study the cost of pigments we find an established tendency to use colour fast high colour value dyestuffs to produce pastel shades and bright colours. When using these pigments of high cost it is essential to produce concentrated dispersions or stainers in which they are dispersed to their ultimate particle size if the maximum colour value is to be obtained. This can only be done if they are processed separately in such equipment as the Steele and Cowlishaw high speed ball mill or Torrance Microflow mill in which the high energy input results in almost complete dispersion under efficient conditions while permitting the production of necessarily small batches.

Paper industry

The application of dispersion to the paper industry falls into two main divisions; the dispersion of ingredients for paper stock and the dispersion of colours and stains.

Dispersion of ingredients for the paper stock

Firstly the dispersion of clay. This is usually a simple operation which is conducted in relatively low speed mixers with chemical additives which, in the aqueous phase, aid dispersion. Addition of retaining agents to the stock then flocculates the clay in the paper pulp and prevents it being lost on the wire during paper making.

Secondly the dispersion of colours and stains. This again is normally considered to be a very simple operation. Most of the materials are received in a pre-dispersed filter cake form and are diluted on site in low energy mixers before addition to the paper stock.

Dispersion of surface coatings

This is one dispersion operation which has undergone considerable changes within recent years. Both the machinery used and the techniques of operation have changed during this period. These changes started when it was shown that a previously complicated multistage process could be condensed into a single unit adopting high shear techniques. When this had been established the simplicity of the formulations employed and the large volumes of materials required soon led to use of automated or semi-automated equipment to feed and discharge the high shear rate equipment so reducing the cost of labour and supervision to a minimum.

The basic formulation consists of a dispersion of clay with other pigments in an aqueous system containing protein adhesives and latex emulsion. The introduction of the Kady or Hynetic mill, the Hydisperser and the Dispersatron to the industry condensed the production of a paper coating into a single stage operation involving the dispersion of the clay and the solution of the protein *in situ* followed by the addition of the latex emulsion. As all the equipment used is underdriven to avoid aeration of the product, the open top tank design lends itself to automated loading and control and most modern paper plants are based on this approach. This does not at first sight seem relevant to actual dispersion techniques but it is an interesting example of a cost saving trend which is being followed by other similar industries when high efficiency dispersion equipment of suitable design is available.

Pharmaceuticals

It is difficult to provide this industry with a truly representative example. Applications of dispersion range from the dispersion of dyestuffs in a sucrose solution for the production of coloured sugar coating, to dispersion of oxides in oil emulsions for use in ointments. One of the more interesting examples is, however, the production of tooth paste as it is one of the few really large scale dispersion processes performed. The constituents from the purely physical point of view are pigment, extender, mucilage, soap or detergent, solvent and flavouring. The process is conducted in many different ways to produce a product of basically similar physical characteristics. In many cases heavy duty low shear rate equipment is used at below the yield point of the slightly dilatant shear-thinning system. This means a low capacity high labour and power cost plant with considerable difficulty in handling the paste-like material. Rheological examination of the material indicated that it could be processed in high shear rate equipment in larger batches using less labour and power to produce a product possessing the same terminal rheology on standing but having a shear-thinned pumpable consistency immediately after processing.

The degree of dispersion and rate of homogenisation in the more modern technique is better than in the low shear rate system. Operation under vacuum in either case removes any aeration, which in the presence of soap or detergent is a serious problem in any dispersion operation, as the dispersion action disperses the air in tiny bubbles which are almost impossible to remove, and these produce a similar effect to dispersed solid particles as far as the rheology of the product is concerned.

Pigments and dyestuffs

The applications in this field fall into three main categories.

(*i*) Precipitation and stabilisation of new surfaces which is usually carried out in slow speed equipment under strict temperature concentration and *p*H control. A considerable amount of work is currently in progress to define the importance of shear rate as a means of further control at this stage and this may result in a modification to the basic technique.

(*ii*) Modification of particle size is frequently performed after concentration by filter pressing in ball mills, colloid mills or vibratory ball mills. In some cases the shape of the crystal and the type of agglomeration make it possible to use more flexible and readily cleaned large batch capacity cavitation mixers or stator rotor mixers. It is frequently found that manufacturers refer to this operation as *grinding* when in fact they really mean de-agglomeration and careful control of concentration, and wetting or surface active agent addition frequently permits the use of the less complex equipment.

(*iii*) Dispersion of pigment or dyestuff in aqueous or oleoresinous suspension in the presence of surface active or wetting agents for use in specific industrial applications.

Whenever possible the pigment or dyestuff manufacturers produce these systems from the filter press cake before drying as this avoids the introduction of cemented aggregates. When producing oleoresinous systems these are obtained by the well known flushing technique in low shear rate equipment. When producing aqueous dispersions it is more usual to use high shear rate stator rotor equipment which, even without wetting and surface activating agents will produce a free flowing fairly stable system, from apparently semi-solid filter press cake or centrifuge concentrates without dilution.

Plastics

This book is intended to cover the dispersion of solids in liquids; as the majority of plastics processes pass through a liquid stage during dispersion due to the application of shear stress and heat, these processes are considered to fall within our field of interest.

Most flexible pigmented plastics are produced in a liquid state by the dispersion of pigments and extenders in plasticiser and plastic polymer. This process is carried out in a range of equipment depending on the rheology of the system. This varies from Z-bar type sigma bladed equipment to the use of Pug or ribbon-bladed or planetary mixers followed by three-roll milling. This heavy duty equipment is necessary for the complex shear resistant, pseudo-plastic rheology of this type of system. Less flexible systems are now frequently processed in the dry state using underdriven high speed mixers in which the sometimes pre-dispersed pigment is absorbed with plasticiser and stabiliser onto the dry polymer powder at elevated temperatures before being passed to the sigma-bladed mixer or sometimes direct to a two stage pugging extruder. In these cases it is only in the penultimate or final stages that the process enters a liquid state. When a pre-dispersed pigment system is employed this enables the manufacturer to obtain the maximum colour development from expensive organic pigments by three roll milling or ball milling in plasticiser suspension before addition to the higher viscosity systems. This is, however, only possible when sufficient plasticiser is available.

The plastics industry is now making increasing use of printing techniques to decorate their products. The inks used are similar in many respects to the gravure and flexographic inks used in the normal printing industry and are not, therefore, discussed under the plastics heading.

Printing inks

News Inks

These are produced from low cost materials such as the cheapest possible carbon-blacks which are dispersed in the cheapest possible often self-colouring oils in low cost mixers of simple design. When the oils used do not possess adequate adhesives or pigment wetting characteristics additions are included. The competitive nature of this type of work requires the use of the cheapest possible materials and equipment with maximum throughputs and lowest possible labour costs.

Lithographic inks

These are normally produced in a two-stage process involving the pre-mixing or pre-dispersion stage in planetary or sigma-bladed equipment followed by three or even five-roll milling. The high viscosity of the resin and oil used and the high pigment concentrations required make extraction formulation impracticable and the water proofing processing requirement makes the selection of any wetting agent or surface active agent extremely difficult as almost all anionic or cationic additives increase the rate of water penetration of both wet and dried ink film. It is therefore necessary to overcome the lack of affinity of the solid for the liquid by the use of low shear rate/high shear stress/high power input pre-mixing followed by a number of passes through a triple roll mill at high pressure. Each pass is normally performed at progressively increasing pressures to counteract the lubricating effect of the material which becomes progressively more Newtonian as dispersion progresses.

Flexographic and gravure inks

These are representative of solvent based liquid ink systems in modern use.

Pigmented, resin-bonded, usually partially plasticised, chips are produced on heavy duty plant at elevated temperature at which the system is partially fluid. The equipment and techniques usually being basically similar to those used for lithographic inks except for the inclusion of wetting agent or surface active agents which are permissible to a greater extent. On cooling and the removal of shear forces the material becomes solid and is discharged from the process, in the form of a chip of varying size from $\frac{3}{8}''$ to as small as 50 mesh size depending on the method of cooling. These chips are then redissolved or redispersed in solvent and additional resin either by the ink manufacturer or the printer to provide an ink of the required shade and flow characteristics.

As all the chips are based on single pigment systems and are miscible with each other and the diluent, the equipment required should be fairly simple. The solubility of the best chips and the degree of dispersion within them is such that redispersion can be achieved in extremely simple-medium/high shear equipment but on occasions it is found that more complex equipment is necessary due to lack of solubility of the chip. This is sometimes caused by polymerisation and/or oxidation during chip manufacture and can also be caused by preferential adsorption of resin components by the pigment reducing the solubility of the remainder. This type of chip can be dissolved by use of higher powered equipment such as the Hynetic or Kady mill and in some cases it is ground into solution by resin size reduction in a ball mill. Some manufacturers prefer to adopt the higher energy system to offset problems of resistance to redispersion. Chips of large size can also cause problems in simple high speed cavitation equipment as they tend to centrifuge away from the high energy zone of mixing and are only processed fully when the viscosity of the mix is increased by the solution of the remainder. The objects of this apparently complex process are to obtain a source for the speedy preparation of inks without the need for high energy equipment. Also the product obtained from this process, if adequately redispersed, possesses a better gloss and greater colour development than that obtained by the direct method of dispersion from dry pigment in resin solution in a ball mill, Hynetic or Kady mill. This is presumably due to the high energies transmitted to the batch in the process of chip manufacture and due to the presence at that stage of an excess of resin to fully satisfy the pigment surface exposed. The direct process is only now used to a very limited extent when economy of production is necessary for low cost pigment systems in which maximum dispersion would not result in any economy.

Rubber

The two main examples from this industry are both for completely liquid systems, but as in the plastics industry there are other applications for solid in liquid dispersion in which the dispersion is performed in a semi-liquid state due to temperature and shear conditions while the material itself reverts to a solid on cooling and/or removal of shear forces.

The two examples have been selected as they represent the solvent based on aqueous based divisions of the field.

Production of pigmented solvent based rubber solutions for fabric impregnation in the tyre industry

Premasticated partially pigmented rubber is dissolved in either high speed sigma-bladed mixers or stator rotor mixers in suitable solvent.

Additional pigment and filler is then added and due to the inherent affinity of the solution for the pigment and filler and its relatively large particle size and low degree of agglomeration, an acceptable product results with little difficulty.

Latex foam base preparations

The basic ingredients for this process are latex emulsion, sulphur accelerators, fillers and pigments. Each of the materials is pre-ground to a fairly carefully controlled particle size and the dispersion of each in its own base liquor is achieved at this stage. This process is performed in ball mills, vibratory ball mills or attritors depending on the feed size and product size requirements of the application. These pre-dispersions are then further intermixed in either stator rotor mixers or turbine mixers to provide a complete intermixing of the ingredients, and the process then passes to the next stage which can be either the dispersion of a finely divided blowing agent, a material which gives off a gas during the subsequent heating stage, or after the addition of a suitable frothing agent or detergent to an aerator. In both cases the degree of dispersion achieved controls the size and distribution of the pores within the latex product while the adequate dispersion and homogenisation of the remaining reactive ingredients ensures complete reaction during the heating stage and controls the shrinkage, flexibility and resilience of the product.

ECONOMICS OF DISPERSION OPERATIONS

In evaluating the economics of a given dispersion operation we are considering means by which a given quantity of acceptable material can be produced at the required rate at the minimum cost. When considering the dispersion from this aspect we must first ensure that the ingredients provided in the initial basic formula are the best available in terms of price and dispersibility. This is a more complex problem than it first appears, and it is the main difference between dispersion operation costing and the costing of other manufacturing operations such as conveying or filling where the materials do not alter. The material most difficult to assess in this way is the solid phase or pigment and by normal standards the more costly the material dispersed the more rewarding does exact costing consideration become. In most systems the dispersion operation does not alter the liquid phase and this should therefore be purchased in the normal way at the lowest available contract price compatible with a given acceptable quality.

Let us consider the solid or pigment phase in more detail to see the reasons for the above statement. Having read the previous chapters we now understand that the degree of dispersion of the solid phase affects its final properties in terms of colour, rheology and opacity as well as in terms of texture, maximum particle size and ease of application. What many manufacturers of dispersions in industry ignore is that dispersion is frequently incomplete when the product reaches a level of acceptance, and that using adequate dispersion equipment and formulation a greater degree of dispersion can frequently be obtained. In those cases in which colour or opacity is required from the dispersion phase it is therefore possible to reduce the quantity of high-cost pigmentation by increasing its degree of dispersion and the surface exposed to reflected or transmitted light. In cases in which a reaction takes place the speed of reaction and even in some cases the type of reaction can be affected in a similar manner. A further consideration is that the same basic material can often be obtained in more than one form. The raw material manufacturer frequently offers his product in two or more grades at differing prices, and these may vary in dispersibility; and it is frequently found that different equipment is necessary to make the most advantage of the finer pretreated grades. Usually when small outputs of a variety of materials are required it is preferable to use the pre-treated form while large throughputs based on untreated materials can frequently be processed more cheaply than the same product based on the treated grade.

The proportion of the cost of the raw material represented in the difference between treated and untreated grades differs depending on the cost of the basic raw material and the type of treatment. There is usually a significant price difference between a pigment sold in a rough-ground calcined form and one in a surface treated micronised form. If the pigment user can process the untreated grade at a lower overall cost than the pre-processed version then he obviously uses it for preference, but the hidden surface cost factor must also be considered. The type of equipment used for the two types of feed material differ and it is frequently shown that the equipment used for processing untreated material is more suitable for development of additional surface than is the equipment used for processing pre-treated raw materials. The high speed equipment used on the pre-treated grade is not capable of reducing the cemented aggregates or particle size nor of applying sufficient energy to the surface of fine pigments to produce the ultimate levels of dispersion obtainable in equipment in the ball mill or high-powered/low-shear-rate ranges.

Also of importance is a careful study of finished formulations, with a view to the extraction of not only an optimum dispersion formulation but also possibly the extraction of high cost pigment concentrates for subsequent addition after high energy processing in the absence of the lower cost constituents of the formulation. This is not always as straightforward as it sounds. Some manufacturers still prefer to use larger quantities of the high-cost ingredients under less efficient dispersion conditions to avoid the risk of flotation, etc. Both these problems can normally be overcome with careful addition and formulation control.

Due to the flexibility of a dispersion process and the large number of variables it is frequently necessary to make a compromise between two optimised conditions. For example, a complete dispersion of a pigmented paint produces a product of greater opacity or colour value and one of greater stability in terms of surface wetting, but as the product must frequently be stored for an extended period a controlled amount of flocculation is necessary to prevent settlement of the pigment into a hard mass at the bottom of the container. In some cases this would not be a problem because stabilising agents are added, but this is not always possible due to their incompatibility or due to their effect on product rheology. In other industries complete dispersions are not required to be stored for extended periods as they are applied after standing in agitated vessels within the same factory in which they are processed. Hence, full knowledge of a process is required not only in terms of its ingredients and their reactions to the dispersion operation, but also in terms of subsequent use and application, before adequate control of an operation is achieved.

DISPERSION STAGE OPTIMISATION AND FORMULATION

This section is divided into three parts:

(*i*) examples extracted from past experience by which practical experiments can be conducted to assess processing equipment and to optimise processing conditions,
(*ii*) practical examples of the use of rheology in this field and the possibility of employing change in rheology as a means of assessing degree of dispersion and
(*iii*) examination of the methods available for the formulation of mill bases by extraction from final formulae without resorting to practical milling experiments.

Optimisation experiments

Three types of equipment offer different problems when it comes to optimisation experiments. These are:

(*i*) open-top machines in which the change is visible during the dispersion exercise, *i.e.* all high and low-shear-rate mixers, except Colloid mills, attritors and all roll mills,
(*ii*) enclosed equipment such as ball mills and
(*iii*) continuous equipment such as Colloid, Perl and sand mills.

In the case of open top equipment optimisation can be performed in small batch equipment by making adjustments to the formulation during processing. This provides a mean point about which further experiments can be conducted utilising the full range of available variables.

In the case of continuous equipment it is not usual for the high-energy zone to be visible during operation but changes in the composition of the feed material are quickly reflected by changes in the product. If these changes in the product are such that they can be assessed fairly quickly in terms of level of acceptance, then in this case simple optimisation exercises can be conducted in the early stages so reducing the number of experiments required.

With enclosed equipment there is no means of shortening the range of experiments necessary and it is therefore necessary to consider the full range of variables involved and evaluate each in turn if full knowledge of the process is to be achieved.

Take for example a ball milling process for the dispersion of Prussian Blue pigment in a long-oil alkyd system for use as a pigment concentrate in a paint formulation. If we list the variables we have to consider within a full evaluation we find that these are as follows:

(*i*) resin solution concentration,
(*ii*) pigment base volume,
(*iii*) mill base volume,
(*iv*) ball volume,
(*v*) ball size,
(*vi*) milling time under optimised conditions, and
(*vii*) effect of high speed ball milling in a planetary ball mill.

The optimum conditions are illustrated in Figs. 6.12–6.17 in which effects in the high speed planetary mill are shown by a continuous line and those in the ball mill by a broken line. The pigment was milled in the ball mill for 20 hours and in the high speed mill for 2 hours. The optimum conditions are as follows:[2]

(*i*) resin solution concentration 25–30% (Fig. 6.12),
(*ii*) pigment concentration 20–25% for ball mill, 40–45% for high speed mill (Fig. 6.13),
(*iii*) mill base volume 25% for ball mill, 25% for high speed mill (Fig. 6.14),
(*iv*) increase in ball volume is shown to increase efficiency within limits of experiment; this would normally be standardised at 45% apparent volume or 28% actual (Fig. 6.15),
(*v*) ball size. Efficiency of dispersion increases with decrease in ball diameter; this is also controlled by discharge rate (Fig. 6.16),
(*vi*) equilibrium is reached at 60 hours in the ball mill and at 4 hours in the high speed mill (Fig. 6.17). Pigment concentration (a) 50%, (b) 30%, (c) 30%, (d) 20%,

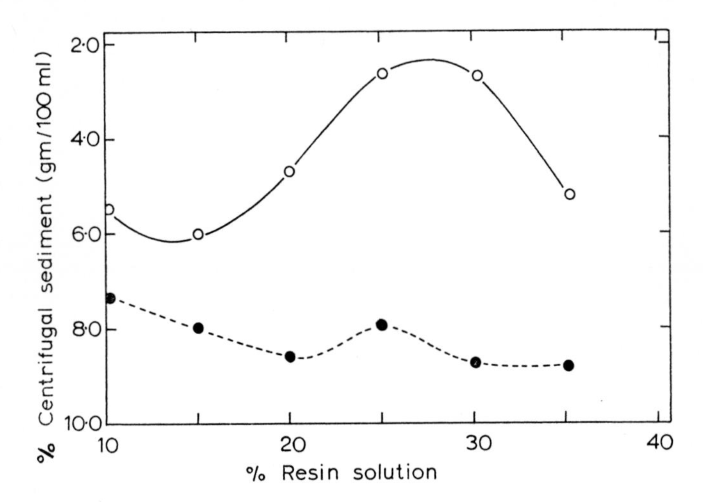

Fig. 6.12. *The influence of resin solution concentration on dispersion efficiency.*

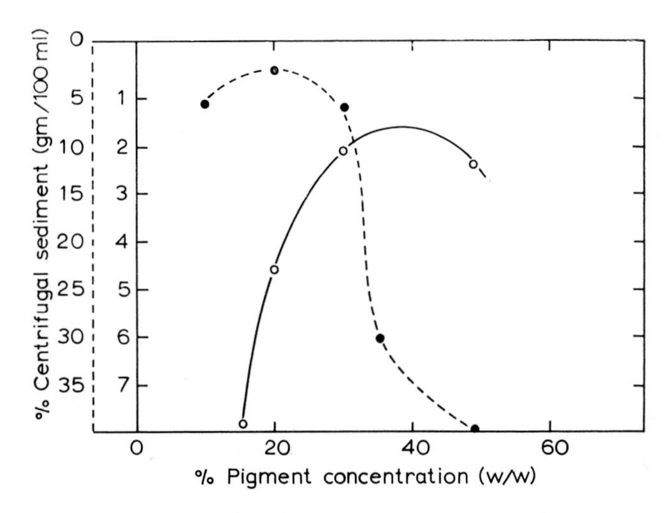

Fig. 6.13. *The influence of pigment concentration.*

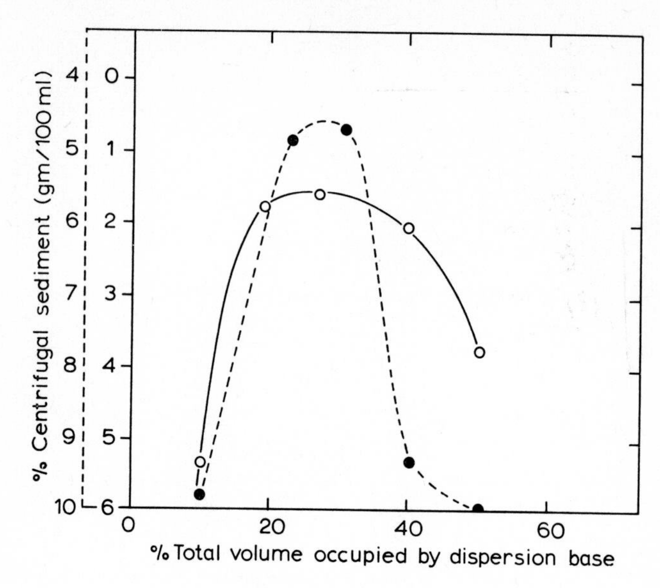

Fig. 6.14. *The influence of charge volume.*

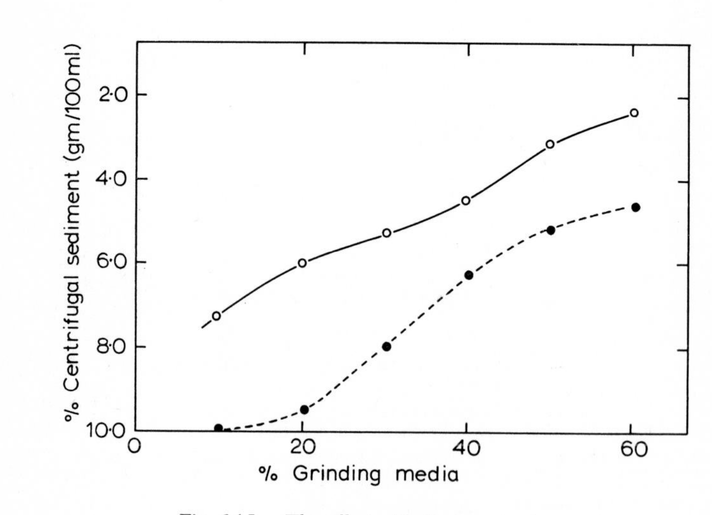

Fig. 6.15. *The effect of ball charge volume.*

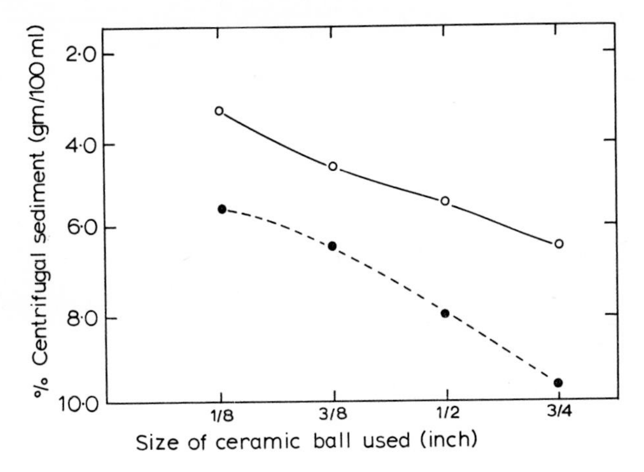

Fig. 6.16. The effect of ball size.

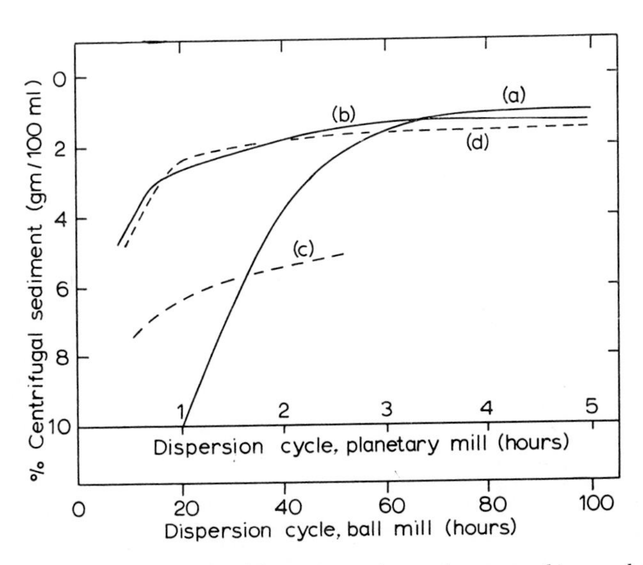

Fig. 6.17. The effect of increased dispersion cycles on the two machines each being used at two different concentrations. Further examination of the samples by other techniques (Fig. 6.3) showed that the colour value of the planetary mill product after 5 hours at 50% was 8% greater than that from the same unit after the same time at 30% which in its turn possessed 17% more colour than the product from the ball mill after 80 hours milling.

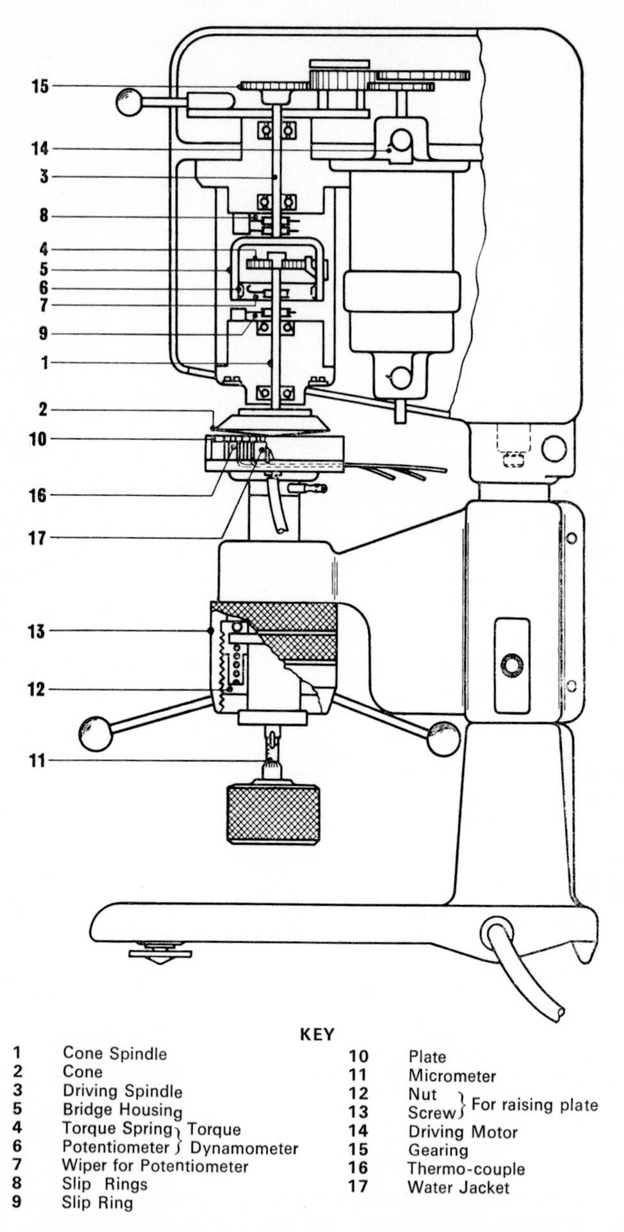

Fig. 6.18. Sectional diagram of a Ferranti–Shirley cone on plate viscometer showing the drive arrangement and the positioning of sensing elements to signal the torque requirement to the galvanometer.

(*vii*) high speed milling permits higher concentrations, higher volume charges and shorter milling cycles.

The testing technique adopted was centrifugal sedimentation and this is not a maximum-sensitivity technique. It is possible to plan a full set of experiments of this type or it is possible to use the *factorial* approach.[3] Both techniques, although extremely sound, require a considerable amount of background work in the laboratory. This can be well worthwhile but also frustrating if only small quantities of the product are required.

Use of rheology

If a sample of an acceptable product is available, together with testing equipment, capable of simulating manufacturing dispersion conditions it is possible to examine it in the laboratory to ascertain if the equipment used is of the most suitable type.

Reference to Fig. 6.3 shows that the apparent viscosity of the system examined on the Ferranti–Shirley Viscometer I (Fig. 6.18) below a shear rate of 160 secs would be extremely high and the behaviour of the product in a low shear rate mixer would therefore be suitable to that equipment. The viscosity of the system at a high shear rate would suggest use of a high shear rate stator rotor machine which would absorb much less power per unit of output. This is one example of the use of rheology in identifying machine usage but this in general proves difficult due to the difficulty in simulating the high shear rate available in some dispersion equipment

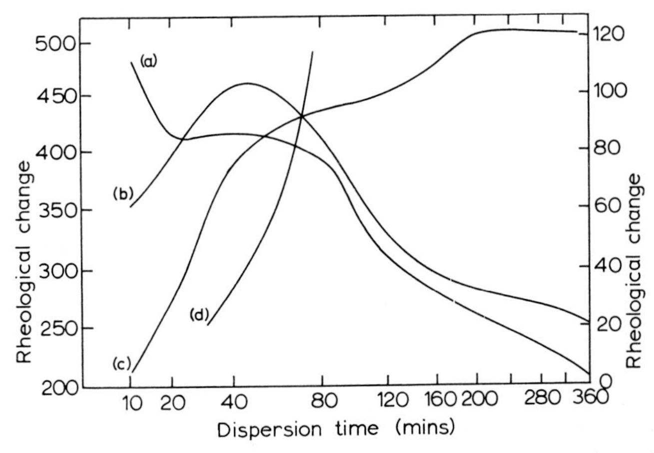

Fig. 6.19. *The effect of dispersion and particle size reduction on the rheology of different systems*; (a) *Neospectra (non-beaded) carbon black,* (b) *Prussian blue,* (c) *microdol in water,* (d) *Neospectra (beaded) carbon black. All systems dispersed or ground for increasing times under optimised conditions.* (*Scale for* (a) *on left hand side, and for* (b), (c) *and* (d) *on right hand side.*)

with information obtainable from laboratory equipment. In certain cases viscometers are designed to simulate low shear rate production equipment quite exactly. An example of this is the Brabender Plastograph which measures and records the torque requirement during the mixing of a small batch of material in a small mixing vessel in which blade speed can be adjusted. The other major difficulty encountered are the changes in rheology of the product during processing,[1] according to the type of process conducted (Fig. 6.19). When such changes are characterised it is shown that grinding or particle size reduction of cemented aggregates tended to produce one type of rheological change and dispersion produces another. It is possible that these changes can be used to indicate continued change in degree of dispersion (Fig. 6.20).

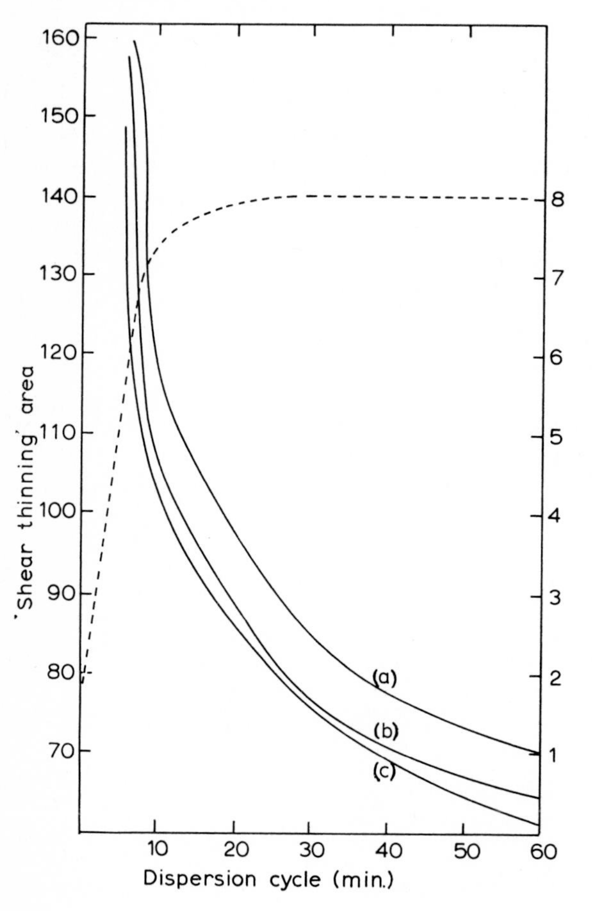

Fig. 6.20. Change in the rheology of the product as a means of detecting significant differences in dispersion below the level of sensitivity of other testing techniques. Continuous lines, (a) 2500 *cc charge,* (b) 2000 *cc charge,* (c) 1500 *cc charge. Broken line, Hegman Gauge reading.*

Extraction from final formulation

In order to avoid excessive laboratory work various techniques have been established to formulate satisfactory mill bases by extraction from final formulation. These depend on either laboratory bench equipment or on the use of the physical constants of the ingredients. Two methods of assessment are oil absorption and flow point, which are used in Pug or U-trough mixing and ball milling to give an indication of optimum dispersion conditions. The formula for formulating dispersion bases in the disc cavitation mixer is[4]

$$F/C = 0{\cdot}90 \cdot \frac{V_s}{145} \cdot \frac{\eta}{40}$$

where F/C is the oil absorption factor for high speed impeller equipment, V_s is the percentage vehicle solids, and η is the viscosity of vehicle in poise at 25°C.

The use of this system has been discussed[4] where it is shown that with an oil absorption of 20 for a titanium pigment the dispersion would be as follows:

Resin Solids (%)	*Oil Absorption*	*Weight of resin (lb)*	*Weight of pigment (lb)*	*Pigment* (% *W/W*)
30	1·58	31·6	100	76
50	1·72	34·4	100	74
75	1·89	37·8	100	72

A further technique for heavy duty mixing was shown to be applicable in that there appears to be a direct relationship between pigment volume concentration and the specific surface of any pigment when optimum conditions are found to coincide with the plastic point. This is shown in Fig. 6.21. Two curves are shown, (*a*) in which the surface area per cc of pigment is plotted against the pigment surface area per cc of resin and (*b*) in which surface area per cc of pigment is plotted against the pigment volume concentration assuming that the pigment is spherical and in cubic packing. The optimum concentration for any pigment within the range of surface area (measured by the same technique) covered by these graphs can be obtained by reading direct providing it is to be dispersed in the same vehicle. A further technique has been established on the basis of the Sonsthaghen formula from which optimum ball mill dispersion base compositions can be evaluated.[5]

The basic formula is

$$V = \frac{W \times S}{BC}$$

where W is the weight of pigment, S is the specific surface area, BC is the

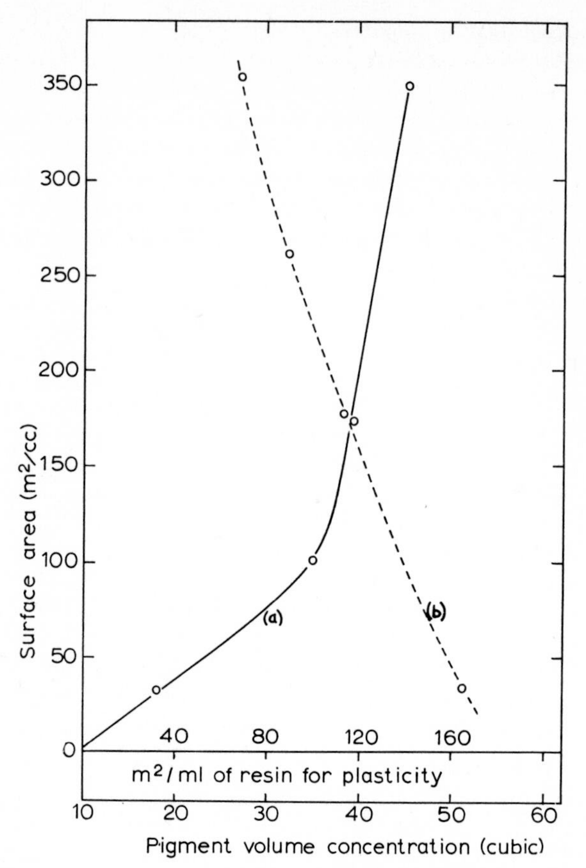

Fig. 6.21. The comparison between pigment surface area and pigment volume concentration at the plastic point *in a Baker Perkins Universal Mixer when the pigments are being dispersed in a long oil alkyd varnish.*

resin solution constant and V is the volume of resin. This has since been modified to include a machine dependent factor which varies according to the type and size of dispersion unit and the type of grinding media employed. The modified formula is

$$V = \frac{W \times S}{BC \times MF}$$

where MF is the machine factor. The machine factors vary from 0·9 for a 6″ diameter laboratory pot mill to 0·99 for a 6′ 6″ diameter horizontal ball mill and also go up to 1·2 for use of steel grinding media (density 7·8 gm/cc) and to 1·45 for a high speed planetary mill.

This formula does in effect modify the original formula according to the hydrostatic energy of the system. The higher the energy level the lower is the resin solution requirement during the dispersion stage. In ball milling a pigment dispersibility factor can be associated with the optimum milling volume,[5] so that if we take a mill having a working volume of 40 % this figure would be multiplied by a factor to modify the working volume according to the dispersibility of the pigment. The factors are:

micronised inorganic pigments	1·00	working volume 40 %,
coarse inorganic pigments	0·8	working volume 32 %,
Prussian Blue and similar pigments	0·7	working volume 28 %,
other organic pigments	0·6	working volume 24 %,
fine carbon-black	0·5	working volume 20 %,

The main problems when utilising any such techniques are

(*i*) the raw materials used tend to vary and frequent physical examinations are necessary in order to ensure that these fluctuations are not reflected in the quality of the finished product,

(*ii*) the formulating technique can never fully replace the complete range of milling experiments, because it is impossible to take into account the many variables encountered during the dispersion exercise as this would make the evaluation of a formulating technique impossible and its use much too complicated, and

(*iii*) the optimum formula when calculated cannot be extracted from the final formula without changing the form of the feed materials.

This last problem frequently arises when resin solutions of low concentration are employed in production when the volume of solvent required at the dispersion stage is greater than that normally associated with a given quantity of resin solution. This is usually overcome by using higher concentration resin solutions in the dilution stage, but sometimes a compromise dispersion formulation has to be used to avoid leaving an addition of high viscosity or even solid resin as the only diluent. When this occurs it is usually found preferable to use a higher resin concentration than the calculated optimum to leave free solvent for the cleaning of the machine and/or viscosity adjustment. The reasons for this are that the technique employed for assessment must reflect not only the type of raw material included in the formula but also the application in which the dispersed product is to be used.

Some examples of assessment of dispersion from the practical standpoint are given below.

(*i*) If a dispersion of silica flour is required in an epoxy resin for use as an electrical insulator then the assessment of its final properties should be electrical rather than those of sieving. As dispersion improves the electrical resistance would be expected to increase as each particle of silica was adequately coated with the resin.

(*ii*) If an absolute dispersion of a costly organic pigment is required for use as a concentrated stainer in the paint industry it would be of little value to examine the product on a Hegman Gauge which possesses a limiting assessment level of approximately 2·5 μ when the pigment has a particle size of less than 0·1 μ. In this case either colour testing or rheology (Fig. 6.20) would be the obvious choice to determine the equilibrium condition.

(*iii*) If a dispersion of pre-treated clay was required for use as a paper coating which was to be applied at 50 μ thick on the paper then the technique of assessment of dispersion would be simply a wet screening of the product through a 350 or 400 mesh screen together with rheology and adhesive strength, as the results achieved in the latter two tests would be of major importance.

As will be appreciated the method of assessment used to compare two samples must be the same and the technique adopted to produce product physical data for use in formulation equations must also be constant. This is due to the extremely fickle nature of fine solids since the result obtained is totally dependent on the technique of assessment employed. Thus when in doubt about the acceptance level of a product the ultimate test must be one of application and however sensitive the testing techniques available the technique adopted must be that which best reflects the application involved. For it is only when this acceptance level has been attained that further investigations can be performed to ascertain if further dispersion can either reduce the cost of the product or further improve its final properties.

REFERENCES

1. I. R. Sheppard and G. Cope, *Rheologica Acta*, **4** (1965) 244.
2. I. R. Sheppard, *J. Oil Colour Chemists' Assoc.*, **47** (1964) 669.
3. D. G. Dowling, *J. Oil Colour Chemists' Assoc.*, **44** (1961) 188.
4. G. R. Lester, *J. Oil Colour Chemists' Assoc.*, **47** (1964) 719.
5. I. R. Sheppard, *J. Oil Colour Chemists' Assoc.*, **47** (1964) 691.

CHAPTER 7

ASSESSMENT OF THE STATE OF DISPERSION

S. H. BELL and V. T. CROWL

INTRODUCTION

The term *dispersion* is used in several senses, leading to some confusion as to its implications. It is used to describe the process of dispersing a powder in a fluid, that is, the incorporation of a dry powder into a liquid medium in such a way that the individual particles of the powder become separated from one another, or form small clusters, reasonably evenly distributed throughout the entire liquid medium. It is also used to describe the resulting product, frequently with some qualification *e.g.* good, poor, fine, coarse, etc. The term *degree of dispersion* may be used to describe the extent to which this objective has been attained, and a powder may be said to be well or poorly dispersed in a liquid medium. Whilst there is considerable confusion of usage, basically these terms are related to the effective size distribution of the powder in the liquid. Frequently, however, some technological properties are used as criteria, *e.g.* in paints, the strength of colour of a pigment developed in the product, or the absence of excessively large aggregates as measured with a Hegman *fineness of grind* gauge.

A further term, *dispersion stability*, describes the extent to which the original degree of dispersion is maintained in the product. A material which is poorly dispersed, *i.e.* with a low degree of dispersion, will frequently show poor dispersion stability due to sedimentation of large aggregates. The fact that a material shows a high degreee of dispersion, however, does not necessarily imply high dispersion stability. Well dispersed individual particles may come together to form *flocculates*, consisting of loosely held clusters of particles as described later in more detail. Such clusters may be separated into smaller units relatively easily by mechanical forces, and may then re-form once the disturbance has ceased. Flocculation is essentially a dynamic process, resulting from an equilibrium between the attractive forces between individual particles and the disruptive effects of thermal motion of the molecules of the liquid phase. A change in the composition of the system, such as that produced by the addition of a surface active agent, may result in a considerable reduction in the extent of flocculation. Similarly, certain agents induce flocculation in a system. Such changes are essentially alterations in the dispersion stability of the system.

In all cases, the two terms *degree of dispersion* and *dispersion stability* are attributes of the particular system. Sometimes the term *state of dispersion* is used to include both these attributes, or in relation to technological criteria involving desirable or undesirable properties of the system in practice. When a powder is described as easily dispersible, the implication is that a high degree of dispersion and good stability may be obtained in the appropriate medium with little mechanical effort.

Dispersions are used in many technologies, with considerable differences in their nature and the technological requirements. Frequently the fluid dispersion itself is involved at an intermediate stage in a process. Thus a liquid paint or printing ink provides a means by which a thin film of a solid pigmented polymer may be deposited upon a substrate, the final product being the dry film or coating. In such cases the degree of dispersion required is governed by the requirements of the end use. For example there should be no pigment aggregates or agglomerates sufficiently large to cause irregularities in the surface of a dried paint film and thus reduce the gloss or spoil some other needed surface characteristic; and the particles should be sufficiently well dispersed to attain the required opacity. Similarly the dispersion stability must be sufficient to give long storage stability without excessive settlement, and to enable the coloured pigments present in the mixture to maintain the desired colour of the surface coating.

In the pharmaceutical industry, uniformity of dispersion and long-term stability are required in order to attain accurate dose rates, and considerable attention has been paid to the degree of dispersion of pharmaceutical materials in relation to assimilation rates of drugs. In other industries such as ceramics and foodstuffs, dispersions may be used soon after production, so that long-term stability is not so important.

The requirements for adequate assessment of dispersion in different technologies are thus very varied. In the sections which follow, the authors have drawn mainly on their experience of the paint and printing ink industries, in which fine particles, mainly below 1 μ in size, are dispersed in many types of aqueous and non-aqueous liquids. The basic principles, however, are common to many technologies.

THE NATURE OF POWDERS

A dispersion may be made by taking lumps of the solid phase and grinding them in the presence of a liquid medium. Such processes are usually highly uneconomic, and natural solid materials, when used, are generally subjected to dry grinding, followed by a stage of classification, to produce dry powders before attempting dispersion in a liquid. Many solid materials to be used as dispersions are produced by chemical means in powder form, but in certain cases materials obtained as an aqueous suspension, *e.g.* as a filter cake, may be used directly to prepare aqueous dispersions without an intermediate drying stage. Direct transfer of such solids from an aqueous to a non-aqueous liquid, thus avoiding the possibility of aggregation or agglomeration on drying, is obtained in the *flushing* process used in the pigment industry.

A powder consists of a collection of particles of various degree of complexity. In the following text, the term 'particle' is used in a general sense. The term 'individual particle' is used to imply a single entity, as opposed to an aggregate, agglomerate, or flocculate, as defined below. Frequently the term 'cluster' is used to indicate a group of individual particles which may involve aggregates, agglomerates or flocculates, acting as a unit in

Fig. 7.1. *Irregular pigment particles of natural materials produced by fracture.*

given circumstances. In a powder produced by dry grinding a solid material, *e.g.* a mineral, there are particles of a range of sizes and shapes; the size range may be reduced by a classification process, but all the particles are of the same type of irregular material produced by fracture (Fig. 7.1). Materials produced by precipitation, *e.g.* most chemical products, organic and some inorganic pigments, contain several different types of particle.

Fig. 7.2. Primary particles of a crystalline pigment. (*Note*: *in this and subsequent figures the primary particles are shown in an idealised form.*)

First, there are the *crystallites*, in which distinct regions of the particles are composed of a single crystal structure. Several crystallites may grow together to form single *primary particles* (Fig. 7.2). Groups of these primary particles may be firmly bound together to form *aggregates* or *agglomerates* (Fig. 7.3).

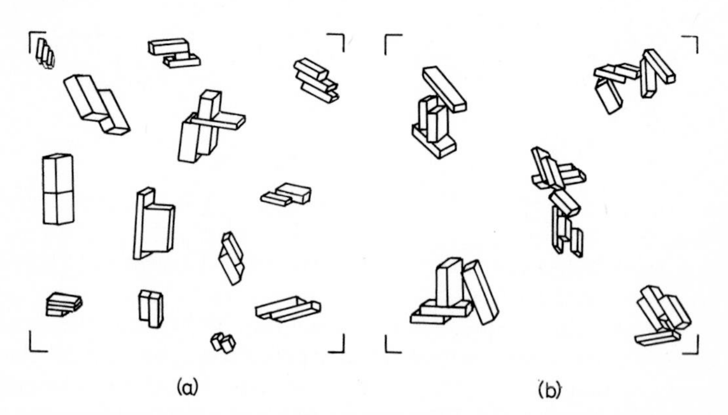

Fig. 7.3. (a) *Aggregates, in which the primary particles are joined at crystal faces.* (b) *Agglomerates, in which a looser structure is formed, joined at the edges or corners of the primary particles.*

A distinction has been drawn between aggregates, in which individual primary particles are joined together at crystal faces, and agglomerates in which the particles touch only at edges or corners, forming a looser, more open structure. The modes of formation of aggregates may vary, but their essential feature is the strength of bonding of the original primary particles into an assembly which behaves as a single particle, and which can only be disrupted by considerable force. With materials produced by precipitation, followed by filtration and drying, the individual primary particles are possibly held together by residual traces of soluble salts; such aggregates

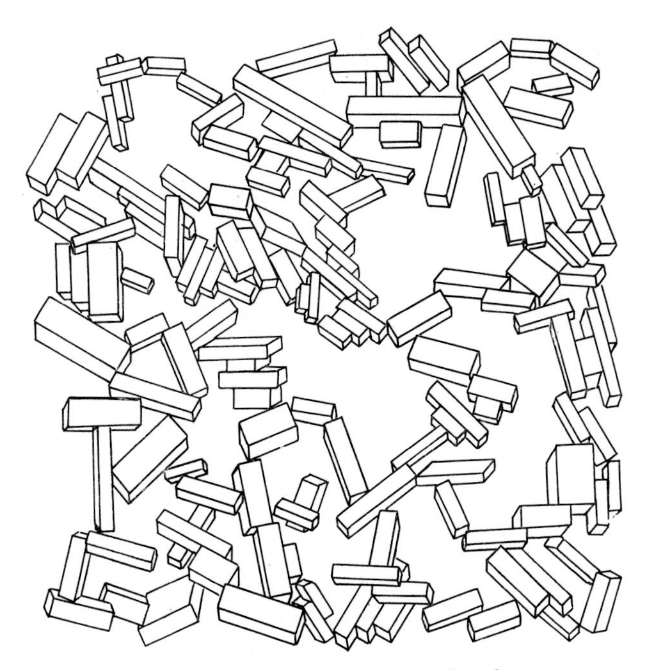

Fig. 7.4. Typical open structure of flocculates.

formed on drying are termed *cemented aggregates*. In some cases, a calcination stage is subsequently applied, and *sintered aggregates* may be formed in which there are considerably stronger bonds between the individual primary particles. In a dry powder macroscopically sized lumps occur, which may generally be easily broken down into individual primary particles, aggregates and agglomerates; these have been designated *air-flocculates* by analogy with similar loose structures in liquid dispersions. Such air flocculates may be easily disrupted in the dry state, but considerable difficulty may be experienced in dispersion in liquid media consisting of solutions of polymers. This is partly due to incomplete displacement of air by the polymer solution, and to selective diffusion of the solvent into the dry mass, leaving a polymer-rich shell around the solid. In paint technology such large clusters are termed *pigment-medium agglomerates*, and are experienced as *nibs*. Such agglomerates may also be produced by other mechanisms, but they are all difficult to re-disperse.

In a dispersion of a powder in a liquid, assuming that there has been complete wetting-out of the powder, varying proportions of primary particles, aggregates, and agglomerates will be present depending on the degree of dispersion attained. In addition there may be *flocculates*, consisting of loosely bound clusters of particles (Fig. 7.4). Flocculation is essentially a dynamic process, and flocculates may continuously form and partly re-disperse as a result of Brownian motion. In this respect, though they may vary considerably in stability, they can clearly be distinguished from aggregates or agglomerates. The overall tendency for flocculates to be present is dependent upon the surface chemistry of the systems, and is greatly affected by the presence of small amounts of flocculating or dispersing agents. A flocculated structure may be set up in a dispersion at rest which may tend to extend throughout the whole volume of the liquid; on mechanical disturbance partial breakdown into smaller flocculates may occur, and on ceasing the agitation the structure may re-form. Processes of this nature are involved in *thixotropy* and some other forms of rheological behaviour of dispersions.

THE PROCESS OF DISPERSION

Three separate stages may be distinguished in the process of dispersing a dry powder in a liquid medium.[1,2]

Wetting.—This is the replacement of the powder–air interface by the powder–liquid interface. For complete wetting to occur all the adsorbed air and other contaminants, *e.g.* adsorbed water, should be displaced, as well as air merely occluded in the pores of agglomerated particles, so that there is direct contact between the liquid medium and the solid interface.

Mechanical disruption.—The powder is thoroughly mixed with the liquid medium and the agglomerates or other large particles present are broken down by mechanical means into particles of the required size, *i.e.* the required degree of dispersion is attained.

Stabilisation.—Conditions are established such that undesired flocculation of the dispersed particles is prevented or minimised. There are attractive forces between particles in an unstabilised dispersion which would normally lead to flocculation. Stabilisation occurs as a result of interactions occuring at the solid–liquid interface, such as the adsorption of molecules or ions from solution, which modify the attraction potential between the particles by the introduction of an electrical charge repulsion or prevent the close approach of particles into the region of strong attractive forces by steric hindrance (entropic repulsion).

These three stages of the dispersion process may occur almost simultaneously, but they are quite distinct in character. It is questionable whether they all proceed to completion in any practical dispersion process. In the dispersion of pigments in paint media it has been shown that release of the last traces of gases entrained by the pigment may take many hours. The degree of dispersion attained in practical paints falls short of complete breaking up of all the agglomerates into individual particles, as is shown by the use of the Hegman gauge in assessing dispersion and by other studies of liquid paint films. Moreover, in the practical concentrated dispersions used in a number of technologies, it is rare for complete deflocculation to occur; indeed this may be undesirable because of the resulting formation on prolonged storage of hard sediments, difficult to redisperse. A partly flocculated system, on the other hand, will form a sediment of larger volume consisting of loosely packed material which is more easily redispersed, and may therefore be preferred in practice.

TECHNOLOGICAL PROPERTIES OF DISPERSIONS

Many important technological properties of dispersions are dependent upon the degree of dispersion attained and the state of flocculation of the dispersion in the conditions of end-use, and it is in the assessment of these properties that the concepts of *good* or *poor* dispersions are frequently used.

What, however, is meant by *good* dispersion and how far can it be expressed quantitatively as to the extent of separation of the particles in relation to the required properties? There has recently been a renewal of interest in this topic, with the development of improved methods of investigation. With dispersions such as paints and printing inks, it is necessary to use electron microscopy in order to obtain the necessary resolution of the pigment particles. For many years the technique used was to prepare thin microtome sections of dried paint films, for direct examination by transmission electron microscopy. By tilting the specimen stage in the electron microscope it was possible to prepare stereoscopic pairs of pictures, which revealed whether apparent clusters of particles were in fact single entities or separate particles or clusters at different levels in the film sections. Hornby and Murley have recently shown that the normal glass or diamond knives used in microtomes can cut through groups of primary particles, thus dividing them into smaller units and producing erroneous results.[3] Kämpf, Liehr and Völz[4] used a preparation technique based on oxidising away the resin by active oxygen, leaving the clusters of pigment particles undisturbed, after which the free pigment is embedded in a matrix, *e.g.*

of polyvinyl alcohol or gelatine, followed by coating with an evaporated platinum/carbon film for examination in the transmission electron microscope. Solvent etching to remove the medium from paint films has been used by Wilska,[5] followed by examination in the scanning electron microscope.

It is not possible, at present, to examine liquid paints directly in the electron microscope, to see whether the clusters in dry paint films existed as flocculates in the liquid paint, or were produced during the film drying stage. In a recent development at the Institute for Electron Microscopy at the Technische Hochschule, Graz, the liquid paint or ink is allowed to fall dropwise through a region cooled by liquid nitrogen. This 'freezes' the dispersion almost instantaneously, so that the individual particles or clusters may be expected to remain unchanged from the liquid state. The solidified droplets are then sectioned on a refrigerated microtome; the sections are exposed to ozone to partly oxidise away the medium; and then coated with carbon to form a replica. The remaining polymer is dissolved with a solvent and the pigment can be removed if required with Caro's acid. The resulting electron micrographs show the state of dispersion of the pigment in the original paint or ink.[6]

Numerous investigations have shown that the state of dispersion, even in paints of accepted good quality, falls far short of full separation of the individual single particles in the dried film, but there may be well-distributed small clusters of particles, apparently as the stable units. Indeed it has been found that the distribution of sizes (number of primary particles in each cluster) is reproducible for a given combination of pigment and medium, and is related to the opacity of thin paint films,[3,7] and to gloss.[4]

It would be expected that such clusters of particles would be of an equilibrium size range determined by the balance of all the forces operating. In paints where the liquid medium consists of a mixture of polymer molecules of varying degrees of complexity, selective adsorption of certain components may occur, so that the medium associated with the pigment may differ from the remainder of the medium in its degree of polarity or molecular weight range. Recent work by Jettmar[8,9,10] using refined techniques of specimen preparation for electron microscopy, has shown that this is indeed so. Certain pigments are associated with the boundaries of, for example, high molecular polymer regions of the paint film, as shown in Fig. 7.5, while others may concentrate elsewhere in the film.

Thus, in practical paints of good quality, the pigment whilst not in the form of individual particles evenly distributed through the liquid paint or dry film, is usually present as pigment-medium units which behave as separate entities of high stability giving the necessary paint properties. In this sense *good dispersion* is frequently used to mean that a good paint is obtained—good for its purpose, rather than that there is maximum separation of particles.

In certain applications, however, the properties which may be desirable, such as a controlled degree of flocculation giving certain required rheological properties, may in other contexts be highly undesirable. It is therefore necessary to review the effects of varying states of dispersion on technological properties. Much of the following section is of general applicability in a wide range of technologies.

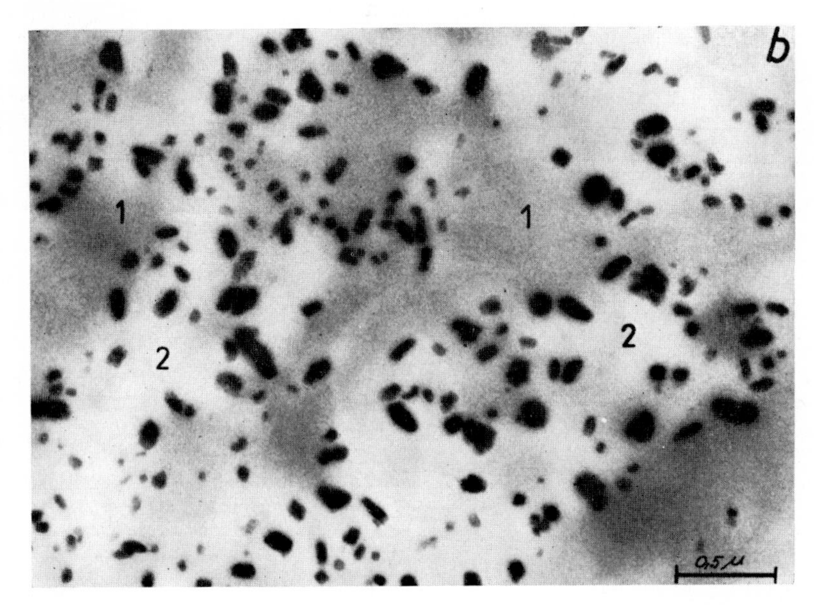

Fig. 7.5. Electron micrograph of phthalocyanine blue pigment particles at the interface between the high molecular, non-swelling alkyd polymer (1), *and the low molecular swelling regions* (2). (*By courtesy of Dr W. Jettmar.*)

Sedimentation behaviour

In the more dilute dispersions of solids in liquids, sedimentation behaviour is of considerable practical importance, and even in some concentrated dispersions such as paints sedimentation may be troublesome, particularly where there is formation of hard sediments, difficult to redisperse.

Stokes Law for the sedimentation of an isolated single particle in a liquid related the rate of fall, v (cm/sec) to the diameter d (micron) of the particle, the viscosity of the liquid η (poise) and the difference in specific gravity between the solid and the liquid $(\rho - \rho_0)$

$$d^2 = \frac{18\eta v}{(\rho - \rho_0)} \times 10^8 \tag{7.1}$$

For a given liquid medium, this relation can be simplified to

$$V = Kd^2(\rho - \rho_0) \tag{7.2}$$

The rate of sedimentation is proportional to the density difference, and to the square of the particle diameter.

Where flocculation has occurred the flocculates may effectively behave as single large particles, and fall at a considerably greater rate. The effective density of the flocculates, including the entrapped liquid phase, may be significantly lower than that of the solid phase and there may be irregular shapes, but these effects are usually small compared with that of the large increase in size. Consequently a flocculated dispersion will in general sediment considerably more rapidly than a deflocculated system, but the nature of the sediment may be very different. The deflocculated particles on sedimentation will pack closely into a compact layer at the bottom of the vessel, as shown in Fig. 7.6a, difficult to redisperse. The

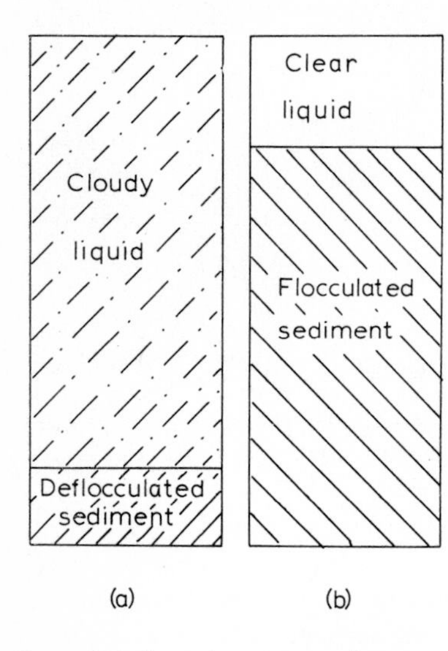

Fig. 7.6. (a) *Sedimentation of a deflocculated material into a tightly packed sediment of small volume.* (b) *Sedimentation of a flocculated material with a loose sediment of large volume.*

flocculated system may, however, form a loosely packed structure of considerably greater volume, as in Fig. 7.6b, since the gravitational force on the particle will be insufficient to break down the structural condition of the flocculates, and then the sediment will be easily re-incorporated into the dispersion by light stirring or shaking. The relationships for the two extreme types of system can be expressed as in Table 7.1.

In practice, there can be various intermediate forms of behaviour. In the concentrated dispersions used in many technologies, the resulting close packing of the particles results in hindered settling, so that a looser type of structure may be formed even where flocculation is not extensive. However, larger particles, agglomerates and aggregates may be able to reach the bottom of the vessel on prolonged standing, particularly if they are of materials of high specific gravity, with the resulting difficulty of re-incorporation into the dispersion.

Sedimentation problems are sometimes troublesome in dispersions consisting of mixtures of powders with differing specific gravities, *e.g.* many paint systems incorporating a mixture of organic and inorganic pigments.

TABLE 7.1

SEDIMENTATION BEHAVIOUR

State of Dispersion	*Rate of Sedimentation*	*Sedimentation Volume*	*Nature of Sediment*
Deflocculated	Slow	Low	Hard, difficult to re-disperse
Flocculated	Rapid	High	Soft, easily re-dispersed

In such mixed systems the function $d^2(\rho - \rho_0)$ for each pigment may be calculated. Where similar values are obtained, separation on settlement is unlikely to take place, but where large differences occur, differential sedimentation is highly probable although other factors may be involved. The suspension and settlement of pigments in paint media has been reviewed by Patton.[11]

Rheological behaviour

The rheological behaviour of dispersions[12,13] is of considerable importance in a large number of industries, and indeed many of the popular terms used to describe consistency are derived from the properties of particular types of material, *e.g.* watery, creamy, buttery, pasty, dough-like, etc. Rheological properties are of importance in the production of dispersions where certain types of mill may operate satisfactorily only within a limited range of flow properties, but more particularly in the use of the finished products. A few examples will suffice; in the forming of ceramic objects involving the rheology of the clay pastes; in the application of paints and printing inks to substrates; in the forming and moulding of filled plastics and rubber; in the flow of dispersions of solids in a wide range of process industries; in the use of everyday materials such as ointments, cosmetics and toothpaste; and in the handling of foodstuffs such as doughs and ice-cream, all are dependent upon flow properties.

Consider two parallel plates immersed in a liquid, a distance x apart. A force F is applied tangentially to one of the plates so that it moves with a velocity v in a line parallel to the other plate. A velocity gradient is set up between the two plates, intermediate layers moving with intermediate velocities. The velocity gradient dv/dx (an incremental change of velocity

dv corresponding to an incremental change in thickness dx) is constant for a Newtonian liquid and is referred to as the shear rate D. Since velocity v may be expressed in cm/sec, and distance d in cm, the shear rate D has the dimensions of reciprocal seconds. The total force acting tangentially on the moving plate is F, normally expressed in dynes, and the area of the plate in A cm^2. This force per unit area is the shear stress τ ($= F/A$), expressed in dyne/cm. The viscosity η is the ratio of the shear stress τ to the shear rate D

$$\eta = \frac{\tau}{D} \tag{7.3}$$

or

$$\eta = \frac{F}{A}\frac{dx}{dv} \tag{7.4}$$

The unit of viscosity is the poise, the dimensions of which are g/cm sec. Where the shearing force is due to the weight of the liquid, as in a flow cup, it is convenient to use another viscosity parameter. In these circumstances F is proportional to ρ so that for a given head of liquid the rate of shear depends on η/ρ. This ratio is termed kinematic viscosity ν and is expressed in stokes. The dimensions of kinematic viscosity are cm^2/sec.

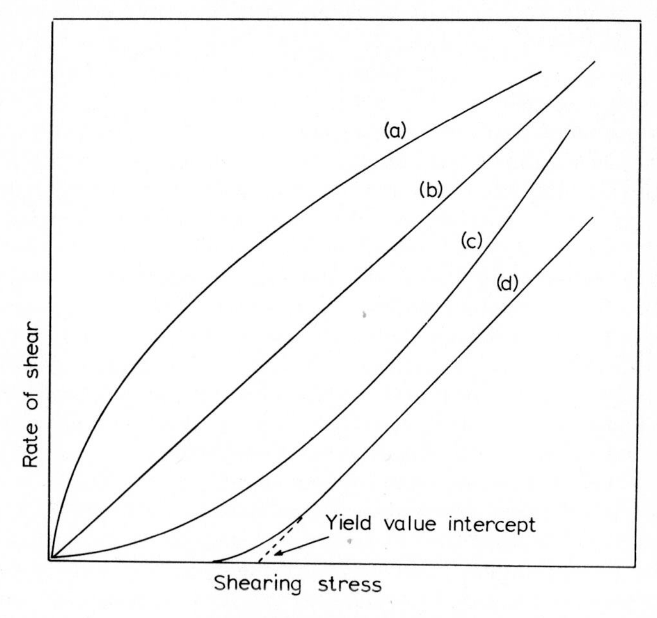

Fig. 7.7. Basic type of rheological behaviour of dispersions. (a) *dilatant,* (b) *Newtonian,* (c) *pseudoplastic,* (d) *plastic.*

Where the relation between the shear stress and the shear rate is linear, the rheological behaviour of the systems is described as *Newtonian flow*, and the slope of the line through the origin is the coefficient of viscosity η. Wide variations from this simple Newtonian behaviour are shown by many materials, including dispersions, and a classification of the different basic type of flow behaviour that can occur is shown in Fig. 7.7. It will be seen that in all cases other than Newtonian liquids the viscosity (= shear stress/shear rate) is dependent on the rate of shear.

A Newtonian liquid flows under any applied force, however small. Materials exhibiting *plastic flow*, however, require a certain minimum shear stress before flow commences; this minimum stress is designated the yield value f. The coefficient of plastic viscosity U is given by the relation

$$U = \frac{\tau - f}{D} \tag{7.5}$$

which applies to the linear portion of the curve above the yield point. This type of flow is also called Bingham flow, and is shown by some dispersions of finely divided solids in liquids such as paints, printing inks, clay pastes used in ceramics, etc. Bingham[14] suggested that in a suspension showing plastic flow the individual particles touch each other, in a flocculated array, and that in order for flow to commence it is necessary to apply sufficient stress to break some of these inter-particle bonds (Fig. 7.8). During shear an equilibrium is set up between the breaking and re-forming of these bonds, so that at a constant rate of shear an equilibrium value for the shear stress is attained. Increasing the rate of shear produces a different equilibrium condition, and so a definite relation between the shear stress and rate of shear is attained, termed the plastic viscosity U.

Certain materials exhibit *pseudoplastic flow* (Fig. 7.7) in which there is no initial yield value, and where with increasing rate of shear, an apparent decrease in viscosity occurs, the shear stress/rate of shear curve being convex to the shear stress axis. Many dispersions and some high-polymer solutions exhibit such behaviour, which can be interpreted in the latter case in terms of interactions between the deformable polymer molecules, analogous to these occurring in a flocculated dispersion of solid particles, which are similarly partly broken down under shear.

The fourth type of behaviour shown in Fig. 7.7 is *dilatant flow*; the apparent viscosity increases with increasing shear rate, the shear stress/shear rate curve being concave to the shear stress axis. This type of behaviour is shown by deflocculated dispersions of pigments and other powders at high volume concentrations. At these high concentrations of the solid phase, the particles are closely packed and disturbance by shear introduces irregularities into the packing, with bridging effects occurring between the particles. Since the packing becomes looser, the total volume of inter-particle space becomes greater, and the liquid present is no longer sufficient to fill the space between the particles; consequently the lubricating effect of the liquid, which enables the particles to slide over one another, is lost. This effect of the change in volume relationships can be observed as a loss of gloss of the surface of a concentrated deflocculated pigment paste when stirred.

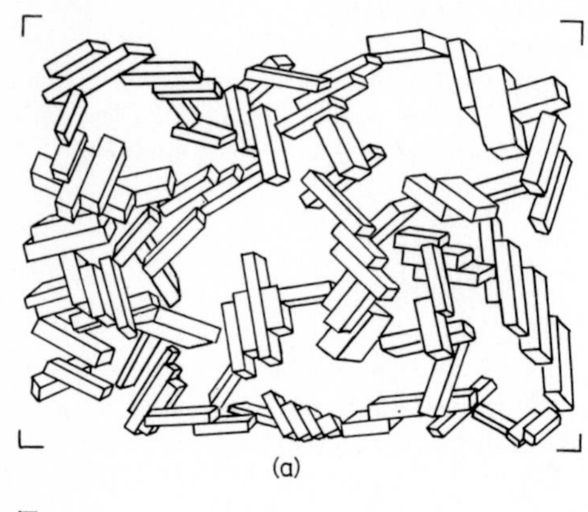

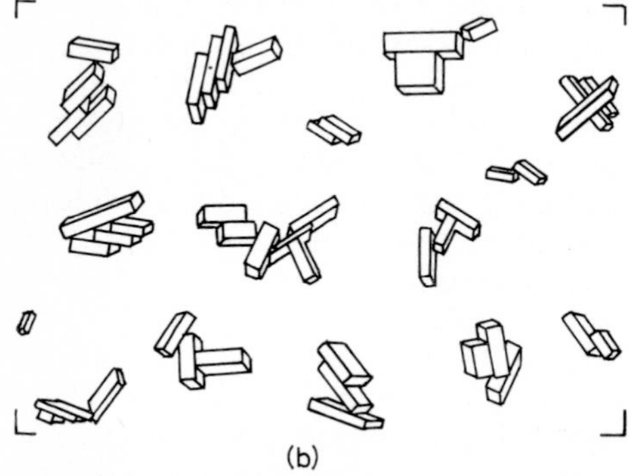

Fig. 7.8. (a) *Thixotropic system at rest.* (b) *Thixotropic system under shear, showing the break-down of the structure into smaller flocculates.*

A further type of behaviour exhibited by certain systems, is *thixotropic flow*. In this case the existing structure of the system is broken down on stirring, and does not re-form immediately, *i.e.* a time factor is involved in the behaviour, which may range from seconds to hours. Figure 7.9 illustrates the type of flow curves which may be obtained using a rotational viscometer with such systems. The curve obtained at decreasing shear rates does not coincide with the original curve obtained at increasing rates of shear, and one measure of thixotropy is the area of the loop between the two curves. This, however, depends largely on the particular circumstances of the type of instrument used and the rate of measurement.

In terms of the mechanism involved, thixotropy is related to plastic and pseudoplastic flow, with the additional effect of time-dependence of structure formation. Where the time effect is short, normal rheological measurements may not show the occurrence of thixotropy, a curve showing plastic or pseudoplastic flow being obtained on rotational viscometers. Numerous types of systems show some form of thixotropic behaviour including some polymer solutions and aqueous and non-aqueous dispersions. Another form of behaviour is *rheopexy*, or shear thickening, in

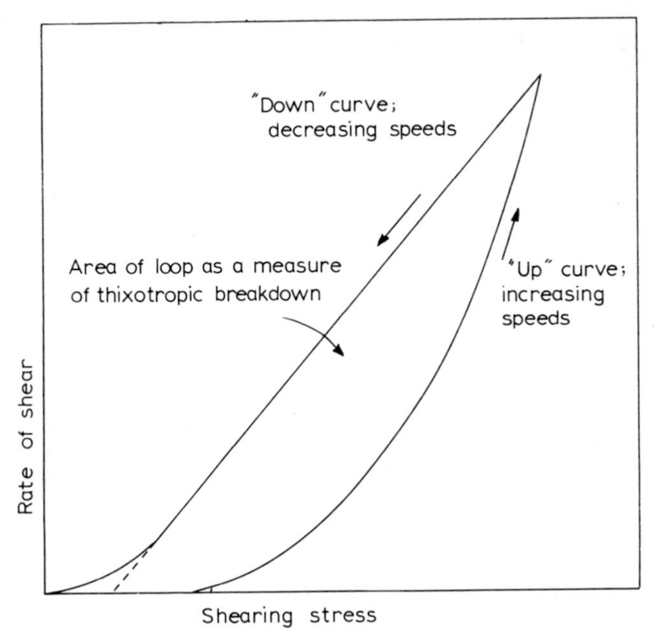

Fig. 7.9. Rheological behaviour of a thixotropic system as measured with a rotational viscometer. The area and nature of the thixotropic loop depends upon the type of instrument and the circumstances of measurement.

which thixotropic materials tend to thicken again on continued movement.[15] There is, however, considerable confusion in the use of this term; it has recently been suggested that a certain minimum shear rate is necessary.[16]

These two properties of dispersions, sedimentation behaviour and rheology, governing overall stability and ease of handling, are fundamental in almost all technological applications of dispersions. A number of other properties are of significance in particular industries. Thus the state of dispersion of pigments affects the optical properties such as colour and opacity of surface coatings of various types. The state of dispersion of carbon black and other fillers in filled rubber and plastics compositions such as tyres, affects the elastic properties and wear resistance, determining the suitability of the product for a particular use.

METHODS OF ASSESSMENT OF THE DEGREE OF DISPERSION AND DISPERSION STABILITY

The many methods used to study the properties of dispersions, range from rather crude tests used in production to assess the degree of dispersion attained, to more elaborate laboratory techniques suitable for academic or technological studies of flocculation. The more common methods of investigation are described below. In some cases the measurements relate primarily either to the degree of dispersion attained or to the dispersion stability, while in others the distinction is not so clear cut. Considerable care should be exercised in assessing the results of certain types of measurement where both factors are involved, *e.g.* sedimentation experiments, or colour development in paint systems with mixed pigmentations.

Control tests for degree of dispersion

With some dispersions, simple sieve measurements may be satisfactorily used to assess the amount of coarse material. A known weight or volume of the dispersion is passed through the appropriate sieve (using a suitable liquid to wash through the dispersed material) and the solid retained by the sieve is weighed. For control purposes it may suffice to use a single sieve, the aperture of which bears some relation to the sizes of particles of importance in the end-use of the product. This technique may also be used with fine sieves, *e.g.* 240 BS mesh, to follow the course of dispersion of fine materials such as pigments containing small amounts of larger sized aggregated material, using a modification of the standard method for determination of residue on sieve.[17] The information obtained relates only to the amounts of material above the sieve sizes, under the conditions used.

Another procedure giving information about oversize material is the use of a wedge gauge, as in the paint and allied industries. The Hegman gauge is of this type, having a wedge-shaped depression about $\frac{1}{2}''$ wide and $5''$ long cut into a polished steel block. The wedge may be $0{\cdot}004''$ ($\sim 100\ \mu m$) deep, tapering along its length linearly to zero depth, with the gauge conventionally marked with a linear scale 0–8 (or sometimes 0–10), so that 0 corresponds to $100\ \mu m$ depth of channel. Gauges are also available with the depth of channel marked directly in microns. In the printing ink industry, finer gauges tapering from $0{\cdot}001''$ ($\sim 25\ \mu m$) to zero are used. A small amount of the dispersion is placed in the deep end of the wedge (held in a horizontal position) and drawn down the gauge by a steel scraper the edge of which is rounded to a radius of 0·01″. The thin wedge of paint remaining in the channel is viewed immediately at grazing incidence, and the point is observed at which a number of large particles project through the glossy film of the dispersion. Individual isolated particles projecting in the deeper part of the gauge are neglected, and the point at which some 4 or 5 particles in a line project through the film is usually taken as the gauge reading. In spite of the somewhat arbitrary definition of the reading, good agreement can be obtained amongst experienced observers, and this type of gauge is used widely in the paint and printing ink industries for process control. It does not measure the overall degree of dispersion, but only the large individual particles, agglomerates or aggregates remaining in the millbase. Nevertheless, as milling is extended to reduce the number of such agglomerates or aggregates sufficiently, it may be assumed that the general degree of dispersion of the finer particles has also been improved.

It is frequently assumed that the depth of the gauge at the point at which the reading is taken corresponds to the size of the largest individual particles, agglomerates or aggregates present in the dispersion. This assumption may not be strictly correct, since in the process of moving the wedge along the gauge, the liquid would be expected to shear midway between the bottom of the wedge and the base of the moving scraper, *i.e.* midway between the top and bottom of the wedge. The thickness of the film of paint left on the gauge would thus be half the indicated depth of the wedge. However, depending on the rheological properties of the paint, some material may flow back under the wedge as it is drawn down the gauge. In addition it is necessary for the particles to project a finite distance above the level of the liquid paint surface to be observable. The size of particle registered on the gauge may therefore be somewhat less than the depth of the gauge at the point of reading.

Appreciable wear of the gauge occurs with regular use and one gauge should be reserved for occasional use as a control standard against which gauges in regular use can be checked. Typical Hegman gauge readings (on the 0–8 scale) for some paints are: high gloss industrial finish $7\frac{1}{2}$ +; gloss decorative finish 7–$7\frac{1}{2}$; undercoats or flat paints 4–5.

Particle size analysis

As dispersion of an agglomerated powder in a liquid medium takes place, the size distribution of the powder changes. Various methods of particle size analysis may be used to follow the changes in size distribution, but there is no method which can be applied directly to a concentrated system without danger of modifying the dispersion. Various stages of dilution, sometimes extensive, are required to produce the system used for measurement, calling for considerable care in dilution procedures to avoid flocculation of the dispersion which could lead to erroneous results. Stepwise dilution procedures with addition of further amounts of resin solution or of surface agents as required are preferable to simple dilution to the required concentration with solvent in one operation, unless there is good evidence that the latter operation does not induce flocculation. In general, particle size analysis yields meaningful results only in highly dispersed systems, and attempts to measure the degree of flocculation of a dispersion can only be significant if the actual measurement is made in the system as it is. Any change in the system such as dilution may result in the establishment of a fresh equilibrium state of flocculation. In addition the techniques of certain methods of size analysis, *e.g.* spreading a dispersion on a microscope slide for optical microscopy, or the flow of a highly diluted dispersion through the orifice of the Coulter Counter, introduce conditions of shear which could lead to re-dispersion of flocculated material. The various methods of size analysis applicable to the study of dispersions will be described in the following sections, but it is important when quoting the results to state the method of determination and the specimen preparation technique.

Sieving

The normal ranges of wire mesh sieves are too coarse to be used in the study of most dispersions (a 240 BS mesh has a nominal aperture of 63 μm), and, as has already been indicated, they are only of value in establishing the freedom from undispersed coarse aggregates or agglomerates. The recent development of electroformed micro-mesh sieves with nominal apertures extending down to 10 μm[18,19] and below[20] suggests, however, that sieving techniques may become more generally applicable in the study of dispersion processes.

Sedimentation methods

Gravitational sedimentation techniques are well established for size analysis and determination of particle size distributions.[21,22] The results of sedimentation analysis are calculated using Stokes Law (eqn. (7.1)) to relate the free falling velocity of isolated particles to the diameter of spheres of the same density with the same rate of fall. The dispersions must by sufficiently dilute for free falling conditions to apply, and a maximum concentration of 1 % by volume has been recommended for some forms of apparatus, and for other methods a concentration not exceeding 0·2 % by volume is specified.[21] Therefore in the study of the degree of dispersion of many practical systems extensive dilution is necessary, preferably in a stepwise fashion to minimise flocculation. The tendency towards flocculation should be examined, *e.g.* by microscopic observation.

A second criterion for the application of Stokes Law to sedimentation behaviour is that conditions of viscous flow apply. The upper limit of size of particle which can be studied is given by the magnitude of the Reynolds number

$$\frac{vd\rho}{\eta} \times 10^{-4}$$

a dimensionless quantity which should not exceed 0·2. At higher Reynolds numbers turbulent conditions apply and Stokes Law is inapplicable. At the lower end of the size range, as d decreases, the time of fall through a given distance increases with the second power. Consequently long sedimentation times are required for particles of 1 μm and below; this presents difficulties in avoiding minor temperature fluctuations which could give rise to convection currents in the apparatus. For many substances of moderate density (3–4) the lower limit of size which can be safely measured is about 2 μm. The lower size limit is decreased for particles of greater density.

Various experimental techniques can be used.[21–25] Sedimentation methods may be either incremental, as in the Andreasen pipette, or cumulative, as in sedimentation balances, the Wiegner tube, and various forms of sedimentation column. For certain purposes, photosedimentation may be used, in which the concentration of solid at the measuring point is determined photometrically. For a more detailed consideration of the theory and experimental methods employed the literature references quoted above should be studied.

Sedimentation methods of size analysis give results directly in weight size distributions, so that no problems of conversion from number size distributions occur. Although many dispersions may contain material below the lower limit of size quoted, of about 2 μm, size analysis is still possible since the total amount of material originally in suspension is known, and may be used as the 100% level, the amounts of material above the various sizes being directly determined.

Centrifugal sedimentation may be used for size analysis of dispersions where the size range is mainly below 2 μm. Various forms of apparatus have been described,[26–30] in some cases of which a two-layer technique is employed. In this latter technique a small amount of the dispersion is introduced into the inner surface of an annulus of liquid in a rotating disc-shaped vessel, and at suitable time intervals the concentration at a fixed radial distance, or the amount of material remaining within a fixed radial distance is determined. Using the two-layer technique the mathematical analysis of sedimentation is greatly simplified, although a new method of operation and calculation of results for homogeneous sedimentation in a disc centrifuge, which overcomes some of the mathematical difficulties, has been published.[31]

Two problems occur with the two-layer technique. First, it is necessary to ensure that flocculation does not occur on dilution of the concentrated dispersion by the spin fluid, and secondly, it is essential to avoid streaming of the dispersion in bulk under the influence of centrifugal force. This latter problem has been overcome by the use of an intermediate buffer layer, which is gently agitated to form a concentration gradient at the inner surface of the annulus formed by the spin fluid.

The two-layer technique has been used to study the changes in particle size distribution of phthalocyanine pigments occurring on milling in alkyd media,[32] and more recently an extensive series of investigations has been made by Carr and his co-workers, using the commercially available ICI/Joyce Loebl disc centrifuge. These show that the instrument is capable of giving reproducible results with aqueous organic pigment pastes[33] and non-bleeding organic pigments in lithographic inks.[34] More recently Carr[35,36] has examined the changes in particle size distribution of organic pigments occurring during milling in alkyd resin solutions, in relation to the development of tinting strength of the millbase. In all this work the importance of careful stepwise dilution of the millbase has been stressed. A method of using a modified ICI/Joyce Loebl instrument as a photocentrifuge for size analysis of organic pigment dispersions has also been described by other workers.[37]

In a more empirical method undiluted dispersions are centrifuged for standard times, in a normal laboratory centrifuge, and the amount of larger sized material accumulated at the bottom of the tube is used to follow the progress of the dispersion process.[38] The results are arbitrary, since at the concentration used the free-falling conditions of Stokes Law do not apply.

The literature on sedimentation analysis has been comprehensively reviewed by the Particle Size Analysis Sub-Committee of the Society for Analytical Chemistry.[22]

Optical microscopy

Size analysis of dispersions can be made by counting on the optical microscope, taking care to ensure that large particles are not driven to the edges of the area under the cover glass so that typical fields are examined. The technique of microscopic counting and the precautions necessary for statistical accuracy in the counting operation have been described in detail in BS 3406, Part 4.[39] The lower limit of size which can be measured accurately with the optical microscope is about 0·5 μm, and consequently it will not be possible to include them in the count. The presence of particles below the limit of resolution of the optical microscope, about 0·25 μm, will be shown as a coloured or partly opaque background.

Sizing is generally performed by comparison of the projected areas of the particles with a series of circles in an eyepiece graticule, frequently in a $\sqrt{2}$ progression of diameters. The number size distributions are converted into weight size distributions assuming that particles over the whole size range are of similar shape. This assumption may not always be correct, *e.g.* where acicular particles are broken down to a more regular shape by milling, and special counting techniques should be used in such cases. In converting a number size distribution into a weight size distribution, the relatively few large particles present become of considerable importance, and it is essential that a sufficiently large overall number of particles should be counted for the results to have statistical significance.[39]

Size analysis by optical microscopy is tedious and time-consuming. A number of automatic counting devices have been developed, frequently based on the flying-spot principle, in which the field is scanned by a spot of light produced on a cathode ray tube, and the interception of the beam of light by particles in the field enables a counting and sizing operation to be performed. In a recently developed apparatus of this type, the Metals Research Quantimet,[40,41] a built-in computer enables a variety of particle size distribution data to be obtained directly. This instrument has been used to determine the particle size distribution of agglomerates of pigments in plastics.[42] Another similar instrument is the Millipore Π M C Particle Measurement Computer.

In another type of apparatus, the specimen slide on the microscope is oscillated over a path to cover the whole field, and the moving images of the particles traverse a slit behind which is a photocell detector. Such automatic procedures greatly simplify counting, but it is necessary to exercise considerable care in the preparation of slides for counting, since factors such as a high density of particles in the field, and overlap or contact between adjacent particles, which can be dealt with satisfactorily in visual counting, may cause errors in automatic counting procedures.

A further technique which has been used to examine the state of dispersion of pigments such as titanium dioxide, red iron oxide, and carbon black in paint or ink films is x-ray contact microradiography. A paint film is drawn down on a fine grain photographic emulsion and then exposed to 'soft' x-rays. The film is developed and then examined under an optical microscope. This technique has a resolution of about 0·5 μm, and avoids the depth of field and scattering problems encountered with normal optical microscopy.[43,44]

Electron microscopy

Where the particles are mainly below 1–2 μm in size, electron microscopy must be used. The techniques of specimen preparation are highly specialised, and because of the small field of the micrograph it is necessary to scan large areas of the specimen to ensure selection of typical fields. One method is to deposit the diluted dispersion on to nitrocellulose or evaporated carbon films, which are then cut into small sections and picked up on the grid for insertion into the specimen holder of the electron microscope.

Dilute dispersions may be prepared using ultrasonics in a low viscosity liquid, or by the use of high shear in a viscous liquid, *e.g.* linseed stand oil, followed by careful stepwise dilution.

Various methods of deposition of the dispersions have been used including spraying droplets, spreading, application with a rubber roller, and floating non-aqueous dispersions in thin films on the surface of water and picking up from underneath on the coated grid. Liquids cannot be examined in the electron microscope, and so the prepared specimen must be allowed to dry before examination. Where the liquid phase of the dispersion is non-volatile, it may be removed by immersion in a volatile solvent. This technique had been found to leave the particles of the dispersion attached to the carbon or nitrocellulose film. Care should be taken to avoid flocculation on drying and the various stages involved in the preparation of the specimens should be followed with the optical microscope. Many other techniques of specimen preparation have been developed for special purposes.

The electron micrographs may be examined as prints or as projected transparencies, and the size analysis may be made as in optical microscopy, using a large transparent graticule,[45] or other sizing device.[46] Certain types of automatic counting and sizing apparatus, *e.g.* the Quantimet, may also be used with transparencies or prints. In all counting methods there are problems of deciding whether particular features consist of clusters or separate individual particles, but reproducible results can be obtained which give some measure of the state of dispersion. With comparatively uniformly sized primary pigment particles (*e.g.* titanium dioxide) it is possible to determine with some accuracy the number of individual primary particles grouped together to form a cluster, by the use of stereoscopic pairs of electron micrographs. Thus a distribution can be obtained, giving the number of primary particles occurring as singles, or in groups of two, three, or more particles.[7] These techniques have the great advantage of being directly applicable to practical paint films, and it has been found possible to relate the cluster number distributions to properties such as opacity and gloss.[3,4]

In academic investigations of dispersion stability, where extremely low concentrations of the dispersed solid are used in the measurement of flocculation rates, the particle concentration may be determined by counting methods using ultramicroscopic techniques.[47] Other methods are based on light scattering for determining the particle concentration. The results, however, may bear little relation to the behaviour of dispersions in industrial practice.

The Coulter Counter

This instrument[48] operates on the principle of measuring electronically the changes in electrical resistance occurring as individual particles of a very dilute dispersion in an electrically conducting liquid pass through a fine orifice (20–400 μm diameter). As the dispersion is drawn through the orifice a series of electrical pulses is produced, the magnitude of which corresponds to the volume of each of the individual particles. These pulses are counted and sized by a pulse height analyser, the various settings of which each correspond to an equivalent spherical diameter. In a sizing operation, a series of cumulative figures are obtained as the size range is traversed.

The instrument is extremely speedy in operation, an individual determination at one size setting in which over 50 000 particles may be counted taking only a few seconds. In more sophisticated versions of the instrument, a whole range of sizes can be counted with one passage of the dispersion through the orifice. The instrument has a high degree of repeatability, and this, coupled with the speed of operation, makes it very suitable for the examination of dispersions taken as milling proceeds. In particular, the large number of particles counted results in greatly increased accuracy at the upper end of the size range, a factor of importance in conversion of the results into weight size distributions.

The dispersions examined in the Coulter Counter are extremely dilute, with 10^6–10^7 particles per ml. At this low concentration even completely unstabilised dispersions will flocculate only slowly. However, at higher concentrations, the flocculation rate of unstabilised dispersions is considerably greater, and it is in the intermediate stages of dilutions of a paint millbase that there is a risk of flocculation if there is insufficient resin or other dispersing agent present. At all stages of dilution the dispersion should be examined microscopically to ensure that flocculation is absent.

The instrument may be used with aqueous or non-aqueous liquids. The necessary electrical conductivity is normally achieved by using solutions of inorganic salts as the electrolyte. In choosing a suitable electrolyte system for examination of non-aqueous dispersions, *e.g.* in resin solutions, the compatibility of the electrolyte with the diluted resin must also be studied. A stepwise dilution procedure is advisable in which further medium or dispersing agent is added to the millbase, followed by successive dilutions with the solvent followed by the electrolyte solution.

The lower limit of size for Coulter measurement is probably above 0·6 μm in aqueous electrolytes; in non-aqueous electrolytes, because of their higher specific resistance, the limit is above 1 μm. Although particles below these limits cannot be sized, the instrument can be used to determine the cumulative amounts of material above these sizes when the total weights of material in the dispersions are known or can be determined.

The changes in size distribution of red iron oxide pigments ball milled in alkyd solutions in white spirit are shown in Figs. 7.10 and 7.11, determined using the Coulter Counter. The millbases were diluted first with alkyd solution, then with white spirit, and finally in successive stages with the electrolyte, a 2·4% solution of ammonium thiocyanate in methyl ethyl

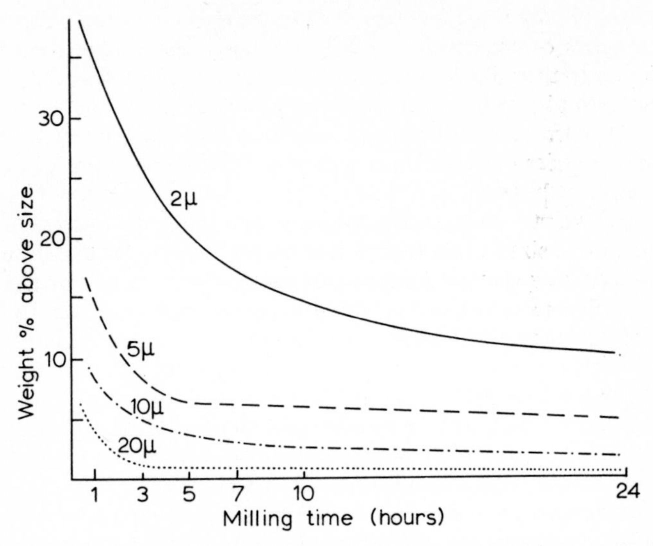

Fig. 7.10. *Dispersion of a natural red iron oxide pigment in alkyd solution by ball milling. Weight percentage residues above* 20, 10, 5, *and* 2 μm *at various times of milling.*

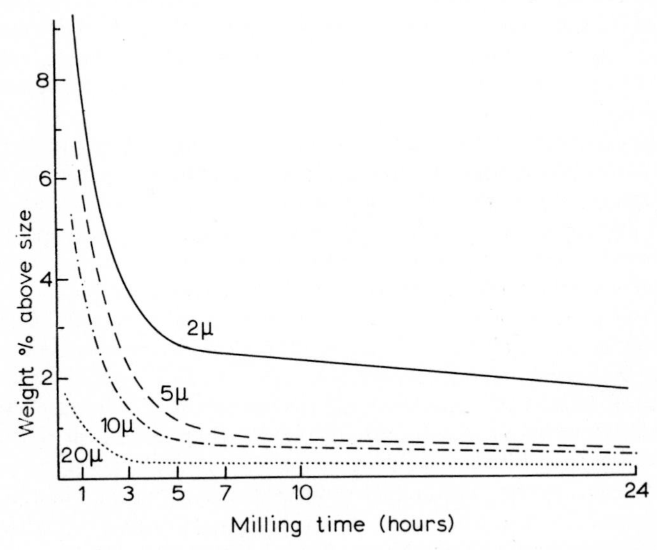

Fig. 7.11. *Dispersion of a micronised natural red iron oxide pigment in alkyd solution by ball milling. Weight percentage residues above* 20, 10, 5 *and* 2 μm *at various times of milling. (Note enlarged vertical scale.)*

ketone, for examination.[49,50] The results, shown as the percentage residue above certain sizes at various time of milling, indicate the considerably more rapid dispersion and greater freedom from larger sized particles of the micronised compared with the normal grade of red iron oxide pigment. Care is necessary in relating results obtained in such diluted systems to practical millbase concentrations, but it has been found that there is good correlation between such measurements and practical paint criteria, such as Hegman gauge readings and the development of pigment tinting strength on milling. The use of the Coulter Counter for the size analysis of organic pigment dispersions of the multi-purpose tinter type has been reported,[51] but difficulties due to flocculation of the organic pigments were found to occur in the aqueous electrolyte used.

Sedimentation behaviour

As indicated above, the sedimentation behaviour of a dispersion is dependent upon a number of factors including the degree of dispersion and dispersion stability. The sedimentation volumes of a number of pigments have been shown to be related to the dielectric constants of the suspending liquids, lower sedimentation volumes resulting from high dielectric constants.[52] Sedimentation tests are easily carried out, and yield directly information of value concerning the technological uses of the dispersions. However, more detailed interpretations of the results, particularly in systems of complex rheological properties and those containing mixtures of dispersed solids, may require considerable thought. Sedimentation processes are of importance in a number of industries, particularly as a means of solid/liquid separation, and an extensive bibliography has been published.[53]

Sedimentation tests, as a means of studying the state of flocculation, are normally performed with the undiluted dispersion, but paint media are generally of high viscosity, and in order to accelerate testing, diluted systems are sometimes examined or observations are made at elevated temperatures. Considerable care is necessary to ensure that the results are meaningful, and this demands experience in interpretation.

The sedimentation rate of isolated particles is proportional to the second power of the diameter, and to the density difference between the particle and the surrounding medium (eqn. (7.2)). Consequently, where flocculation occurs, the rate of fall of the flocculates is considerably increased unless the flocculates are so loose and entrain so much medium that their effective density is lowered considerably. In most dispersions of technological interest with solids concentrations greater than 1% by volume, some form of hindered settling takes place, which in the extreme case may be regarded as a passage of the liquid phase upwards through the mass of downward moving interacting particles. The simple Stokes Law conditions do not then apply; large dense particles will fall most rapidly, and fine low density particles will tend to remain in suspension, but the behaviour of the main part of the dispersion will be dependent upon the nature of the structure in the dispersion, *i.e.* the degree of flocculation, as much as on the particle size and density of the individual particles. Consequently sedimentation behaviour has been extensively used in the study of flocculation.

Two separate but interrelated factors are involved, the rate of sedimentation, and the sedimentation volume (Table 7.1). In very dilute dispersions where Stokes Law conditions apply the rate of sedimentation may be used as a measure of the size of the flocculates, provided that the material has previously been completely dispersed by mechanical means. The normal techniques of sedimentation analysis may be used,[54] but it should be noted that the effective density of flocculates is lower than that of the solid material. Little value, other than in qualitative terms, can be attached to experiments in which powders are merely shaken up with liquids and allowed to sediment.

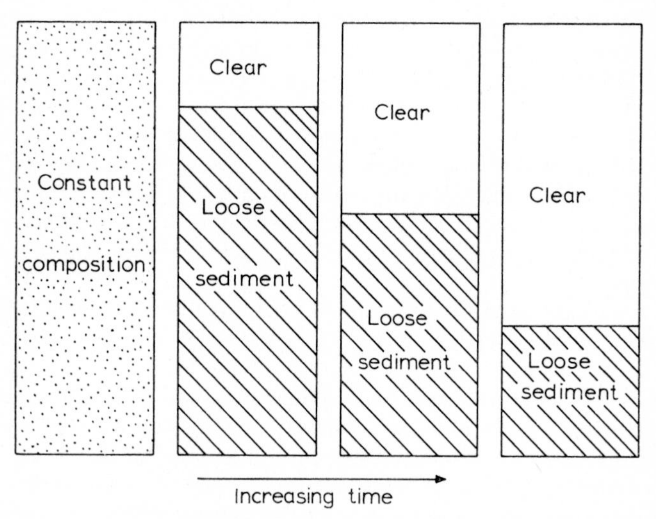

Fig. 7.12. Sedimentation behaviour of a concentrated flocculated dispersion.

In more concentrated dispersions, measurements of the rate of sedimentation cannot be related directly to flocculate size. In a completely flocculated dispersion the dispersed phase will settle rapidly in bulk, leaving a clear layer of liquid, and the rate of fall of the top surface of the sediment is dependent upon the flow of the liquid through the packed bed (Fig. 7.12). The composition of the sediment is substantially uniform throughout its depth.

The behaviour of practical concentrated dispersions such as paints is more complex, and two basic patterns of sedimentation behaviour are illustrated in Figs. 7.13 and 7.14 in each case from a completely homogeneous dispersion. In a partly flocculated concentrated dispersion, especially if the size range is narrow, the type of behaviour is shown in Fig. 7.13. Coarse material will commence to form a layer of sediment at the bottom of the vessel and above this will be a layer of variable composition. As sedimentation proceeds, a clear layer of liquid will be formed above the sedimenting layers, and this will gradually extend downwards. The sediment will be loosely packed and of large volume. In a completely

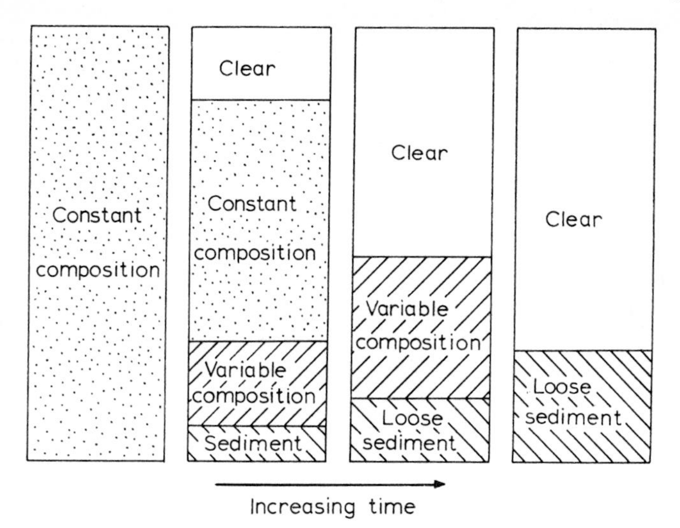

Fig. 7.13. Sedimentation behaviour of a concentrated partly flocculated dispersion.

deflocculated dispersion, as shown in Fig. 7.14 no clear layer is formed for some very considerable time. A layer of sediment commences to form, and above this is a layer of variable composition, with no clearly defined upper boundary. As sedimentation proceeds, the layers descend through the vessel. The final sediment formed is compact and of small volume, and above it the cloudy liquid phase still contains finer particles, which will eventually settle over a long period. This type of sediment is very difficult to re-disperse, by comparison with the loosely packed material shown in Fig. 7.12. This type of sedimentation behaviour is most frequently shown by deflocculated dispersions containing lower concentrations of materials with a wide spread of particle size.

Measurement of sedimentation volumes thus yields information about the degree of flocculation of the dispersed phase, a high volume corresponding to increased flocculation. Sedimentation may take a considerable time to come to equilibrium, and the determination of the final sedimentation volume is consequently difficult. Figure 7.15 illustrates the asymptotic nature of the changes in (*a*) the upper level of sediment in a partly flocculated dispersion, and (*b*) the sediment building up from the bottom in a deflocculated dispersion. Wolff[55] has suggested the use of the half-value time, *i.e.* the time at which the change in sedimentation volume has reached half its final value, as a measure of the degree of flocculation. The final value can, if necessary, be estimated by extrapolation.

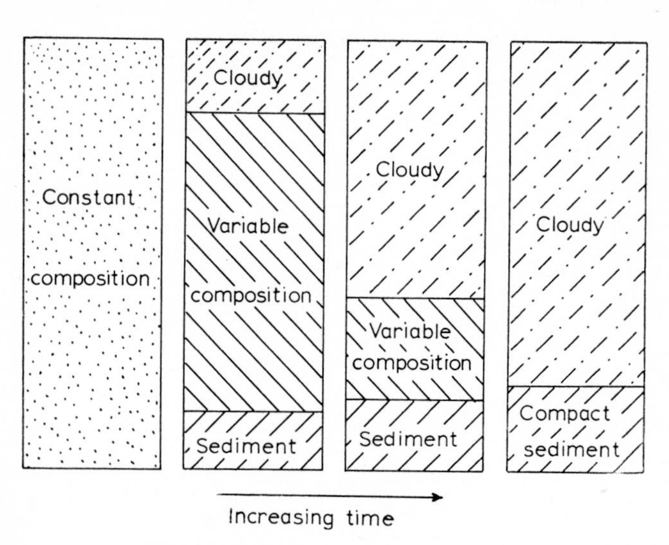

Fig. 7.14. *Sedimentation behaviour of a concentrated deflocculated dispersion. The sediment formed is difficult to redisperse.*

Practical measurement of sedimentation effects in concentrated dispersions such as paints presents problems. Most test procedures involve some means of thrusting a probe downwards through the sample under controlled conditions, to determine the nature of the sediment. Patton[11] has described a simplified gauge in which increasing force is applied by adding weights to a test probe. Landon[56] has described an x-ray absorption technique for the study of sedimentation in paints.

A factor of considerable practical importance in paint technology, in dispersions of mixtures of pigments, is the occurrence or absence of co-flocculation, *e.g.* of white and coloured pigments. Co-flocculation or similar dynamic behaviour of flocculates of the individual pigments serves to prevent any colour separations under certain conditions of paint application. Sedimentation tests of partly diluted dispersions can yield information of value, particularly where there are appreciable differences in density between the pigments. Where co-flocculation occurs, a uniformly coloured sediment layer is obtained, in spite of differences in density, while in the absence of co-flocculation, separate coloured layers may be formed.[57,58] Care is required in the interpretation of such data, for particles of one pigment may be swept along by particles of another, in either a gravitational or an electric field. This phenomenon has been termed *syresis.*[59]

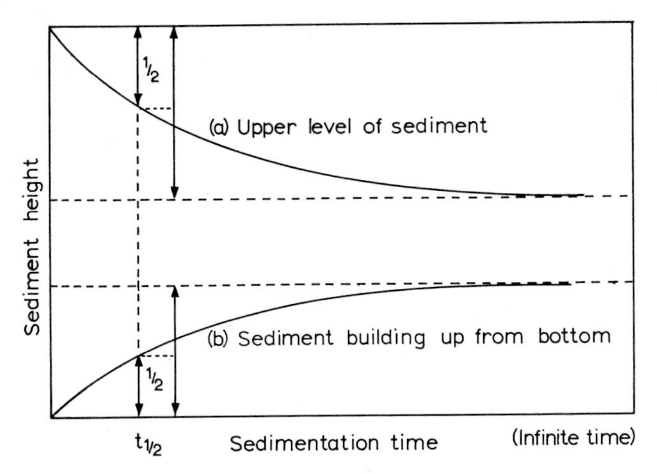

Fig. 7.15. *Sedimentation curves illustrating the asymptotic nature of changes in the height of the sediment with time.* (a) *Upper level of sediment in a partly flocculated dispersion,* (b) *build up of sediment from the bottom in a deflocculated dispersion. The* half-value *times may be used as a measure of the degree of flocculation.*

Rheological properties

The basic types of rheological behaviour of dispersions have been outlined above, and the extensive literature has been reviewed by Fischer.[12] Rheological measurements are widely used to characterise dispersions, to study their fundamental properties, and to assess their suitability for particular technological purposes. A knowledge of their rheological behaviour is needed for almost all dispersions in order to handle them satisfactorily; in many cases the observations may be simple and empirical, *e.g.* time of flow on inverting a standard container or assessment of the type of flow from a palette knife. Many forms of viscometer are used and there has been a tendency in certain fields to characterise materials by determination of a simple rate of flow at a given shear stress, with the use of the term *apparent viscosity*; this is the quotient obtained by division of the shear stress by the rate of shear, with the implied assumption that the flow curve is linear, passing through the origin, *i.e.* that Newtonian behaviour is postulated. Such measurements may be suitable for routine manufacturing control but they do not adequately characterise rheological properties nor are they capable of more detailed interpretation of the flow behaviour in terms of mechanism.

More advanced methods are necessary for adequate investigation, and a rotational viscometer is frequently used. There are considerable advantages in the use of a cone and plate instrument, *e.g.* the Ferranti–Shirley viscometer, or the Weissenberg rheogoniometer, where the rate of shear is constant over the whole of the radial distance at a given rotational speed. The range of rates of shear over which measurements is made will depend upon the requirements of the technology of the particular industry concerned. Thus extremely high rates of shear are involved in the application of many printing inks, and in the brushing of paints, while very low rates of shear are involved in *levelling* or *sagging* of liquid paint films during the setting after application.

The Einstein equation

$$\eta = \eta_0 (1 + C\varphi) \tag{7.6}$$

where φ is the volume fraction of the dispersed solid and C is a constant, relates the apparent viscosity η of a dispersion to the viscosity of the liquid phase η_0. This equation applies only to low concentrations of idealised spherical particles. Normally dispersions show some form of plastic or pseudoplastic flow, dependent upon the concentration of solid particles. Casson[60] developed an equation for the rheological properties of dispersions of particles flocculated in chain-like groups, of the form

$$F^{\frac{1}{2}} = C_0 + C_1 D^{\frac{1}{2}} \tag{7.7}$$

where F is the shear stress and D the rate of shear, and C_0 and C_1 are constants depending on the system. This equation was applied successfully to printing ink dispersions.

The rheological properties of a paint millbase change during the process of pigment dispersion, and may be used to characterise the various stages of dispersion. Jefferies[61] showed that as milling proceeded, the curves of apparent viscosity against shear rates changed their shape giving an indication of the progress of dispersion. Cope and Boulton[62] using the concept of thixotropic area, *i.e.* the area between the ascending and descending rheological curves, showed that behaviour was highly specific. In some cases the thixotropic area increased, and in other cases it decreased with increased time of milling. A distinction was drawn between the processes of *grinding* and *dispersion*, as typified by different types of change of rheology as milling proceeded.

Considerable further investigation is required of the rheological changes occurring during dispersion processes, including investigations of the properties of the systems by other means, in order that the rheological properties may be interpreted in terms of the structure of the dispersion including the extent of flocculation.

Optical properties

The optical properties of pigment dispersions depend upon the state of dispersion, though not necessarily linearly over the whole range. With coloured pigments an increase in the degree of dispersion results in an increase in the strength of tint developed on admixture with a stable white pigment dispersion, giving a widely used method of evaluation of the dispersion of coloured pigments used as tinters in paints. The tinted paints are examined either as wet or dry films, either by visual comparison or instrumentally using a tristimulus colorimeter with the appropriate filter in the absorbing region of the spectrum. For more critical evaluation, full spectrophotometric curves are used, and the apparent strength of colour assessed by comparison of the heights of the absorption peaks. It is assumed that the state of dispersion of the white pigment in the paint is unaffected by the addition of the tinting paste.

There is no satisfactory theoretical treatment of the relation between the degree of dispersion of the coloured pigment and the optical properties of the system, but it has been predicted, on the Mie theory, that for pigments within certain ranges of particle size and refractive index, the colour strength should be approximately inversely proportional to particle size.[63] In other cases a maximum is reached at a certain low particle size. It would be expected that the strength of tint developed would be related to the effective area of surface of the coloured pigment optically exposed in the system, *i.e.* inversely related to the extent of aggregation, agglomeration or flocculation of the coloured pigment, and this is known to be qualitatively true. The tinting strength of a series of carbon black pigments, dispersed with zinc oxide in linseed oil, has been shown to be proportional to the specific surface area of the carbon black calculated from electron microscope size analysis.[64] Similar relationships have been found with iron oxide and ultramarine pigments.[65] In some cases, the development of strength of tint can be correlated with the degree of dispersion of the coloured pigment in the tinting paste, as judged by Hegman gauge readings, but, particularly with finer size organic pigments, tinting strength continues to increase on further milling *beyond the gauge*.[66] This would be expected, since in relation to the actual size of the primary particles of organic pigments, the Hegman gauge is an extremely coarse tool. Recent work by Carr[35,36] using the ICI/Joyce Loebl disc centrifuge, has shown that for some organic pigments the colour strength (expressed as an absorption/scatter ratio K/S) rises slowly with decreasing particle size, but with others, *e.g.* phthalocyanine blue, the rate of increase becomes increasingly steep as the particle diameter falls below 0·35 μ. Similar effects were found for phthalocyanine blue by Hauser, Herrmann and Honigmann.[67] This has considerable practical importance, in that a small amount of flocculation of finer sized phthalocyanine blue particles could produce a very marked optical effect in reducing the strength of tints.

Many practical problems occur in paint films with mixed pigmentations which may result in changes in the optical properties of the systems, due mainly to differences in the degree of flocculation of the individual pigments, or to some form of pigment segregation occurring during the later stages of film formation. In the liquid paint a coloured pigment may be flocculated to a certain extent, and differing methods of application, *e.g.* by brushing or spray application (with high shear conditions) and by dipping (with low shear conditions), may cause differences in the relative degrees of flocculation of the pigments in the film, with consequent colour changes. Two other effects may occur.

Flooding is a differential separation of the mixed pigments through the depth of the film, the surface layer being either enriched or impoverished of one of the component pigments. Many factors may be involved, including the rate of evaporation of the solvent, the effective viscosity of the medium under the conditions of drying, and particularly differences in density of particles or flocculates of the component pigments, where gravitational effects may occur.

Flotation in a paint film is also a differential effect where a regular or irregular pattern is produced in the film, which in films dried horizontally may consist of hexagonal shaped Bénard cells with vertical sides running down into the thickness of the film. With evaporation of the solvent from the surface of the paint film as it dries, a regular circulatory motion occurs within each cell, bringing the solvent-rich medium to the surface. Associated with this circulatory motion there can be segregation of the different pigments, and the cell edges can become coloured differently from the main surface of the film. In some cases it has been found that the cell edges are free of pigments, and dark in appearance, emphasising the surface pattern. When the films are dried in an inclined or vertical position the cells may be elongated to give regular or irregular striations. The effects are more highly developed in thicker films.

Factors suggested as being involved in Bénard cell formation, include convection, due to differences in temperature caused by solvent evaporation, localised differences in surface tension, and differences in effective particle or cluster sizes between the pigments. The function $d^2(\rho - \rho_0)$, where d is the particle diameter, and $(\rho - \rho_0)$ the effective density of the particle in the medium for the different pigments has been shown to be of importance in determining whether flotation effects will occur.[68] As already indicated flotation and similar defects may be minimised by some degree of co-flocculation of the white and coloured pigments in tinted paints.[57,58]

The effect of the state of dispersion of pigments on opacity has received less attention. The problems involved in the application of a theoretical treatment of opacity have been reviewed by Orchard,[69] and, more recently, Dowling and Tunstall,[70] in an elaborate statistically designed investigation, have shown that many factors besides the nature and extent of pigmentation of the film can contribute to opacity, for example, highly reflective substrates could greatly increase hiding power. It seems reasonable to assume that if a pigment is fully dispersed, the number and extent of pigment–medium interfaces available for the reflection and refraction of light will be greater than if the pigment is aggregated or flocculated; consequently the opacity should be greater. In a pigmented film the pigment particles are seldom completely separated, and, even if the primary particles are uniform in size, there is a variety of cluster sizes. This has already been referred to (pp. 273–4) in relation to work by Jettmar[8,9,10] and by Hornby and Murley.[3] The latter authors made it clear that it was not possible to decide, on the basis of available evidence, whether the clusters were present in the liquid paint, or whether they are formed on drying the film. However, opacity is essentially a property of the dried film and unlike colour has no significance in the bulk liquid paint.

A paint film may contain a variety of individual particle sizes and a range of clusters of different sizes and degrees of internal packing. For fairly large particles the scattering power of a given volume of film containing a certain volume percentage of pigments can be taken as roughly proportional to the surface area of the particles present, increasing steeply with decreasing particle size. For particles smaller than the wavelength of light the amount of scatter is much lower. Between these two extremes there must be a particle size giving maximum scattering for any given ratio of the refractive index of the pigment to that of the medium. The optimum size for scattering is smaller for blue light than for red, and there can be differences in colour of transmitted and reflected light with certain dispersions, *e.g.* of fine particle zinc oxide.

It has been predicted theoretically and confirmed in practice that the scattering of light by pigment particles is not greatly reduced by the presence of neighbouring particles at separations exceeding one particle diameter (corresponding to a pigment volume concentration in the film of about 20%). But in matte paint of ink films the particle separation may be much smaller and scattering may be reduced considerably. Again, even in a film with a pigment content of less than 20% by volume flocculation may result in high local concentrations and some reduction in scattering efficiency.

Fischer[71] has shown that there is a relation between the opacity and the degree of flocculation of dispersions of titanium dioxide in linseed oil with various additives to control flocculation, but it has also been found that the opacity of pigmented films with one medium changes very little with variations in the time of milling. However, the same pigment, milled

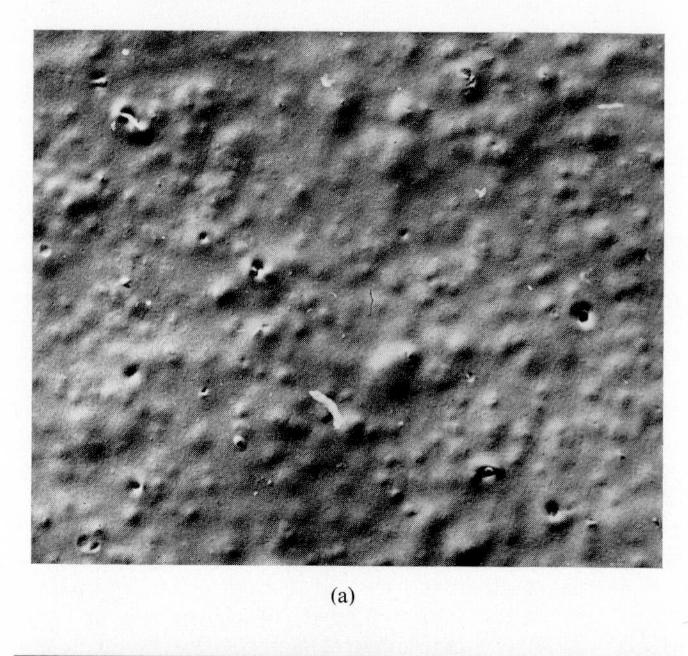

(a)

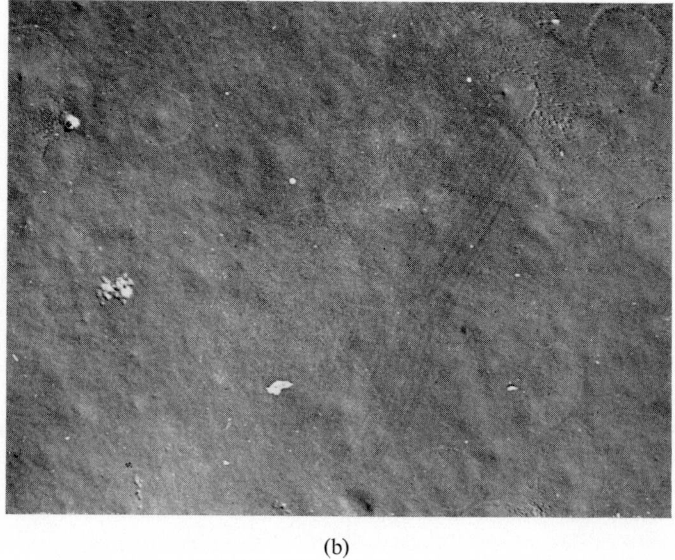

(b)

Fig. 7.16. Electron micrographs (× 5800) *of replicas of surfaces of paint film of* (a) *low and* (b) *high gloss.*

to a constant Hegman gauge reading, shows marked differences in opacity in different media.[72]

The *gloss* of pigmented paint films is a surface phenomenon and is related to the degree of dispersion and the extent of flocculation of the pigments. Large individual particles, aggregates or agglomerates can project through the smooth film surface interfering with the specular reflection characteristic of a high gloss finish. Carr[36] has recently examined the effect of particle size of organic pigments on gloss, and showed that as the mean size was reduced there was a steady increase in specular gloss, until a mean particle diameter of about 0·3 μm was reached, after which further reduction in particle size produced no alteration. This size is of the order of the wavelength of light. However, further reduction in particle size produced a decrease in the extent of gloss haze.

In recent papers, interference microscopy and goniophotometry have been used to measure gloss haze.[73,74] The gloss of carbon black pigmented film has been related to the degree of dispersion of the pigment in the film, as measured with the Quantimet.[75] The quantitative relation between gloss and dispersion in titanium dioxide pigmented paints containing silica flatting agents has been examined by Dunderdale and his colleagues,[76] using the concept that a surface defect, *e.g.* caused by an oversized particle or cluster of particles, produced a finite area of zero reflectance on the surface. The appearance of the surfaces of films of low and high gloss is shown in the electron micrographs in Fig. 7.16(a) and (b).

High gloss is frequently associated with a clear surface layer of medium free from pigment, above a more even layer of pigment particles, as in Fig. 7.17. Such clear layers are extremely thin, and can therefore be easily disturbed or broken through by large pigment particles or clusters of particles. Thus the progress of milling increases the evenness of dispersion and decreases the number of larger particles, resulting in an increase in gloss. This is true even though such milling may not have major overall effects on the pigment distribution throughout the film such as may affect light scattering and hence opacity. Also, as the film shrinks on ageing and the surface layer is attacked by weathering, the particles project through the surface and some may become completely exposed.

Miscellaneous methods

Electrical properties of dispersions are affected by the state of dispersion. The electrical conductivity of flocculated dispersions may be greater than that of similar deflocculated systems, due to direct contact between the individual particles in the continuous network of particles. Thus it has been shown that a deflocculated dispersion of carbon black in organic solvents has no measurable conductivity, while undisturbed flocculated dispersions allow appreciable currents to flow.[77] Disturbing the flocculated structure greatly reduced the conductivity. Several studies have been made

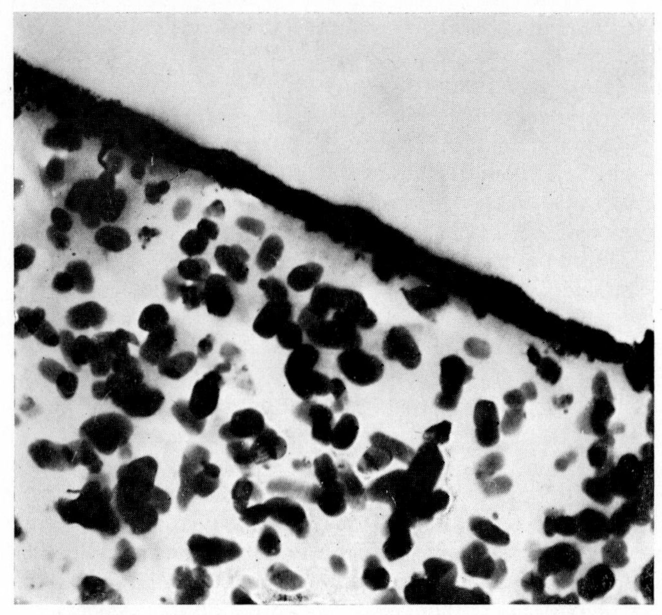

Fig. 7.17. *Electron micrograph* (× 15,000) *of the upper part of a section cut through a high gloss paint film from the surface to the underside. The surface of the film has been silvered before sectioning, to show the demarcation between the medium and the embedding resin. A layer of medium containing relatively few particles can be seen above the body of the film.*

of carbon black in natural and synthetic rubber media[78–80] and in mineral oils.[81]

The dielectric properties of dispersions are also dependent upon the degree of flocculation. Bruggemann[82] and Voet[83] have developed equations relating the dielectric constants of the dispersion to those of the components and the volume fraction of the dispersed phase. Voet also developed an approximate formula incorporating a form factor for the dispersed phase; a change in the observed form factor would indicate flocculation. Oesterle,[84,85] has studied the variation in dielectric loss with frequency for a number of pigment dispersions. He showed that in certain cases maxima occurred in the curves at low frequencies, and developed a theoretical treatment in terms of electrical model systems. By this means the pigment–medium interfacial behaviour of pigmented paints could be classified into two distinct types of wetting behaviour; in 'pseudo-wetting' external forces caused the close approach of pigment and medium which intrinsically had no tendency to interact, while in 'effective wetting' there was maximum interaction due to adsorption or chemical interaction between the pigment and the medium.

THE NATURE OF PRACTICAL DISPERSIONS

As indicated earlier, variations in the state of dispersion are of considerable practical importance in a number of technologies. The examples which follow are drawn from the paint and printing ink industries, but they serve to illustrate certain general features.

Pigments have been dispersed in paint media for many hundreds of years, and although labour-saving dispersion techniques are now in use, and the pigments and media have changed considerably, the essential required characteristics of the product are unchanged. What then is the level of dispersion attained in practice, and how does this affect the properties of the final product, the solid, pigmented film?

A paint or printing ink consists of a solution of a polymer or a resin or oil suspension or emulsion in which the pigment is dispersed. There are many types of paint or ink but in most cases when the paint is applied to a substrate the first change is simple evaporation of solvent to leave a film of pigmented polymer. With air drying or stoving finishes, reactions involving oxidation or cross-linking take place to convert the polymer into an insoluble, inert, hard film. Certain pigmented lacquers do not undergo such chemical reactions, the process being complete on evaporation of the solvent. The common feature is the conversion of a liquid dispersion to a solid film of pigmented polymer with relatively high pigment volume concentration and a wide range of particle sizes including many below 1 μm in diameter.

It is difficult to observe visually the state of dispersion of pigments in liquid paints, without introducing some artificial constraint to the system. It is not at present possible to examine liquid compositions in the conventional electron microscope, and even with a 500 kV electron microscope dry films thicker than 2 μm could not be examined due to focussing difficulties,[3] but it is known from other methods of examination that some degree of flocculation occurs, and may be essential for the development of many of the required properties. However, it is possible to examine the state of dispersion or flocculation of pigments in dried paint films by making microtome sections for examination by optical or electron microscope. A normal single coat of paint dries to form a film some 25–40 μm thick, and sections cut from surface to underside of such films are more easy to prepare and examine than those of printing ink films, which may be only some 1–5 μm thick on a paper substrate. The technique, particularly for specimens for electron microscopy, demands considerable skill, but the results are rewarding, as can be seen from the electron micrographs of sections of typical paint films shown in Figs. 7.18 and 7.19. One feature which is common to all of the film sections examined is the presence of clusters of pigment particles; fields consisting of individual pigment particles evenly dispersed are only very rarely seen.[86]

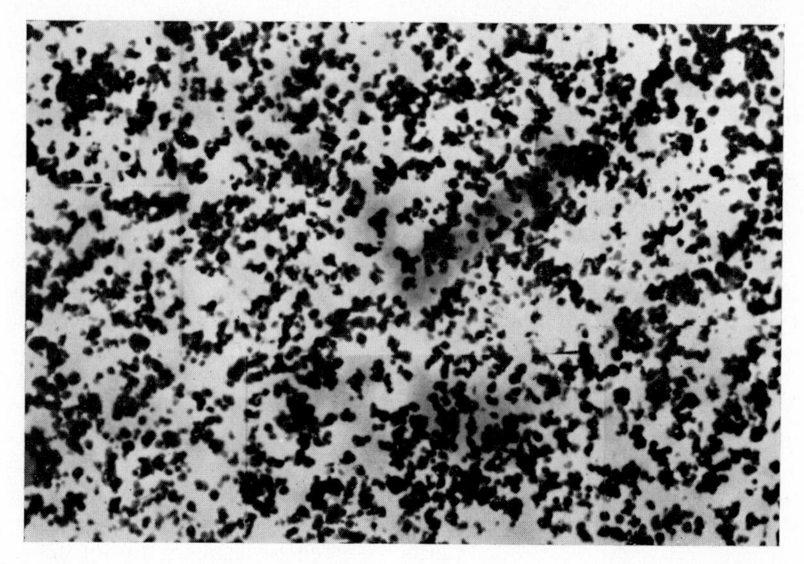

Fig. 7.18. Electron micrograph (× 5400) of a paint film section showing good practical dispersion in a solvent-based alkyd paint. The distribution of the pigment particles is far from even.

Fig. 7.19. Electron micrograph (× 4000) of a paint film section showing relatively poor dispersion in a water based polyvinyl acetate emulsion paint.

Other methods of examining the pigment distribution in paint films, *e.g.* by electron microscopy of replicas of fractured surfaces of films[87] and by various techniques of removal of the medium from between the pigment particles, which have previously been described (*see* p. 273), have also shown the presence of clusters of pigment particles; seldom is there complete separation of individual particles. Whether this cluster formation exists in the liquid paint, or whether is occurs during the drying process, is at present impossible to ascertain with certainty; the evidence from simple optical microscopic examination of liquid paints, and from the more elaborate and sophisticated methods of electron microscopy of frozen liquid paints and inks (*see* p. 274) tends to suggest that such clusters do occur in the liquid paint, possibly in a state of dynamic equilibrium, and that they may persist in the dried film. To some extent this might be expected as a result of the relatively high pigment volume concentrations occurring in paints and inks.

Underlying all studies of dispersion assessment is the question what is a 'particle' and what is the true significance of 'particle size'? Electron micrographs reveal that even 'individual particles' of pigments with no obvious evidence of clustering may themselves be made up of numerous fine particles or crystals. There are some pigments, including titanium dioxide and china clay, of which individual particles can be clearly identified by freedom from fine structure or by shape, but in many cases identification of individual particles is indeterminate, and a true 'size distribution' in the traditional sense is impossible.

What is most often observed is a size distribution of particles, clusters, etc. which may vary according to the medium used, the pigment concentration and the methods of preparing and observing the specimens. With appropriate selection of conditions, such observations are relevant to the probable situation of the pigment in the paint or other product as used. It is significant that in recent years much more attention has been paid directly to the assessment of cluster sizes in recognition of their practical importance in relation to product behaviour.

It cannot be too highly stressed that if the results are to be meaningful in practice the methods of assessing dispersion must be selected in relation to the character and conditions of use of the product or the nature of the stage of production which it is derived to investigate.

REFERENCES

1. V. T. Crowl, *J. Oil Colour Chemists Assoc.*, **46** (1963) 169.
2. T. C. Patton, *Paint Flow and Pigment Dispersion*, Interscience, New York and London, 1964, 216.
3. M. R. Hornby and R. D. Murley, *J. Oil Colour Chemists Assoc.*, **52** (1969) 1035.
4. G. Kämpf, W. Liehr and H. G. Völz, *Farbe u. Lack*, **76** (1970), 25, 127, 1105.

5. S. Wilska, *J. Paint Technology*, **43** (1971) 65.
6. K. Merkle and W. Herbst, IX Fatipec Congress Report, (1968) Sect. 2, 65.
7. A. E. Comyns and R. D. Murley, *Schweizer Archiv.*, **31** (1965) 390.
8. W. Jettmar, VII Fatipec Congress Report, (1964) 343.
9. W. Jettmar, IX Fatipec Congress Report (1968) Sect. I, 68.
10. W. Jettmar, *J. Paint Technology*, **41** (1969) 559.
11. T. C. Patton, *J. Paint Technology*, **38** (1966) 387.
12. E. K. Fischer, *Colloidal Dispersions*, John Wiley, New York, 1950, 147.
13. T. C. Patton, *Paint Flow and Pigment Dispersion*, Interscience, New York and London, 1964, Chaps. 1–7.
14. E. C. Bingham, *Fluidity and Plasticity*, McGraw Hill, New York, 1922, 228.
15. G. W. Scott-Blair, *Elementary Rheology*, Academic Press, London, 1969, 33.
16. K. M. Oesterle, *J. Oil Colour Chemists Assoc.*, **51** (1968) 1007.
17. British Standard 3483. Methods for Testing Pigments for Paints (1962).
18. F. J. Colon, *Chemistry and Industry*, (1965) 263.
19. H. W. Daeschner, E. M. Seibert, and E. D. Peters, A.S.T.M. Special Technical Publication 234 (1959) 26.
20. H. B. Carroll and J. B. Akst, *Rev. Sci. Instruments*, **37** (1966) 620.
21. British Standard 3406. Methods for the Determination of Particle Size of Powders. Part 2, Liquid Sedimentation Methods, (1963).
22. Society for Analytical Chemistry, Particle Size Analysis Sub-Committee. Determination of Particle Size, Part I. A Critical Review of Sedimentation Methods (1967).
23. Clyde Orr and J. M. Dallavale, *Fine Particle Measurement,* Macmillan, New York, 1959, Chapter 3.
24. R. D. Cadle, *Particle Size Determination*, Interscience, New York, 1955, Chapter 7.
25. R. D. Irani and C. F. Callis, *Particle Size*, John Wiley, New York, 1963, Chapter 5.
26. C. Slater and L. Cohen, *J. Sci. Instruments*, **39** (1962) 614.
27. E. Atherton and D. Tough, *J. Soc. Dyers and Colorists*, **80** (1964) 521, **81** (1965) 624.
28. M. J. Groves, B. H. Kaye and B. Scarlett, *Brit. Chem. Eng.*, **9** (1964) 742.
29. M. W. G. Burt, A.W.R.E. Report 0/76/64 (1964).
30. M. H. Jones and T. R. Manley, *Analytical Chem. Acta*, **38** (1967) 143.
31. R. D. Murley, *Nature*, **207** (1965) 1089.
32. J. Toole, J. S. F. Gill and R. C. Tainturier, VII Fatipec Congress Report (1964) 289.
33. J. Beresford, *J. Oil Colour Chemists Assoc.*, **50** (1967) 594.
34. G. A. Lombard and W. Carr, *Printing Technology*, **13** (1969) 148.
35. W. Carr, *J. Paint Technology*, **42** (1970) 696.
36. W. Carr, *J. Oil Colour Chemists Assoc.*, **54** (1971) 1093.
37. G. Eulitz and K. Merkle, *J. Oil Colour Chemists Assoc.*, **54** (1971) 196.
38. I. R. Sheppard, *J. Oil Colour Chemists Assoc.*, **47** (1964) 669.
39. British Standard 3406. Methods for the Determination of Particle Size of Powders, Part 4, Optical Microscope Method (1963).
40. M. Cole, *Microscope*, **15** (1966) 148.
41. C. Fisher and M. Cole, *Microscope*, **16** (1968) 81.
42. M. J. Smith, *Microscope*, **16** (1968) 123.
43. M. D. Garrett and W. M. Hess, *J. Paint Technology*, **40** (1968) 367.
44. L. J. Venuto and W. M. Hess, *Amer. Ink Maker*, **45** (1967) No. 10, 42, No. 11, 30.

45. V. T. Crowl, Proceedings Particle Size Analysis Conference, Loughborough, Society for Analytical Chemistry 1966, 36.
46. F. Endter and H. Gebauer, *Optik*, **13** (1956) 97.
47. R. H. Ottewill and D. J. Wilkins, *J. Colloid Sci.*, **15** (1960) 512.
48. Society for Analytical Chemistry, Particle Size Analysis Sub-Committee Determination of Particle size, Part 2, Electrical Sensing Zone Methods (Coulter Principle) 1971.
49. V. T. Crowl, *J. Soc. Dyers and Colorists*, **81** (1965) 545.
50. V. T. Crowl, *Verfkroniek*, **40** (1967) 318.
51. J. Beresford, W. Carr and G. A. Lombard, *J. Soc. Dyers and Colorists*, **81** (1965) 615.
52. H. Weissberg, *Official Digest*, **34** (1962) 1154.
53. J. B. Poole and D. Doyle, *Solid–Liquid Separation*, HMSO London, 1966.
54. V. T. Crowl and M. A. Malati, *Discussion Faraday Society*, **42** (1966) 301.
55. R. Wolff, *Kolloid Z.*, **150** (1957) 71.
56. G. Landon, *J. Oil Colour Chemists Assoc.*, **46** (1963) 555, **47** (1964) 44.
57. F. K. Daniel, VII Fatipec Congress Report (1964) 280; *J. Paint Technology*, **38** (1966) 309.
58. V. T. Crowl, *J. Oil Colour Chemists Assoc.*, **50** (1967) 1023.
59. H. H. McEwan and S. J. Gill, *J. Oil Colour Chemists Assoc.*, **52** (1969) 46.
60. N. Casson, *Rheology of Disperse Systems*, Pergamon, London, 1959, 84.
61. H. D. Jefferies, *Rheologica Acta*, **4** (1965) 241; *J. Oil Colour Chemists Assoc.*, **45** (1962) 681.
62. C. Cope and A. J. Boulton, *J. Oil Colour Chemists Assoc.*, **47** (1964) 704.
63. A. Brockes, *Optik*, **21** (1964) 550.
64. F. Frith and R. A. Mott, *J. Soc. Chem. Ind.*, **64** (1946) 81.
65. V. T. Crowl, D. L. Tilleard and L. G. Kieran, Paint Research Station Research Memorandum 333 (1964); *Verfkroniek*, **40** (1967) 318.
66. H. G. Cook, *J. Oil Colour Chem. Assoc.*, **48** (1965) 17.
67. P. Hauser, M. Herrmann and B. Honigmann, *Farbe u. Lack*, **76** (1970) 545.
68. S. H. Bell, *J. Oil Colour Chemists Assoc.*, **35** (1952) 573.
69. S. E. Orchard, *J. Oil Colour Chemists Assoc.*, **51** (1968) 44.
70. D. G. Dowling and D. F. Tunstall, *J. Oil Colour Chemists Assoc.*, **54** (1971) 958, 1007.
71. E. K. Fischer, *Colloidal Dispersions*, John Wiley, New York, 1950, 145.
72. S. H. Bell, *British Ink Maker*, **8** (1965) 21.
73. H. Becker, H. Noven and H. Rechmann, *Farbe u. Lack*, **74** (1968) 145; Kronos Information leaflets 29E and 30E.
74. R. D. Amberg, *J. Oil Colour Chemists Assoc.*, **54** (1971) 211.
75. W. M. Hess and M. D Garrett, *J. Oil Colour Chemists Assoc.*, **54** (1971) 24.
76. J. H. Colling, W. E. Craker, M. C. Smith and J. Dunderdale, *J. Oil Colour Chemists Assoc.*, **54** (1971) 1057.
77. C. M. McDowell and F. Z. Usher, *Proc. Roy. Soc.*, **A131** (1931) 409, 564.
78. F. E. Amon and C. J. Brown, *Amer. Ink Maker*, **19** (1941) No. 11, 25.
79. L. H. Cohan and J. F. Mackay, *Ind. Eng. Chem.*, **35** (1943) 806.
80. P. E. Wack, R. L. Anthony and E. Guth, *J. Applied Phys.*, **18** (1947) 456.
81. A. de Waele and G. L. Lewis, 2nd. Int. Congress Surface Activity 1957, Vol. 3, 21.
82. D. A. G. Bruggemann, *Ann. Physik.*, **24** (1935) 636.
83. A. Voet, *J. Phys. and Colloid Chem.*, **51** (1947) 1037; **53** (1949) 597.
84. K. M. Oesterle, VI Fatipec Congress Report (1962) 341.
85. K. M. Oesterle and K. Müller, *Paint Technology*, **27** (1963) No. 6, 16.
86. S. H. Bell, VII Fatipec Congress Report (1964) 74.
87. W. Funke, *J. Oil Colour Chemists Assoc.*, **50** (1967) 942.

CHAPTER 8

DISPERSION OF INORGANIC PIGMENTS

H. D. JEFFERIES

INTRODUCTION

Classification

Inorganic pigments, including the carbon blacks, are notable for the great number of commercially available variants of each basic type. These variations may be only physical, as in the case of natural dolomites or calcites ground to different degrees of fineness, or in ratio of constituents, as in the case of lithopones of different zinc sulphide content, or physico-chemical, as in the case of the various grades of post-treated titanium dioxide, whose pigmentary behaviour is influenced by both the primary particle size and the quantity, composition and physical structure of the coating oxides. For information about the chemistry and technology of commercial pigments, the reader is referred to the brief bibliography at the end of this chapter and to the technical information provided by the various suppliers, which by its nature is normally more up-to-date than the standard texts.

Some appreciation of the variety of chemical compounds involved can be obtained from the following list:

(i) White (high refractive index): anatase and rutile titanias, zinc sulphides, zinc oxide, lead hydroxycarbonate (white lead), co-precipitated zinc sulphide-barium sulphate (lithopone), basic lead sulphate.

(ii) White (low refractive index): calcium carbonate (whiting, natural or synthetic calcite), silica (natural or synthetic), magnesium-calcium carbonate (dolomite), hydrated aluminosilicates (china clay), barium sulphate (barytes, blanc fixe), magnesium silicate (talc), potassium aluminosilicates (mica).

(iii) Black: natural graphite, carbon blacks, black iron oxide (natural and synthetic).

(iv) Reds, browns: mercuric sulphide (vermilion), lead oxide (Pb_3O_4, red lead), cadmium sulphide, cadmium sulpho-selenide, lead molybdo-chromates, red iron oxide (natural and synthetic), copper oxide, complex ferromanganic, *etc.* oxyhydrate (umber), hydrated ferric oxide (ferrite browns).

(v) Orange, yellow: lead molybdo-chromates, basic lead chromate, hydrated ferric oxides (ferrite yellows), alkali-zinc chromates.

(vi) Green: chromium oxide, co-precipitated lead chromate and alkaline ferri-ferrocyanide (struck chrome greens).

(vii) Blues: complex alkaline sulpho-silicates (ultramarine green/blue/violet), alkaline ferri-ferrocyanides (iron blues, Prussian and milori blues, *etc.*).

(viii) Metallic flakes and powders: lead flake and powder, zinc powder, copper powder, stainless steel powder and flake, aluminium flake, copper bronze flake, copper–aluminium–zinc 'gold' bronzes, nickel flake.

The above list is far from comprehensive, and the chemical compositions are only indicative. For example, the 'lead chromate' pigments are produced in shades ranging from primrose to orange by adjusting the ratio of lead to chromate salts. In other cases, such as the synthetic iron oxides, small amounts of other elements are added to favour a particular crystal form and/or to aid control of growth to the desired primary pigment particle size.

Although the basic hue is determined by chemical composition and crystal structure, the tone of synthetic inorganic pigments can be varied very considerably by adjustment of the mean particle size and size distribution. Consequently the state of dispersion of a pigment, by determining its effective particle size distribution, affects the observed colour of a pigmented object, such as a paint film or sheet of plastic.

The achievement of an adequately stable pigment dispersion requires, in addition to some initial mechanical dispersion process, an interaction between the pigment (disperse or solid phase) and the liquid (continuous phase) in which it is dispersed. The interaction most often consists of adsorption on the pigment surface of material from the continuous phase, so that the volume ratios of the disperse and continuous phases are modified, as well as the compositions of the phases. A side-effect of this is that the consistency and rheological characteristics of the suspension are altered to some degree. Further complications ensue because the initial pigment dispersion, if used in paints or inks, is almost always mixed with other materials, some of which may be readily adsorbed by the pigment. The nature of these is outlined below.

Pigments in paints and inks

The formation of a dispersion of pigment in a liquid is the basic step in manufacture of ink or paint, within which groupings cosmetics may be included. The oldest known paintings are those in the caves at Lascaux, so that the preparation of a pigment dispersion, to be used as a surface coating, has a good claim to be the only one of man's early technological achievements to have survived almost unchanged.

The very ancient craft of paint manufacture has a corpus of empirically acquired knowledge. This in itself has been a deterrent to expenditure on scientific investigation of the dispersion process and the disperse state, but the most effective deterrent has been the complexity of a typical surface coating. An ordinary liquid paint or ink will usually contain the following components:

(*i*) *Solid phase.*—A chemically and physically heterogeneous mixture of white, black or coloured materials, organic and/or inorganic.

(*ii*) *Liquid phase.*—Usually a mixture of a material capable of film formation and a volatile diluent.

(*a*) Film-former (or binder). A mixture of compounds, either polymeric or capable of polymerisation.
(*b*) Volatile (solvent). Either water, or one or more organic liquids chosen partly for their ability to dissolve the binder to give the desired solids content and viscosity and partly because their rates of evaporation facilitate application of the coating.

(*iii*) *Driers.*—One or more soaps, formed from polyvalent metals (usually lead, manganese, cobalt, or zirconium) and organic acids of sufficient molecular size and suitable structure to *solubilise* the cation in the binder.

(*iv*) *Anti-skinning agent.*—Usually an oxime or phenolic material, intended to prevent premature oxidative polymerisation (skinning) of the paint or ink during storage. These materials also inhibit drying after application of the paint or ink so that a balance must be achieved between volatility and effectiveness.

(*v*) *Dispersing or wetting agent.*—These may be metallic soaps, such as zinc naphthenate, or other surface-active materials of natural or synthetic origin. Driers can and do act as pigment dispersants. Many 'dispersing' agents in fact produce a degree of flocculation of the pigment (*see* (vi), below).

(*vi*) *Rheology-adjusting agent; the avoidance of hard sediment.*—In the manufacture of a paint or ink, the degree of dispersion attained will have depended on many factors, not all amenable to control, but in the final packed product there is likely to be a proportion of *oversize* pigment. The oversize may consist of agglomerates or aggregates which were not broken down during manufacture, or flocculates which formed after the milling (pigment dispersion) stage. In certain cases the primary particles of pigment may be too large for a stable dispersion to be made. Whatever the cause, the formulator has to face the fact that some portion of the pigment is likely to settle out if the product is stored long enough; the sediment, when it forms, may be either a *disperse type* or a *flocculant type*.

A disperse type sediment is produced by particles originally in a well-dispersed condition; on settling these pack closely, forming a hard, tough layer, like a stiff clay. The sediment is very difficult to re-disperse.

A flocculate is an entity created in a suspension when previously separate particles come into contact with each other, chiefly at corners and/or edges, forming a skeletal structure whose gross, or effective, volume may be more than twice that of the total solid volume of its component parts. A paint or ink, in which most of the pigment is flocculated, usually has a high structural viscosity when the system is at rest. When subjected to a shearing stress, flow does not occur until a certain minimum stress (the yield value) is exceeded, after which the stress/strain relationship may, or may not, become linear.

The rate of sedimentation of particles, and in particular flocculated particles, in a liquid, such as paint, is not determinable from Stokes' Law, and the viscous resistance to gravitational settling of flocculated pigment may prevent settling, or the flocculates may settle slowly, forming a voluminous sediment beneath an upper layer of substantially unpigmented medium. The voluminous sediment is soft and can be reincorporated in the medium quite readily.

The formulator usually opts for the readily redispersed flocculant sediment, whilst hoping that flocculation-induced structural viscosity will prevent settling altogether. It is for this reason that many paints contain anti-settling agents which are actually pigment flocculating agents. This type of anti-settling agent, however, must be distinguished from those additives which are expected to achieve their object by inducing a marked degree of structural viscosity in the liquid portion of the paint. These are either materials such as hydrogenated castor oil derivatives, metal soaps (such as aluminium stearate) or specially treated montmorillonite-type clays which create a gelatinous structure in the medium, or a thixotropic alkyd resin formed by reacting a polyamide with an alkyd during resin manufacture to give a material in which hydrogen bonding creates structural viscosity.

The above list is by no means exhaustive since large numbers of materials are offered to the paint or ink formulator as a cure or palliative for every possible defect of his product.

This brief outline of the principal components of a paint (and some types of ink) has placed most emphasis on the minor components that may be unfamiliar (to those not paint or ink technologists) but which have marked effects on the state of dispersion of the pigments. In the following pages the major components and their interactions are discussed.

Pigments: manufacturing processes

Inorganic pigments may be synthesised, involving one or more chemical operations, or natural materials that have been physically processed. The chief chemical processes and some of their products are the following:

(i) Oxidation (in the gaseous phase): *e.g.* fume-process basic lead sulphate, zinc oxide, chloride-process titanium dioxide.
(ii) Reduction (in the gaseous phase): *e.g.* carbon blacks.
(iii) Precipitation (from solution): *e.g.* lithopone, chrome yellows, Prussian blue, sulphate-process titanium dioxide.
(iv) Calcination: *e.g.* ultramarine blue, synthetic iron oxide.

The physical processes chiefly involved are:

(i) Leaching or washing, to remove soluble matter. Often carried out concurrently with wet grinding of natural pigments.
(ii) Calcination, to obtain particles of pigmentary size. This may involve fusion of crystallites (*e.g.* lithopones and sulphate-process titanias) or sintering of particles into an agglomerate of pigmentary size (*e.g.* certain synthetic iron oxides).
(iii) Comminution, either by wet or dry grinding.
(iv) Size classification. There are two chief methods, wet, often used in sequence with water-leaching and grinding of natural pigments, and dry, by entraining particles in a gas stream.

In practice, all inorganic pigments except the carbon blacks usually undergo a final grinding process, since calcination to pigmentary size inevitably produces a percentage of oversize material. The pigment may be ground dry, in swing-hammer or Hardinge-type ball mills or wet, usually in ball mills of various types, either batch or continuously operated. Some pigments are given a final milling in a microniser or equivalent jet-mill; the effect of this on pigment properties is discussed later.

A number of inorganic pigments consist of a core surrounded by a layer of different composition. The thickness of the outer layer varies according to the purpose of the coating and the intended use of the pigment. Certain pigments for use in anticorrosive paints consist of a relatively thick layer of the 'active' component on an inert core, whilst the particles of after- or post-treated titanias have a comparatively thin coating of inorganic oxides, usually of low refractive index. Still thinner inorganic coatings may be applied to the lead chromate based pigments, in order to reduce their photosensitivity. In the case of the carbon blacks, the smaller the particle size, the greater is the proportion of the pigment weight that consists of complex organic and metallo-organic compounds. These, however, are deposited on the pigment surface as an integral part of the pigment particle formation, not applied at a later stage.

In order to increase the ease with which the pigment particle is wetted by liquid, a coating, usually of about monomolecular thickness, of an appropriate surfactant compound may be applied. Those most frequently encountered are either amines, amine-soaps or polyhydroxy compounds. Treatment with silicone compounds is also effective, one approach being the exposure of a coated titanium dioxide to methyl polysiloxane vapour, with the object of achieving a hydrophobic pigment surface by covering it with methyl groups, linked to the inorganic coating on the pigment through the silicon atom. The effect of these coatings on pigment behaviour is discussed later.

Particle size, opacity and colour of inorganic pigments

The optical properties of a typical paint film cannot be expressed in a rigorously derived mathematical form. Most modern work on this subject is based on the theory developed by Mie in about 1906, but even using computers the calculations can only be made for grossly over-simplified conditions or for hypothetical light paths. The subject has been reviewed fairly recently by Brockes[1] and, with special reference to opacity, by Orchard.[2]

The theory divides the optical properties of a pigment into two components, the colour strength (tinting strength) and the opacifying (hiding) power. These components are considered as functions of basic pigment properties, namely:

Particle size
Light scattering coefficient ‘S’
Light absorption coefficient ‘K’
Refractive index ‘n’. (For a rigorous treatment, the three values of ‘n’ for the three axes of the crystal must be used.)

The theory of opacity most often used is that developed by Kubelka and Munk.[3] This allows the simplifying assumption that for white pigments the value of ‘K’ can be neglected. Conversely, for black pigments the value of ‘S’ is small. For coloured pigments the value of ‘K’ is of major importance, but in the case of inorganic pigments, many of which have high refractive indices, ‘S’ is not necessarily negligible.

For inorganic coloured pigments, the colour strength (tinting strength, or staining power) increases with light absorbing capacity (increasing ‘K’) whilst the opacity (hiding power) increases with the light scattering power (increasing ‘S’), reaching a maximum at a particle diameter of about 280 nm, since scattering is independent of hue. In this latter respect the coloured and white inorganic pigments are similar and sharply differentiated from organic pigments, whose refractive indices are, in general, much lower.

Opacifying power is only one of several factors that influence the manufacturer's choice of mean particle size for a given pigment. These other factors are related to the pigment colour.

White pigments

A white pigment ideally absorbs no radiation and therefore is colourless in its massive form. The whiteness and opacifying properties, that result from the scattering of incident radiation by multiple reflection and refraction by particles of pigments, are therefore functions of the ratio of the refractive indices of the pigment and its surrounding medium, and of the particle size. The refractive index is an intrinsic property but the particle size can be deliberately chosen. Whatever the actual size of the primary particles, the state of dispersion in the final pigmented product determines the effective particle size and thus the optical properties.

The dependence of opacifying power on the ratio of the pigment/medium refractive indices causes a number of white pigments to opacify only media of low refractive index; in oily, resinous or plastic media they are translucent or transparent. These low refractive index pigments are usually referred to as 'extenders' or 'fillers', and their chief use is for the adjustment of rheological properties of suspensions and the mechanical and optical properties (gloss) of paint and plastic films. They can be used as opacifying and whitening agents in conjunction with low refractive index binders (*e.g.* glue, casein or egg-white) or very highly pigmented ('underbound') films containing too little high refractive index binder to substantially reduce the opacity derived from light scattering at pigment/air interfaces.

White pigment particle size

The greater the number of particles in a unit volume of film, the more effective is the scattering of the radiation entering the film, provided that the particle diameters are not less than half the wavelength of the incident radiation. The optimum particle size for opacification under a monochromatic illuminant is thus readily determined, but this is not the case for illumination by white light. The visible spectrum, which constitutes 'white light', ranges from about 400 nm to 720 nm, whilst the human eye has a peak response at about 560 nm, in the green portion of the spectrum. There is therefore no single optimum particle size for a white pigment and the apparent optimum size is dependent on the pigment content (opacity) and thickness of the pigmented film.

Light reaching a highly pigmented completely opaque white film is scattered and reflected in the upper layers of the film. If the pigment particles range from about 200 nm to 1000 nm the film will appear white, unless there is a 'clear layer' of substantially unpigmented medium at the surface of the pigmented layer. (In this latter case, interference effects may affect the apparent hue, even if the medium is completely colourless.)

If the pigmented film applied to a light absorbing (black) ground is not completely opaque, then the apparent colour and opacity of the film becomes a function of pigment particle size. As an example, we may consider two films, with equal pigment contents and respectively containing uniform particles of 200 nm and of 360 nm diameter, so that there are 5·832 times as many small as large particles per unit volume of film. The refractive index of any pigment increases inversely with the wavelength of the incident radiation, so that blue light is always scattered more efficiently than red.

Some incident light passes through each film and is lost, so the apparent hues of the films are somewhat grey, *i.e.* less bright, than a fully opaque white film containing either pigment. The greater number of small particles, efficiently scattering the short wavelengths, more than compensate for any loss of brightness due to failure to reflect red light; in addition, two or more small particles, close together, may function as a large particle and interfere with the longer wavelengths of incident radiation. The overall effect is that the film containing the small particles appears bluer and brighter than that containing the large (that appears yellower and duller (greyer) relative to the other).

The practical significance of this effect is important in the design of white pigments that are to be used at low concentration in relatively thick films, for example in plastics or rubber. For this kind of application, a pigment with a mean particle size of about 150 nm is found much superior to one with a mean size of 220 nm or more, such as is found most suitable for use in oleo-resinous paints. (The optimum particle size for a general-purpose pigment is also influenced by the effect of particle size on the hue of mixtures of white and coloured or black pigments. This is discussed later.)

Postulated models of pigmented film structure

The complex mechanism of opacification has caused various workers to postulate models of film structure in order to determine the hypothetical optimum white pigment concentration. The basis is usually that the uniform spherical particles of a hypothetical pigment are considered as located concentrically with larger spheres arranged in some particular way. The packing fraction or solid volume/gross volume ratio of any given array of uniform spheres is independent of the actual sphere diameter (*see* Table 8.1).

In a given array, it can be assumed that each pigment particle, of diameter $d = \lambda/2$, is located about the centre of a hypothetical sphere of diameter $D = \lambda$ so that each pigment particle is separated from each neighbour by a gap $\lambda/2$ wide. The ratio of the volume of each pigment particle to that of its concentric sphere is given by

$$\left(\frac{d}{D}\right)^3 = \left(\frac{1}{2}\right)^3 = \frac{1}{8}$$

Thus for various arrays of spheres, the optimum pigment concentration is always an eighth of the solid volume fraction of the chosen array. The appropriate array to choose is arguable; the closest possible packing with a packing fraction of 0·74 is a highly artificial concept. More appropriate would seem to be either the experimentally determined closest random packing, with a packing fraction of 0·64 or, for a given pigment, the packing fraction calculated from its oil absorption (BSS method) paste composition by the method described later. In any case, the initial assumptions are so gross and the hypothetical structure so far divorced from the reality that any apparent agreement between the experimentally determined optimum pigment volume concentration and that calculated in this way can only be coincidental.

TABLE 8.1

DIMENSIONS AND SOLID VOLUME FRACTIONS OF UNIFORM SPHERES IN CLOSE-PACKED ARRAYS

	No. of Contacting Spheres	*Dimensions of Unit Volume*	*Solid Volume Fraction*
Rhombohedral	12	$d \,.\, (\frac{2}{3})^{\frac{1}{2}}d \,.\, (\frac{3}{4})^{\frac{1}{2}}d = d^3/(2)^{\frac{1}{2}}$	0·740
Rhomboidal	8	$d \,.\, d \,.\, (\frac{3}{4})^{\frac{1}{2}}d = (\frac{3}{4})^{\frac{1}{2}}d^3$	0·605
Cubic	6	$d \,.\, d \,.\, d = d^3$	0·524

Black pigments

An ideal black pigment completely absorbs incident radiation of any wavelength; such a pigment does not yet exist. Actual black pigments vary in the completeness with which they absorb incident light, a fact that has a more marked influence on the undertone, the hue of a mixture of white and black pigments, than on the mass-tone, the hue of the pure colour.

Light absorption is a function of molecular and crystal structure as well as particle size. Very finely divided metal particles are usually relatively black, if the particle arrangement and surface structure do not allow metallic reflection effects. This is true even for metals that are coloured in the massive state, such as gold. Most commercial metallic pigments are, however, more or less dark grey in hue, or deliberately made in a form that retains metallic reflection properties. The chief truly inorganic black pigment is the black oxide of iron, either natural magnetite, a mixed ferro-ferric oxide usually containing some silica and alumina, or the synthetic product which is similar but may contain additional compounds added to control crystal form and particle growth.

The most important black pigments are those based on graphitic carbon, produced by controlled oxidation of organic matter. Traditionally, the highest quality carbon blacks of maximum jetness were made by the channel process, but increasing cost of the raw material (natural gas) and restrictions on atmospheric pollution have favoured the trend to the furnace process. Improved manufacturing methods have made the colour of the best furnace blacks equal to the best channel blacks, whilst the nature of the furnace process facilitates production of blacks with specific properties.

Mass-tone and undertone

The smaller and more uniform the particle size of carbon black, the greater is its jetness and colour (tinting) strength. Conversely, the larger the particles the washier (greyer) is the mass-tone of the black. Particle size is difficult to define in the case of carbon blacks because of their tendency to form reticulated chains of primary particles; this tendency increases inversely with primary particle size and it is doubtful whether the very small particle blacks are ever completely dispersed in normal use.

The small particle blacks are too small to interfere with the longer wavelengths of the visible spectrum, so that when mixed with white pigment the resultant grey has a marked red-brown tone. Conversely, the larger particle size blacks, with fewer particles per unit weight or volume, do not interfere with the shorter wavelengths and when mixed with white pigment give greys with a 'clean, blue' tone. This effect of particle size, it should be noted, is the reverse of that which occurs with white pigments; thus a small particle white with a large particle black gives a clean, blue undertone and a large particle white with a small particle black gives a red-brown, 'dirty', undertone. 'Small' and 'large' in these cases are relative terms; for white pigment a particle diameter of about 200–230 nm would be small and about 300–400 nm large, whereas the effective particle sizes of small and large carbon blacks are about 10–30 nm and 100 nm respectively.

Coloured inorganic pigments

The fundamental theory of colour of inorganic compounds is much less developed than that for organic pigment dyestuffs. A major reason for this is that whilst the mechanism, absorption of all but a fairly narrow band of visible wavelengths, is the same for both, the colour of inorganic compounds is also dependent to a much greater degree on crystal structure and particle size. Distortion of the crystal lattice by small amounts of other elements may cause marked variation in hue, as may the degree of hydration of a particular ore. The range of hues of the iron oxide pigments, from yellow through red to black, provides examples of these effects.

The defining characteristics of a coloured pigment are its hue and lightness or saturation. The curve of spectral reflectance from the pigment illuminated by white light normally shows a peak in one portion of the spectrum, corresponding to the major hue of the pigment. Some nominally 'pure' pigments have secondary peaks, whilst purples have concave curves with reflectance increasing at both the blue and red ends. The lightness of the colour is defined as the total fraction of the incident light that is reflected, regardless of wavelength, whilst the saturation (chroma or intensity) is defined as the intensity of the colour compared to a neutral grey of equal lightness. Colours with sharp peaks, reflecting substantially completely light of particular narrow bands of wavelengths, the spectral colours, are the most saturated.

The mass-tone of a coloured pigment is defined as the hue of the pigment undiluted with white, and often refers implicitly to the colour of the pigment in an oil or resinous matrix. The undertone is the hue of the pigment when diluted with white or applied to a white substrate in a very thin layer (as may be done with printing inks).

The hue of the mass-tone and the undertone of inorganic pigment may differ considerably, but both are influenced by the mean particle size of the primary particles in a fashion similar to the black pigments. Decreasing particle size causes a shift, especially in undertone, towards the red, increasing size towards the blue. An example is provided by a commercially available range of ten synthetic red iron oxides, said to differ only in primary particle size, ranging from 100 nm, 'strong yellow cast' to 1000 nm, 'strong violet cast'.

The staining power, or tinting strength, of pigments increases inversely with particle size, whereas the scattering power, being independent of colour, reaches a maximum at a particle size that varies according to the pigment concentration and thickness of the pigmented film. The size usually quoted is 280 nm, a half of the wavelength at which the human eye is most sensitive. The scattering effect is more important in the case of inorganic than organic pigments, since many of the former have high refractive indices, a primary cause of the frequently observed higher opacity of inorganic pigments compared with organic pigments of similar hue.

Purity of shade is proportional to the narrowness of the band of reflected light; in the case of inorganic pigments the width of the band may be affected by the particle size distribution. Except for some pigments whose tone is achieved by sintering primary particles to give the necessary pigment particle size, the milling processes used in paint manufacture do not have a significant effect on pigment particle size. The effect on hue of the state of dispersion of the pigment is most often observed in tints in the form of such effects as flooding and floating.

Dispersion media

The liquids in which inorganic pigments are dispersed can be classified as follows:

(i) Thermally liquefied solids.
(ii) Solutions of surface-active compounds, either aqueous or non-aqueous.

(iii) Liquid film-forming materials.
(iv) Solutions of film-forming materials (mostly non-aqueous).

The majority of the first class of dispersions are of pigment in thermoplastics, where the dispersion is achieved by mechanical action and maintained by the high viscosity of the medium, that prevents pigment flocculation. Often, the material is rapidly cooled directly after dispersion has occurred, as for example in injection moulding using polystyrene granules previously dry-tumbled with pigment. This type of dispersion will not be discussed in any detail since the major influence factors are mechanical design and rate of cooling (rate of viscosity increase) of the polymer. These are dealt with in detail in a number of books.[4]

The second class of dispersion medium is the subject matter of Chapter 4 and any extended discussion would be repetitious. In passing, it should be noted that most inorganic pigments catalogued as self-dispersing in water contain some water-soluble dispersant, most often a sodium phosphate or similar material. The division between some of the surfactants and the media of the third and fourth types is not sharp; many of the latter are dependent for their dispersing power on the presence of small amounts of surface-active material, such as free fatty acids in vegetable oils.

The third class of media includes material such as linseed oil in various forms, low viscosity alkyds and silicone oils and resins. In many cases, media that at first sight appear to belong to this group actually belong to the second class, because their intrinsic pigment dispersing powers are small and some 'wetting agent' is always added.

The fourth class includes the great majority of pigment dispersing media for use in non-aqueous paints. These are mostly alkyd resins, modified alkyd resins or oleo-resinous varnishes in the case of air-drying paints. Many baking enamels are based on a pigment dispersion in alkyd, to which is added a thermosetting amino resin, the principal alternative to these alkyd/amino resin media being the thermosetting acrylic resins.

Of the various media those whose behaviour as pigment dispersants have been most closely studied are linseed oil and alkyds.

Linseed oil

Linseed oil, being a natural product, varies according to the climatic conditions during growth and to soil conditions. The raw oil has limited compatibility with most synthetic resins; it contains a quantity of mucilaginous material, the 'linseed oil foots' that must be removed if the oil is to be thermally processed. Foots removal can be by prolonged storage (possibly the origin of the term 'stand oil') which was the process used by the 16th century Dutch artists. The more usual procedures are treatment with acid or alkali. The first-named, not unreasonably, gives an oil with a higher acid value that is a better pigment dispersant due to its greater content of free fatty acids and it is for this reason that the acid value of oil used for the determination of the 'oil absorption' value of a pigment must be that specified. An increase or decrease of one or two units of acid value can produce a corresponding variation of 10–20% in the oil absorption. A monomolecular film of the oil, assuming a molecular weight of 875 and a density of 0·933 g/cm^3, has a thickness of about 1·56 nm, a surface area of about 1 nm^2 per molecule and weighs about 1·46 mg/m^2.

Linseed stand oil, or linseed lithographic varnish, is made by thermal polymerisation of linseed oil. The viscosity of a stand oil may be from 2–1000 poise and therefore should always be clearly specified. The molecular weight distribution is usually fairly broad and both the number average and weight average values should be determined as a guide to the degree of heterogeneity.

Alkyd resins

Alkyd resins are formed by the polycondensation of a polyhydric alcohol (*e.g.* glycerol or pentaerythritol) with a polycarboxylic acid (usually phthalic) in the presence of a proportion of monocarboxylic fatty acid. The latter may be a vegetable fatty acid (FA) or a mixture of FA and rosin (abietic acid), in which case the alkyd is usually described as 'oil-modified'. Either drying (*e.g.* linseed), semi-drying (*e.g.* soya bean) or non-drying (*e.g.* castor, coconut) oils may be used as the fatty acid source, the drying/non-drying properties of the oil being transferred to the alkyd. If the fatty acid chain is unsaturated then vinyl compounds, most often styrene or alpha-methyl styrene (vinyl toluene), may be introduced.

Oil-modified alkyds are generally classified as short-oil (30–50% FA), medium-oil (50–57% FA) or long-oil (>57% FA) types. Solubility in white spirit (which is usually about an 85/15 w/w mixture of aliphatic and aromatic hydrocarbons) requires at least 55% FA content, solubility in only aliphatic hydrocarbons rather more. Short-oil alkyds are usually dissolved in xylene or a similar type of solvent, whilst some very short (30–35% FA) alkyds require a solvent mixture of xylene with isopropanol or butanol.

Alkyds are complex mixtures, owing to the variety of secondary reactions that can occur. Thus, especially in long-oil alkyds polymerisation of the fatty acids usually takes place, as well as some formation of esters of glycerol or pentaerythritol. The alcohols may etherify to some degree, especially if the 'monoglyceride' process is used, in which the vegetable oil is heated with glycerol, at a molar ratio of 1:2, in the presence of a catalyst. Technical grade pentaerythritol may initially contain some di-, tri- or poly-pentaerythritol. Ring closure during alkyd manufacture is theoretically possible, but if it took place to any significant extent would almost certainly lead to gelation of the resin.[5]

The classification of alkyds in terms of oil-length, based on the weight per cent of fatty acid in the base resin, diverts attention from the molar ratios of the constituents. For example, a typical long-oil alkyd may be described as a pentaerythritol (P/E) esterified alkyd, containing 24% phthalic anhydride (P/A) and 62% soya fatty acids. In terms of molecular ratios the ratio of P/A:FA:P/E is about 3:4:3·15, or in terms of functional groups, 6:4:12·6. In this form the excess of hydroxyl over carboxylic groups is apparent. Similarly, a glycerol esterified short-oil alkyd, with 48·5% P/A and 31·0% coconut FA has molecular and functional group ratios of about 9:4:10·5 and 18:4:31·5 respectively. The ratio, 1·43, of hydroxyl groups to carboxylic is even greater in this case.

These values are characteristic, since it is usual to allow an excess of 5–10% by weight of polyol over the stoichiometric amount, because this is found to prevent an excessive rate of increase in viscosity, with its attendant risk of gelation during manufacture. The final alkyd resin always contains some 'free', unreacted, hydroxyl and also carboxyl groups which, through hydrogen bonding, make a considerable contribution to the melt viscosity of the resin. This can be shown by reacting the carboxyl groups with diazomethane, and the hydroxyl groups with acetic anhydride. The effect on the melt viscosity is very great, but on solution viscosity is much less, and on the intrinsic viscosity virtually negligible.

Study of alkyd resin structure is usually carried out by solvent fractionation of the alkyd,[6–9] but the results obtained depend very much on the solvent used. The basic alkyd contains molecules varying in shape, size and degree of polarity; thus polar solvents will fractionate by polarity, rather than size or shape, and conversely for non-polar solvents. Comparative results of fractionating with different solvents are given by Brett[9] from whose paper Table 8.2 is taken.

The conventional expression of acid value (AV) and hydroxyl value (HV) as mg KOH/gram resin allows the weight of resin per carboxyl or hydroxyl group to be calculated, using the expressions

$$\frac{56\,000}{\mathrm{AV}} = \text{grams resin per } -\mathrm{COOH} \text{ group}$$

or

$$\frac{56\,000}{\mathrm{HV}} = \text{grams resin per } -\mathrm{OH} \text{ group}$$

The weights of resin per $-$COOH or $-$OH group for the resin fractions listed in Table 8.2 are given in Table 8.3.

From the tables it can be seen that use of a polar solvent mixture gave fractions in which the acid value decreased initially but was thereafter independent of increasing molecular weight, so that the weight of resin per carboxylic group was substantially constant for fractions 4–7. In contrast, with a non-polar solvent, acid value increased with the molecular weight

TABLE 8.2

Alkyd	Extracting Solvent	% Wt.	Acid Value mg . KOH/g	Hydroxyl Value mg . KOH/g	Oil Length %	Mol. Weight Number Av.	Mol. Weight Weight Av.
			(a) Benzene/methanol fractionation				
Original			17·2	34·6	70	1 940	59 000
Fraction 1		23·2	40·8	49·4	69	1 030	7 900
2		14·1	17·1	41·1	74	1 500	8 000
3		22·5	10·2	26·5	70	2 540	22 000
4		11·7	8·7	26·1	69	8 500	46 000
5		7·7	8·6	28·9	69	9 800	49 000
6		14·3	8·7	28·9	68	17 000	400 000
7		6·5	8·2	47·6	67	35 000	Contained some gelled material
			(b) Petroleum ether fractionation				
Original			17·2	34·6	70	1 940	59 000
Fraction 1	Pet. ether b.p. 40°C	34·8	15·6	22·6	80	1 300	10 000
2	40–60°C	9·1	17·2	35·8	73	1 500	22 000
3	60–80°C	18·1	18·4	35·9	66	2 470	30 600
4	80–100°C	18·9	19·3	37·2	63	3 900	61 400
5	Residue	19·1	20·1	38·6	62	4 060	171 000

and the weight of resin per carboxylic group correspondingly decreased. The hydroxyl values in general followed the pattern of the carboxyl groups, if the value for the benzene/methanol fraction 7 is ignored. (Brett suggested that this might be due to methanol retained in the highly viscous resin.)

Comparison of the number average and weight average molecular weights listed in Table 8.2 shows that the fractions are extremely heterogeneous, regardless of the type of fractionating solvents used. This extreme heterogeneity is a basic characteristic of alkyd resins and makes it very difficult to apply the solvent parameter concept to them in a meaningful way. All in all, the evidence from studies of alkyd structure indicate that only the broadest conclusions can be drawn about their behaviour in pigmented systems such as paints or inks. Discrepancies between the findings of various workers are at least as likely to be due to their use of different alkyds as to any other single cause.

Solubility of alkyd resins

A typical modified alkyd resin contains compounds ranging from residues of the initial ingredients up to polymers with a molecular weight of

500 000 or more. Since the monocarboxylic fatty acids are condensation chain-stoppers, the average and maximum molecular weights tend to increase inversely with the fatty acid content. The variety of components causes alkyd solubility to be greatly influenced by the mutual solubilities of similar components and, to a lesser extent, by the mutual insolubility or incompatibility of the most dissimilar compounds. This latter effect may become important if the stability of the resin solution is marginal and certain constituents are effectively removed by adsorption on the surface of pigment.

TABLE 8.3

WEIGHT OF RESIN PER —COOH AND —OH GROUP, CALCULATED FROM DATA IN TABLE 8.2

	Benzene/methanol fractionation				*Petroleum ether fractionation*			
	Mol. Weight Number Av.	*Mol. Weight Weight Av.*	*Weight per —COOH Group*	*Weight per —OH Group*	*Mol. Weight Number Av.*	*Mol. Weight Weight Av.*	*Weight per —COOH Group*	*Weight per —OH Group*
Original alkyd	1 940	59 000	3 260	1 620	1 940	59 000	3 260	1 620
Fraction 1	1 030	7 900	1 370	1 130	1 300	10 000	3 590	2 480
2	1 500	8 000	3 280	1 360	1 500	22 000	3 260	1 565
3	2 540	22 000	5 490	2 115	2 470	30 600	3 045	1 560
4	8 500	46 000	6 440	2 145	3 900	61 400	2 900	1 505
5	9 800	49 000	6 520	1 940	4 060	171 000	2 785	1 450
6	17 000	400 000	6 440	1 940				
7	35 000	—	6 840	1 175				

The fundamental requirement for mutual solubility (spontaneous mixing), according to Hildebrand and Scott,[10] is that the compounds should have similar cohesive energy densities, but it was found that to apply this concept to polymer/solvent combinations required the hydrogen bonding tendency of the solvent be taken into account. A practical method of applying these concepts to some resins and solvents used in the surface coating industry was described by Burrell,[11] since when various workers have refined the original method, generally by subdividing Hildebrand's 'solubility parameter' (the square root of the cohesive energy density) into dispersion, polar and hydrogen bonding components. All the better known methods are however based on the theory of regular solutions proposed by Hildebrand[12] and Scatchard[13] and applied to polymer solutions by Flory[14] and Huggins.[15] This assumes that there is no strong association between molecules, which may be valid for dispersion and polar forces but is probably wrong where hydrogen bonding forces are significant. A method for overcoming this problem has been proposed by

Nelson *et al.*[16] which utilises infrared frequency shifts to calculate a 'net hydrogen bond accepting index'.

Determination of this hydrogen bonding index for an alkyd would require the prior separation of the constituent molecules of the resin. This is not feasible, yet alkyd resins are known to have their melt viscosities increased by hydrogen bonding. This effect largely disappears when the resin is in solution, but figures quoted by Burrell[11] for a group of alkyds and soya bean oil suggest that the original solubility parameter concept is not of much use in studying the interaction of alkyd resins with solvents or solvent combinations.

An alternative approach is the study of resin solution viscosities to determine the effect of solute/solvent interaction on the mean equivalent hydrodynamic volume of the polymer molecules.[11] Such data would provide a useful check on data obtained from studies of the effect of the solvent on alkyd adsorption on pigment surfaces, where selective adsorption of certain resin components may occur. The findings of both the solution viscosity and the adsorption experiments are valid only for the resin used; no general rules or equations can be derived from them.

Solvents

The chief functions of the solvent in a pigment/film-former/solvent combination are to adjust the viscosity of the suspension and, at a later stage, to evaporate at a suitable rate. The choice of solvent may be subject to restrictions on odour, toxicity and flammability and, increasingly in the future, by regulations similar to the now well-known Los Angeles County Air Pollution Control District Rule 66, that forbid the use of solvents whose vapours undergo certain chemical changes in the atmosphere.

The dissolving of a film-former involves a solute/solvent interaction, one effect of which is to influence the configuration of the film-former molecules. This in turn can affect the configuration and/or amount of film-former molecules adsorbed on the pigment surface and consequently the degree and stability of the pigment dispersion. Also, the solvent itself may be capable of dispersing or flocculating the pigment, though the correlation between pigment behaviour in binary (pigment/solvent) and ternary (pigment/film-former/solvent) systems is not readily predictable.

Solvent selection has been shifted from a wholly empirical to a qualitatively predictable basis chiefly by the proselytising of Burrell[17,18] on behalf of Hildebrand's[10,12] *solubility parameter* concept. The original simple technique suggested by Burrell used three groups of solvents, representing poorly, moderately and strongly hydrogen bonded types, each group containing solvents covering about the same range of solubility parameter. The solubility parameter of a resin was determined from the limiting values of the solvents in which a stable solution was obtained. Various refinements of the method have been proposed, involving division of the solubility parameter into three components, representing (London) dispersion, dipole and hydrogen-bonding intermolecular forces. The methods employed by the authors to calculate these forces vary in the soundness of their theoretical foundations; in every case the greater work involved in using the improved method seems largely unrewarded. An excellent review of many of these papers and other aspects of the concept has been provided by Burrell.[18] A critical review of the shortcomings of the basic concept has also been made by Hildebrand.[19]

The solubility parameter permits a rational explanation of a number of apparently anomalous solubility effects, and allows qualitative predictions about solubility in various solvents. The concept is not very helpful in explaining the discrepancy between the postulated influence of the solvent on polymer solution viscosity at low and at high solute concentrations.

It is postulated[20] that in a 'poor' solvent each polymer molecule occupies the minimal volume, its equivalent hydrodynamic volume is therefore minimal and the increase in viscosity due to the presence of a solute is at a minimum. In a 'good' solvent, the polymer molecules are expanded and the converse effects are produced. These relationships have been confirmed for dilute solutions of polymers having fairly narrow molecular weight distributions.[21] From experiments with solutions of a short oil alkyd, solvent/viscosity relationships that were substantially the reverse of these were reported and alkyd micelle formation in weak solvents tentatively suggested as a possible cause.[22]

The absence of a soundly-based theory of liquid structure is a severe handicap to all attempts to explain solution behaviour in a fully acceptable fashion. The two approaches to the theoretical explanation of liquid structure, by treating the liquid state as an extension of the solid and of the gaseous state respectively, have been set out in some detail by Kimball;[23] the cases of a plastic solid, such as a polymer, mixed with a relatively small volume of solvent and of a solvent containing a relatively small volume of polymer respectively provide an example of each system.

In the particular case of alkyd resins, as mentioned earlier, the heterogeneous nature of the resin must have a significant effect on its behaviour in solution. In the general case, association between solute and solvent molecules would increase the effective volume of the disperse at the expense of the continuous phase, an effect that would be exaggerated at higher solute concentrations. The formation of polymer micelles or aggregates cannot be ruled out, especially near the 'throw-out' point, at which resin molecules coagulate and precipitate from solution; such aggregates could immobilise a large proportion of the continuous (solvent) phase within themselves, with a resultant marked increase in solution viscosity. Useful papers have been published by Shaw and Johnson,[24] Bobalek *et al.*[25] and, in particular, Reynolds and Gebhart,[26] who dealt with solution concentrations of practical interest.

The influence of the solvent on the pigment in a suspension seems largely dependent on the presence or absence of a film-forming material, as well as on the pigment dispersing capacity of the film-former. The interaction of inorganic pigments and solvents was studied by Weisberg,[27] who used dielectric constant as the characterising value for the solvents and also took into consideration the influence of the trace electrolytes inevitably present in a commercial pigment produced by a precipitation process. He was unable to confirm the specific sedimentation volume (SSV) effect postulated as a characterising constant of a pigment by Dintenfass,[28] but his results showed that in general the sedimentation volume decreased inversely with the dielectric constant of the solvent, with certain exceptions. The latter were provided by the behaviour of the pigments in the solvents with the lowest dielectric constants; the pigments formed dense approximately spherical aggregates. This effect can also be produced very easily by exposing a typical chalk-resistant grade of rutile titanium dioxide to a high humidity atmosphere, then trundling the high moisture-content pigment in benzene. The pigment rapidly forms small indented aggregates, reminiscent of miniature golf-balls. (Trundling the moist pigment in ethanol or butanol produces a relatively stable dispersion, as might be expected.)

Attempts to correlate pigment dispersion with solvent parameter have not been very successful. Hansen[29] attempted to determine a solvent parameter value for pigments, but was not very successful in obtaining reproducible results. Another attempt[30] to correlate pigment dispersion with solvent parameter was not very successful; the influence of intermolecular association of the solvent molecules appears to be greater than that of the solvent parameter, and this aspect of solvent structure seems more readily characterised by dielectric constant.

Sorensen[31] was able to show a quantitative relationship between viscosity of certain printing inks and the hydrogen bonding potential of the vehicle. The relationship varied in form between inorganic pigments (titanium dioxide, chrome yellow) and two groups of organic pigments and also between film-formers (calcium resinate, Versamid 930 polyamide resin and SS grade nitrocellulose). This differentiation between inorganic and organic pigment behaviour was also observed by the Toronto Society for Paint Technology.[30]

It seems likely that the effect of solvents on pigment dispersion behaviour is more greatly influenced by internal bonding forces between the solvent molecules than by the solvent parameter itself. Pending the development of a rigorously founded theory of the structure of liquids, a quantitative theory of solution properties will be difficult to formulate. For ordinary dispersions, made from commercial materials, the influence of trace impurities, batch to batch variations and, in the case of pigments, the variation of adsorbed moisture content with atmospheric humidity, may greatly reduce the practical value of a quantitative treatment when it becomes available. None of the elaborations so far suggested seems to offer a truly significant advance over the original method of using the solvent parameter proposed by Burrell[11] in terms of information obtained for work involved.

CHARACTERISATION OF INORGANIC PIGMENTS: PHYSICAL AND CHEMICAL ASPECTS

It is quite true that the only characteristics of a pigment of importance to the user are those that may be manifested in the final product, whether printed ink, dry paint film or plastic moulding. On the other hand, it is impossible to quantify the factors that determine the success or failure of the pigment in a particular application unless the initial physical and chemical characteristics of the pigment are known. Of these, probably the most important are the particle shape, size and size distribution, since these have a major influence on the rheological characteristics of the pigment suspension. Also, comminution of pigment particles during the dispersion process is unlikely to be achieved by any machine other than the ball mill. Knowledge of the pigment size distribution is therefore necessary, so that before large scale manufacture of a pigment dispersion is undertaken, the probability of obtaining a dispersion of suitable quality, using the proposed equipment, can be assessed.

Assuming that the pigment can be satisfactorily dispersed, the stability of the dispersion will in most cases depend on the adsorption on the pigment surface of components of the liquid phase. This adsorption can occur only on a surface accessible to the adsorbate and will be influenced by the physical structure and chemical composition of that surface.

Particle shape

Particles of inorganic pigment may be fragments of naturally occurring minerals or the products of a chemical process. In the former case the variety of shapes and the range of sizes is likely to be larger than in the latter, but in both cases the accurate description of the particle shape presents difficulties. (Materials prepared by precipitation on an industrial scale are rarely perfect crystals.) A number of workers have attempted to produce systems in which the shapes of the (three-dimensional) particles could be adequately defined in terms of the two dimensions measurable on a projection screen or photomicrograph; one recent study[32] of carbon black involved computer simulation of the flocs that were observed in the electron microscope. Particle size is often estimated from the EM photographs by using a transparent screen on which is marked a horizontal row of circles, increasing in diameter from left to right, whilst below each circle is a vertical row of ellipses, each one more elongated than the one above it. The sizes are such that, assuming the particle to be parallel to the screen, normal to the line of sight of the observer, and symmetrical about its major axis, then all the particles that can be fitted to one column of the grid are of equal volume. The particle size can then be described in terms of the 'equivalent sphere', that is, the sphere whose diametrical cross-section is the circle at the head of the particular column. For 'structured' pigments made up of fused aggregates and agglomerates of primary crystallites, such as carbon black, the 'equivalent sphere' volume can be compared with that calculated from the number of crystallites in the particle. (*See* later, under Structured pigments; carbon blacks.)

The scanning electron microscope, that can provide stereoscopic pictures of the particles, will assist study of the structure of aggregated and agglomerated pigment structures. One interesting discovery using this instrument has been that particles of chalk whiting are made up from remarkably well-shaped aggregated crystals of calcite.

Particle size and size distribution

The value described as the particle size of a pigment is almost always an average particle diameter derived in one of the following ways:[33]

(i) Median diameter: d_{med}. The mid-point value of the cumulative size distribution curve, *i.e.* 50% of the sample is larger and 50% smaller.

(ii) Arithmetic-mean diameter: $d_1 = \Sigma nd/\Sigma n$ where n is the number of particles in each class d.

(iii) Length-mean diameter: $d_2 = \Sigma nd^2/\Sigma nd$.

(iv) Volume-surface mean diameter: the diameter of a particle with the same ratio of $d_3 = \Sigma nd^3/\Sigma nd^2$ surface to volume as the whole of the sample. (This value is usually obtained from surface area measurements.)

The kind of diameter chosen to describe the particle size is often influenced by the method used to determine the size distribution, but in every case the use of an average value involves the almost always incorrect assumption that all the particles are of the same shape. A fact of greater importance is that the numerical values of the various diameters may vary by a factor of two, or more, for the same pigment.

The practical problems of particle size measurement were the subject of a conference, the record of whose proceedings,[34] and especially the verbatim reports of the discussions, should be studied by those interested in the subject. The problems of measuring surface area by gas adsorption have been discussed by Sing[35] with special reference to the interaction of adsorbate and adsorbent and to the assessment of the effect of pores and fissures. The measurement of the micropore volume from gas adsorption data, using modifications of the t-plot method originally proposed by Lippens and de Boer[36] has been described by Sing[37,38] and Day and Parfitt.[39] Urwin[40] has published data showing the influence of micropores and fissures in the surface of coated titanias on the surface area measured by gas adsorption. Typically, only about 1·5 $m^2 g^{-1}$ out of an increase in surface area of about 4 $m^2 g^{-1}$, caused by coating, would be available to molecules larger than that of water. It is this 'accessible' surface area that should be used in calculating the adsorption per unit area of pigment surface.

The problems of measuring the surface area of fissured or sintered pigments, in a way that gives results allowing calculation of a mean particle size having some relevance to the effective particle size of the pigment in suspensions, have kept interest alive in the Carman[41] gas permeability method of surface area measurement. Kaye and Jackson[42] have reviewed the theory and practice of permeability measurements and drawn attention to Carman's own remarks[43] on the subject, with special reference to the problems created by the presence of a proportion of very fine powder in the material under test. In general, permeability methods cannot be considered as very suited to the measurement of surface area of pigments, as these generally consist of particles too small to form satisfactory plugs with adequate reproducibility.

Apart from the two above-mentioned indirect methods of measuring particle size, all methods of pigment particle size measurement involve the preparation of a dispersion. The methods are discussed in Chapter 7. The amount of work done on the pigment during the preparation varies considerably, both between methods and in many cases between operators using the same method. The determination of particle size thus may become a method of assessing either the comparative efficiency of different methods of pigment dispersion or the comparative ease of dispersion of pigments by a given method. As such, they have their uses, but the results obtained require careful interpretation.

Rugosity, roughness or smoothness factors

There is a need to amplify data such as mean particle size by some indication of the deviation of the particles from sphericity. The deviation may on the one hand be only slight, caused by indentations or roughening of the particle surface, or on the other hand the particles may be far from spherical, rough surfaced and angular. Another problem, for which it is difficult to find a satisfactory solution is provided by pigments such as carbon black and Prussian blue, that consist of agglomerates of fine particles with a considerable internal volume; the structure of the agglomerates may be such that the void spaces are roughly spherical, or at the other extreme they may consist of very narrow pores. Often one sample of pigment may contain both types of structure.

The established methods of characterising such pigment properties require that the pigment size distribution be known to a fair degree of accuracy. Measurement may be by either sieve analysis (not a suitable method for pigments), sedimentation or computation from electron micrographs. From the measured size distribution of the pigment may be calculated either the number of uniform spheres having the same mean diameter, density and total weight as the fraction, or the surface area per unit weight of the pigment.

Robertson and Emödi[44] suggested that the quotient of the specific surface area divided by the surface area of the equivalent spheres should be defined as the coefficient of *rugosity*. In the original work the specific surface was measured by the permeability method of Lea and Nurse,[45] using a modified cell.

Anderson and Emmett[46] proposed that the quotient of the BET nitrogen surface area, divided by the corrected area computed from the electron micrographs, should be termed the *roughness factor*.

The reciprocal of the roughness factor has been termed the *smoothness factor* (or smoothness ratio).

The coefficient of rugosity is reasonably satisfactory only as a means of indicating the surface roughness of individual particles; the problem of determining the specific surface in a more satisfactory manner requires consideration. The coefficient of rugosity is, nevertheless, of practical significance. Discrete particles with a highly indented surface can behave in a disperse system as particles of a size equivalent to the 'shell' which could enclose the particle. The liquid filling the indentations is effectively removed from the continuous phase, so that the effective solid volume fraction in the system is much larger than the calculated value, and the viscosity of the system correspondingly increased. This effect was verified, albeit for particles considerably larger (38–279 microns) than most pigments, by Ward and Whitmore.[47]

The roughness and smoothness factors have each been used to indicate the porosity of carbon black and Prussian Blue agglomerates; they cannot be considered as ideally suited to the job. The roughness factor, for example, when applied to material such as carbon black can reach a value of 2–2·5 without requiring any significant degree of true porosity.

The effect of milling (micronising) of pigments

Optimum performance, in any particular application, is given by pigment particles of a particular size, so the problem of producing material having a fairly narrow size distribution has received considerable attention. Dividing the final product into size classes is one way but not the most economic, since it is likely neither that all the fractions will be in equal demand nor that the most copious fractions will be those in most demand.

In the case of synthetic products careful control during manufacture may ensure that the size of the primary particles approximates to the optimum, but it is difficult to prevent the formation of some sintered aggregates in the case of materials subjected to calcination, or of cemented aggregates if a precipitation process is used. Products made by the comminution of minerals usually have an undesirably wide spread of particle size, and some form of classification should take place concurrently with the grinding process, so that material ground to the correct size is at once removed from the milling zone.

Pigments present particular problems because of the small size of the particles. For optimum ball milling efficiency the grinding elements should be of similar size to the material being ground, a condition which cannot be conveniently fulfilled if the grinding elements and the milled products are to be separated readily. Contamination of the product by material abraded from the mill also occurs. The logical solution to the problem was to devise a mill in which the feed was ground by impaction on itself, contact with the walls of the mill was minimised, the smaller particles separated and removed from the grinding zone, and the larger particles concentrated in the zone in which the milling action was at a maximum.

A number of mills were designed to meet these requirements, of which the best known is the *microniser*, developed at the Dutch State School of Mines. The process is also known as *jet-milling*, or *fluid-energy milling* but, in Europe at least, the term *micronising* is used to describe any process in which milling is carried out by acceleration of the particles of the feed so that, usually in conditions of turbulent flow, they impinge upon each other.

In the case of the true microniser the powder which is to be milled is fed into an annulus, around the outer circumference of which are nozzles through which superheated steam, compressed air or inert gas is injected tangentially to the inner wall. The particles of powder circulate in the high velocity gas stream, centrifugal action ensuring that the largest particles remain near the circumference (in the zone of maximum velocity) whilst ultra-fine material is carried by the escaping gas stream toward the centre and then upward, out of the mill. The maximum velocity and the intersections of the streams of gas from the jets ensure turbulence in the gas stream so that there is very considerable inter-particle contact, which is the only substantial source of grinding action. Where the feed to the microniser consists of markedly flocculated or aggregated material then occluded air or moisture may expand rapidly, when heated in the annulus, and cause some disruption of the floccule. This mechanism is not considered to be of great significance and in any case it is essential that the moisture content of the pigment fed to the microniser be kept low, to minimise latent heat losses. If steam, used to drive the microniser, becomes 'wet', then milling efficiency falls; in some cases the pigment may be pelletised, instead of broken down into particles.

The net effect of micronising the pigment is to increase the ease of dispersion by paint milling machinery, and this improvement is particularly notable when the dispersion process is carried out in mills dependent on high speed stirring action to achieve pigment dispersion (*e.g.* the Kady, Cowles, Torrance Cavitation mills, etc.).

Whilst it is accepted that micronised pigments need less milling, there is little physical evidence to indicate the reason for this. The properties of the micronised pigment, relative to the feed, may be summarised as follows.

(i) The milled pigment particles tend to be more rounded, with smoother surfaces.
(ii) The surface area, measured by nitrogen adsorption is usually lower, and never greater, than the original surface area.
(iii) The particle size distribution usually shows that oversize particles have been broken down, and the finer particles removed (drawn away in the exhaust gases), so that the size distribution is narrowed and the average size reduced.
(iv) Highly acicular pigments tend to fracture across the main crystal axis.
(v) Pigments such as Prussian blue, which are highly agglomerated when fed to the mill, tend to show the most startling size reductions, but are not normally broken down to primary particles.
(vi) Coated pigments, such as fully processed grades of titanium dioxide, show effects (i) to (iii), above, but do not fracture or lose their coating under normal micronising conditions.

A number of papers have been published on the use of organic additives to improve the efficiency of micronising of minerals. Various grades of titanium dioxide, in particular, are now available which are coated, in part at least, with organic compounds, presumably added prior to micronising. Most often used are amines, amine-soaps and polyhydroxy compounds, any of which are readily adsorbed on pigment surfaces. By choosing a compound of suitable structure the organically-treated surface may be made hydrophobic or hydrophilic in character and, in addition, the micronised pigment has a slightly narrower particle size distribution than a 'control' without the additive.

Treated pigment is more readily wetted by an appropriate medium so that a smaller proportion of the pigment milling time is required for wetting and disruption of agglomerates and a correspondingly greater proportion is available for the dispersion stage. However, it is not easy to decide whether the greater ease of dispersion of these pigments is due to more rapid wetting because of the organic material already adsorbed on their surfaces, or to the narrower particle size distribution associated with the use of organic additives when micronising.

It is established that the apparent ease of dispersion of most pigments is improved by micronising, but this improvement is very often measured by the reduction in the time required to obtain a particular 'fineness-of-grind' reading. About 40 000 uniform spheres of 0·25 microns diameter in random close packed array could be contained in a sphere of 10 microns diameter, so fineness-of-grind gauge readings do not give any useful guidance to the actual state of dispersion of particles less than 1 micron in diameter. The gloss of alkyd/amino resin based baking enamels provides a more useful guide, and data obtained in this way suggest that the effect of micronising is a physical one. Given adequate milling, the baking enamel containing an uncoated unmicronised rutile will have the same gloss value as that containing a fully treated (coated and micronised) rutile.[97] Data on the comparative hiding powers of very well dispersed uncoated, unmicronised rutile and fully treated (coated and micronised) rutile in an air-drying long oil alkyd medium have been published.[48] The paint compositions were not adjusted to compensate for the lower content of opacifying pigment in the treated grade (the low refractive index coating oxides do not make a direct contribution to the hiding power) and the plots of hiding power against pigment volume concentration (p.v.c.) showed that after prolonged (7 days) laboratory ball-milling the paints containing the untreated grade were slightly superior in hiding power at the higher p.v.c. values. In the lower p.v.c. range the superior particle size distribution of the treated pigment compensated for its lower content of actual opacifying pigment and the opacities of the paints were very similar. These results indicate that both pigments were dispersed to the same extent in the dry paint films.

It seems probable, therefore, that the principal effect of micronising is the breakdown of aggregates and agglomerates, so that only comparatively weakly-bound agglomerates, formed by the unavoidable flocculation of the pigment in air, have to be disrupted during the dispersion process. The adsorption characteristics of the pigment surface are probably not much affected by micronising, apart from the changes in apparent characteristics caused by the rounding of the particles. Pending the publication of data correlating the effect of micronising, adsorption characteristics and degree of dispersion, the improvement in ease of dispersion associated with micronising seems attributable to the modification in particle shape and size distribution rather than any other cause.

Structured pigments; carbon blacks

The terms 'low structure' and 'high structure' are of rather obscure origin, but seem to have come into use in the rubber industry to describe the relative effect of a reinforcing carbon black on the consistency of the rubber during compounding. The term is now applied chiefly to carbon blacks, but sometimes also to iron (Prussian) blues and materials such as diatomaceous silica. The common characteristic of these pigments is that their particles consist of agglomerates and/or aggregates of very small primary particles. (Diatomaceous silica, fossil plant skeletons, is a special case.) As a result, when suspended in liquid, each particle has an effective hydrodynamic volume much greater than its nominal size, owing to the quantity of the liquid that is immobilised in the interstices of the pigmentary particle.

The most intense studies of the nature of structured pigments have been made on the carbon blacks, especially the rubber-reinforcing (breaking-strain increasing) grades, where it has been shown[49,50] that the characteristics of the vulcanised rubber can be predicted from the physical characteristics of the black used in it.

The most recent studies of the fine structure of the primary particles of carbon blacks[51,52] have shown these to consist of distorted graphitic carbon layers, without any actual crystals being present. These primary units of carbon black occur in either a fibrous (*e.g.* channel and furnace blacks) or spheroidal form (thermal blacks). From these primary units are built up three-dimensional aggregates, that the more powerful modern electron microscopes show to be quite different to the convoluted chain structures formerly proposed.

The highly structured blacks consist of relatively large aggregates that have more 'open' structures and therefore occlude a larger proportion of fluid relative to the volume of black forming the particle. Characterisation of the blacks is carried out either by visual comparison of electron micrographs of the aggregates with a range of standard micrographs, or by an automated scanning method[53] that measures the optical density of the photographs,the darkness of the picture being proportional to the thickness of the aggregate. (This has the disadvantage that a two-dimensional image is being used to characterise a three-dimensional object.) The data is used to characterise the particles in terms of estimated size and number of particles, cross-sectional area and maximum linear dimension. A shape factor is calculated from the relation of the major dimension to the diameter of a circle of equal area. The sphere generated from this circle is described as the *equivalent sphere* and the ratio of the volume of this sphere to the solid volume calculated from the size and number of the particles provides a measure of the 'structure' of the black. Oil and dibutyl phthalate (DBP) absorption values calculated from data obtained in this way show good agreement with the experimentally determined values.[54]

All carbon blacks are of relatively low intrinsic density and have in addition high bulking values (*i.e.* low bulk densities). For these reasons, densified or pelletised forms of these blacks are popular. The spheroidal thermal process blacks and the low structure furnace and channel blacks will densify quite readily, but the particles of the medium and especially the high structure blacks do not readily pack closely together and therefore resist densification. In all cases, the work done on the pigment in the densification process has to be nullified during the pigment dispersion process. Dispersion methods that allow application of shearing forces powerful enough to disrupt the black pellets are therefore essential. Surfactants may assist pellet disruption, but the secondary effects of such materials on the performance of the product in its intended use should be taken into consideration.

The composition of the pigment surface

Information about the composition and nature of inorganic pigments that is relevant to the behaviour, during the dispersion process and subsequently, of the pigment is, not surprisingly, scarce. A large proportion of the information that is available unfortunately refers to pigments made in the laboratory for the investigation, whilst a still greater proportion of the information is reduced in value because of inadequate characterisation of materials; pigment chemists paying little attention to the media, resin chemists treating pigments in similar cavalier fashion. It is unfortunate, too, that in many cases the behaviour of the pigment is greatly influenced, if not controlled, by trace impurities such as residual electrolytes (from the manufacturing process) that cannot be completely leached out. In addition, in normal industrial conditions the surface of a pigment is covered by about a monolayer of water adsorbed from the atmosphere. The fact that most coloured pigments are likely to be blended to maintain their mass-tone and staining power within specified limits is rarely considered even though in the case of such pigments as lead chromes this may involve quite major variations in chemical composition between one batch and another.

Another factor that is not much considered is the tendency for impurities to concentrate in the surface layers as the crystal grows in its mother liquor, or during a calcination process. The surface layers may therefore differ substantially in chemical composition from the bulk, and the trace impurities may have a disproportionate influence on pigment dispersion behaviour. It is possible that by the deliberate incorporation of 'impurities' during manufacture a pigment with the required surface could be obtained without an additional process such as coating with other oxides.

The composition and structure of the surface layers of a pigment also affect other properties besides dispersibility. Clay[55] showed that the effectiveness of an alumina coating as an improver of light fastness of lead chromes is reduced as the proportion of coating oxide relative to base pigment is increased. He also stated that a mixture of the hydrated oxides of aluminium and titanium or silicon is superior to either of the additives used singly. An alumina coated chrome was found to be more resistant to darkening when a mercury-vapour lamp is used than with a tungsten light as a source; a silica on alumina coating gave maximum resistance to mercury light while an alumina on silica coating is superior when tungsten lamps are used. Unfortunately Clay did not indicate whether tests were carried out under conditions of equal total energy input to the films.

Coating with other oxides is most firmly established in the titanium dioxide pigment field. These pigments are made either by the sulphate process from a ferro-titanic ore, usually ilmenite, or by the chloride process from natural rutile. The latter method avoids the massive quantities of by-product (ferrous sulphate) and acidic aqueous effluent of the older process; in addition purification by distillation of the intermediate, titanium tetrachloride, is simpler than the wet methods used in the sulphate process and allows production of material of better colour at acceptable cost. The final products are essentially similar, either anatase or rutile types produced in the sulphate process by fusion of crystallites to pigmentary size in a calciner or by controlled crystal growth in the chloride process. The untreated (unrefined) pigments consist of primary particles with no coating oxides, although most rutiles contain a small amount, about 1% w/w, of other oxide, either zinc oxide, alumina, antimony oxide, or alumina and antimony oxide mixture. These latter oxides are added to encourage growth in the rutile form. They usually cause some degradation of the crystal lattice, zinc oxide in particular causing the particle to be somewhat rounded. This is considered desirable, since it prevents the abrupt discontinuity in surface properties that occurs at the crystal plane junctions of a perfect crystal.[56] Both anatase and rutile have the defective lattice structure associated with semi-conductor and catalytic properties.

Treated, or coated, titanium pigments have a surface layer of oxides, most often alumina, silica, titanium dioxide and sometimes zinc oxide. These coatings are precipitated on to an aqueous suspension of the primary particles, either simultaneously, one at a time, or in various combinations. The titanium oxide coating is often applied first, since it seems to assist the subsequent adhesion of the alumina and/or silica, though this adhesion-promoting effect seems to function equally well when alumina and titania are co-precipitated.

Apart from the amounts and kinds of oxides used to form the coating, and whether they are precipitated as single oxides, mixed oxides, or part singly, part mixed, the physical form and crystal structure of the oxide coating is markedly dependent on the conditions used, all of the following factors being important:

Temperature
Concentration of reagent solutions
Rate of precipitation (*i.e.* speed of addition of precipitating reagent)
Time in contact with mother liquor.

The coating technique used for pigments intended for general application, is designed to give a relatively smooth uniform thickness of coating oxides, which constitute about 5–7 % w/w of the final pigment. More recently introduced types, intended specifically for use in polymer latex based paints ('emulsion paints'), consist of about 85 % TiO_2, 15 % coating oxides, with coatings applied in such a way that they form dense spiky protrusions from the crystal surface.

The objective in coating application is to ensure that the initial precipitated crystallites adhere to the titania particles and then grow in a controlled fashion. Kämp and Völz[57] have published electron micrographs showing the type of result produced by an unsuitable coating technique. Simple accretion of discrete precipitated crystals does not produce a satisfactory coating, but even a relatively smooth, dense coating of the type applied to 'general purpose' pigments is inevitably fissured by the gaps between adjacent coating crystal growths, and is likely to have also some pores in the coating. These pores and fissures cause a disproportionate increase in the BET surface area, typically from about 8 $m^2\ g^{-1}$ to about 12 $m^2\ g^{-1}$ for a coating that, examined under the electron microscope, is uniform, with a fairly smooth surface, and a thickness of about 50 Å. Of the new surface, some 2·5 m^2 is contributed by the pores and fissures, accessible to a water molecule, but not to larger molecules.

The coating oxides, as applied, carry down with them water, either as hydrates or occluded, as well as small amounts of water soluble salts, components of the precipitation reaction. The coated pigment, therefore, has to be washed very thoroughly (but some traces of soluble salts are always left) and then dried, before receiving a final milling treatment, usually micronising or an equivalent process.

As noted earlier, some grades of pigment have on their surface residues of an organic compound added prior to micronising. This varies from 0·5–1·0 % w/w according to the compound and probably forms a monolayer. Another end-treatment involves exposure of the pigment to methyl siloxane vapour, with the object of producing a pigment surface covered in methyl groups, bonded through the silicon atom. This type of treatment is intended to produce a hydrophobic surface with less tendency to adsorb water vapour and with a greater ease of 'wetting' by polymers, such as polyethylene, in the course of a plastic moulding process.

The oxide coating can be heat-treated, causing many of the pores and fissures to close as the coating fuses. Improved 'chalk-resistant' properties are claimed for these pigments, since, in theory at any rate, the semi-conductor titania crystals are sealed away from the surrounding film-forming material, and therefore cannot act as oxidation-catalysts.

A similar 'annealing' effect also takes place at ordinary temperatures. Aided no doubt by the moisture that in ordinary circumstances is inevitably adsorbed on the particle surface, some crystals join together, so that after about a year there is a small but insignificant drop in specific (BET) surface area. Parallel with this effect is a fall in the so-called 'reactivity' of the pigment, measured as the rate of viscosity increase of a paint made from the pigment and an intrinsically unstable varnish. The viscosity increase seems to be due primarily to an adsorption mechanism, the pigment particles acting either as nuclei or as bridges between polymer particles. The fall in 'reactivity' with time may be due to the reduced surface area, a reduction in the number of crystal plane junctions (high activity zones) or to a reduction in surface energy caused by hydration of the pigment surface.

In the absence of organic material, the surface of a coated titanium dioxide will nominally consist chiefly of alumina, silica, or both, although in fact the true surface is most likely to be about a monolayer of water on a surface of hydrated alumina and/or silica. The chemical characterisation of such surfaces has been discussed by Parfitt.[58]

The surfaces of the pigments just discussed are all essentially inorganic, except for those that have a small amount, possibly a monolayer, of some specific organic compound. The chemical and physical nature of the surfaces of carbon black pigments presents a very different picture. The physical form of these pigments ranges from the spheroidal particles of thermal blacks by way of the fairly simply shaped aggregates of the *low structure* to the very complex shapes of the *high structure* blacks. The very small primary particles consist of graphitic layers, but not graphite crystals, so that the unaligned stacking and the distortion of the surface layers provide a very high energy surface. The distorted or bent layers may be associated with dislocations or even atomic vacancies in the lattice. Since the channel and furnace blacks are produced by burning, these blacks have a surface that is partially oxidised (or incompletely reduced) and consists of various oxyhydrocarbons. The thermal blacks, produced by the high-temperature *cracking* of natural gas or acetylene, without combustion, have as a result almost unoxidised surfaces, in marked contrast to the former types.

The physical form of a carbon black is partly dependent on the feedstock, traditionally natural gas for channel blacks and also for some furnace blacks, though a large proportion of modern furnace blacks are made from oil. This difference in the feedstock also affects the chemical composition of the surface. The usual channel black process produces considerable atmospheric contamination and anti-pollution regulations have forced closure of some plants, but some new plants are claimed to be able to meet the strictest requirements. The furnace process is intrinsically cleaner, so that there has been considerable effort to produce blacks by this process that are closer in pigmentary properties to the channel blacks.

One step in this direction was taken when it was found that injection of alkali metal salts into the furnace reduced the flame ionisation potential and resulted in the production of blacks with much less structure, down to the channel black levels. The effect of alkali addition of this type is reflected in the pH value of an aqueous slurry of the black, but this property is also dependent on the volatile content and thus on the manufacturing process.

The volatile content of carbon black pigments is usually expressed as percentage weight loss on heating at about 950°C in the absence of air. It consists of carbon- and hydrocarbon-oxygen complexes and is usually slightly acidic in character. The volatile-free carbon black surface is slightly basic. An aqueous slurry of a thermal black, that has only about 0·5% volatile, has a pH of about 9·5. Oil furnace blacks, with about 1% volatile, give a pH of 8·5. An original process channel black has about 4–5% volatile and gives a pH of about 5. Certain channel blacks are heated to red heat in air as a secondary stage in manufacture. This increases the volatile content to 10% or more and the pH drops correspondingly to 3·5–3·0. The treatment also increases the nitrogen adsorption surface area, but chiefly by the formation of internal pores too small to affect the amount of surface available to paint or ink media.

Rivin[59] analysed the volatile constituents of carbon black and postulated a stepwise oxidation process, from phenols and/or hydroquinones (weakly acidic) through quinones (neutral) to either lactones (neutral) or carboxylic groups (strongly acidic). Further oxidation caused carbon dioxide formation, with pore formation as its consequence. Scott[60] has published figures for the contents of the above chemical groupings in the volatile content of various commercial blacks and Dollimore[61] has reviewed the more fundamental aspects of black surface characterisation.

A generalised summary of the relationship between surface composition and ease of dispersion of carbon blacks is not very useful, since the dispersion properties depend in the first instance on the ease with which the black is disrupted into the aggregates that function as pigmentary particles and this in turn depends on the past history of the black, for example whether it has been densified (pelletised) deliberately or accidentally. The rheology of a carbon black suspension is in turn primarily dependent on the physical nature of the aggregates, that is, on the structure, low, medium or high that determines the volume of liquid occluded by a unit volume of black. The chemical nature of the surface is important only insofar as it affects the speed of wetting and the tendency of the particles, subsequent to dispersion, to adsorb components, from the suspending fluid, that will stabilise the dispersion. The experimental evidence[62] suggests that a relatively high content of acidic 'volatiles' does in fact assist both pigment wetting and dispersion stabilisation.

DISPERSION OF INORGANIC PIGMENTS

In the dispersion of a pigment in a liquid by mechanical action, three processes are involved.

(i) Wetting, in which air is displaced from the surface of and between the particles of the pigment clusters and is replaced by medium, but the form of the clusters may remain unchanged.
(ii) Disruption, by mechanical energy, of the clusters (aggregates and agglomerates) into smaller units dispersed in the medium.
(iii) Flocculation, the process which involves interparticle collisions that reduce the total number of particles.

These processes are not necessarily sequential, all may be occurring concurrently in different parts of the system, but the third process does not become important until the first two are substantially complete. In the absence of some other force, maintenance of the dispersed state will depend on the relative rates of mechanical separation and of flocculation of the particles. These aspects of the dispersion process are discussed in Chapter 1.

If the liquid phase is viscous, pigment mobility and hence rate of flocculation are reduced, a circumstance put to practical use in plastics injection moulding. The pigment is dispersed in the molten plastic by the shearing forces created during the injection process, whereupon cooling in the mould quickly raises the viscosity of the plastic to a level that prevents any significant flocculation of the pigment. For suspensions in more or less permanent liquids an alternative method of stabilising the dispersion is required. This is usually the presence in or addition to the liquid of a surfactant that can be adsorbed on the pigment surface and substantially prevent flocculation. (For reasons mentioned earlier in this chapter, an attempt may be made to induce a very loose, voluminous, flocculated pigment phase, but this will not be considered here.)

For various reasons, some of which are discussed later, the pigment dispersion process is often carried out in a liquid with some degree of intrinsic pigment dispersing power. Wetting of the pigment is aided by low viscosity of the medium and in machines using impacting forces, such as the ball mill, the mode of action requires the suspension to have fairly high mobility together with the highest possible pigment content. In other machines such as the triple-roll mill, where dispersion is by shearing forces, a high-viscosity medium with strong internal cohesive forces is required to transmit the shearing forces produced by the roll speed differential to the pigment. Whilst the method of achieving dispersion varies from one type of machine to another, the efficiency of the process is in every case dependent on the suitability to it of the rheological characteristics of the pigment suspension.

Rheology of pigment suspensions

A liquid subjected to a certain shearing stress will flow in order to relieve that stress; if the stress τ is not too great the flow will be laminar and the rate of shear ε will be a simple function of the applied stress. *i.e.*

$$\tau = \eta\varepsilon \tag{8.1}$$

where η is the viscosity of the liquid.

The presence of suspended particles in a liquid subjected to an applied stress makes the response of the system to the stress more complex; the problems of expressing this behaviour mathematically have been discussed by Oldroyd.[63] The rheological behaviour of suspensions as a class has also been reviewed by Roscoe.[64] So far as the dispersion of a relatively large volume of pigment in a liquid is concerned, the major stages can be enumerated

(i) Agglomerated 'dry' pigment with a volume of internal voids equal or even greater than the volume of liquid. Apparent consistency of system very high.

(ii) Pigment wetted and partly dispersed, but with a large number of clumps of particles in which a substantial proportion of the available liquid is immobilised. Consistency lower than (i) but still high.

(iii) Progressive breakdown of the pigment clumps with release of immobilised liquid causing a further reduction in apparent consistency.

(iv) Pigment particles dispersed but tending to flocculate rapidly if shearing stress is reduced or removed. Characterised by reduction in viscosity (increase in rate of shear) with increasing shear stress. The system may have a *yield value*, a certain minimum shear stress being required to induce flow.

(v) Pigment particles substantially dispersed and adsorbing material from the liquid phase, so that the effective volume of the disperse (pigment) phase is increased and that of the continuous phase decreased. The system is more Newtonian than (iv), but its viscosity slightly higher; if the shearing stress is raised sufficiently, adsorbed material may be stripped off.

Of these stages, the first is often critical, since if the medium fails to wet the pigment and so assist in the mechanical disruption of the agglomerates, the machine may be unable to accomplish this task. It is therefore necessary to ensure that the proportions of pigment and liquid are suitable for the accomplishment of this stage, but not such that after wetting-out, the consistency of the mixture is too low for efficient milling. Appropriate proportions can be estimated from the results of some simple tests.

Oil absorption value of pigments

There are a number of specified methods for the determination of the oil absorption but discussion here will be limited to the British Standard (BS) method. This requires a small amount of pigment, usually 10 g, to be mixed, on a glass or marble slab, with linseed oil of specified acid value, by an operator using a steel spatula and exerting maximum effort. The oil is added dropwise and the end-point is the formation of a just coherent paste that will not smear the slab. The amount of oil required is strongly influenced by the acid value of the oil and the amount of effort put in by the operator, including the thoroughness with which each drop of oil is incorporated before the next is added. The result obtained should be compared with the result obtained, by the same operator, using a sample of a standard pigment, in order to minimise the operator effect. Oil absorption values are usually quoted in millilitres or, more often, grams of oil required by 100 g of pigment, a practice which exaggerates any error by a factor of ten.

Marsden[65,66] has reported the results of an evaluation of the test method. He concluded that the BS end-point, correctly interpreted, was quite sharp but that the amount of oil required to reach the end-point was subject to an operator-effect, because of the work factor. He suggested that the oil-absorption value of a pigment was the sum of two terms:

(i) an oil-coating term, which varies as the specific area of the pigment and as the specific gravity of the oil used. The most important factor, however, is the operator himself, who by rubbing is able to obtain more or fewer touches of particle with particle, thereby greatly affecting the magnitude of the oil-coating term;

(ii) an oil-adsorption term which under normal conditions varies with the nature of the pigment, with the oil used and with the state of division of the pigment. Provided each particle is coated with oil the operator cannot affect the magnitude of this term.

Besides the effect of the amount of work done on the paste during the test, there is also a dynamic, or age, factor, due to the end-point being reached whilst the test is, more or less, still in progress. If the end-point paste is stored either it does not change, or it becomes softer, or it becomes dry and crumbly. The second effect will be observed if the pigment particles can rearrange themselves into a more closely packed condition. Marsden[65] suggests that this effect is particularly likely with materials such as iron oxides, which are readily influenced by the electrical charges induced by rubbing, but the effect is the definitive characteristic of all dilatant systems. The third effect would result if the pigment particles tended to separate as the paste aged; the effect might be caused by increased adsorption of oil on the pigment surface, but penetration of oil into residual voids in pigment agglomerates is an equally likely cause.

Despite the observations of Marsden and other workers, the oil absorption paste is popularly defined as a system consisting of the pigment particles, each with an adsorbed monolayer of linseed oil on its surface, in their closest possible random packing, together with just enough extra oil to just fill the voids between the particles. For a pigment of density ρ_p and nitrogen surface area S m^2 g^{-1}, and linseed oil $\rho = 0{\cdot}933$ the theoretical oil absorption Y g per 100 g pigment can be calculated from the formula

$$Y = 0{\cdot}146S + 0{\cdot}933\left(\frac{100}{\rho_p} + 0{\cdot}156S\right)\left(\frac{1}{F} - 1\right) \qquad (8.2)$$

where $0{\cdot}146S$ g and $0{\cdot}156S$ ml are the weight and volume of a monolayer of linseed oil of specific gravity 0·933 and molecular weight 877. F is the solid volume fraction of the particles in their closest possible random packing. (The spatial relationships of the particles are assumed to be identical in the dry state and in the oil absorption paste, except for the extra centre-to-centre separations necessary to accommodate the adsorbed oil layer.)

The calculated thickness of the adsorbed monomolecular oil layer is about 1·55 nm, and the pigment surface area, S, is related to the volume-surface mean diameter, d_3, in microns, by the formula $S = (6/\rho_p d_3)$. This enables the formula for Y to be transposed to:

$$Y = \frac{0{\cdot}876}{\rho_p d_3} + \frac{93{\cdot}3}{\rho_p}\left(\frac{d_3 + 0{\cdot}0031}{d_3}\right)^3\left(\frac{1}{F} - 1\right) \qquad (8.3)$$

In this form, the dependence of the theoretical oil absorption on pigment density, surface area and packing fraction is clearly shown. The influence of the adsorbed oil layer on the p.v.c. in the oil absorption paste, insignificant for particles of 1 micron (1000 nm) mean diameter, becomes very considerable as the mean particle size diminishes.

The above equations for the theoretical oil absorption value can be rearranged in various ways, but only the pigment surface area and density can be independently determined. For all pigments, the surface area value used should be the value corrected to allow for the area due to walls of fissures and pores, not accessible to oil molecules. Also, especially in the case of very small particles (30 nm or less in diameter) the effect of adsorbed water is not negligible. The surface area is measured on the dried (out-gassed) pigment, the oil absorption on the pigment in its normal state, with at least a monolayer of adsorbed water. The effect of a water layer 0·28 nm thick is not negligible relative to a 1·55 nm thick film of oil. The factor by which an adsorbed oil layer increases the effective volume of the coated particles is the cube of the quotient of the increased and original particle diameters and therefore increases very rapidly as the particle diameter decreases. For very small particles, the effect of the hypothetical adsorbed oil layer on the apparent oil adsorption is dominant. This suggests an explanation of the marked structural viscosity that can be produced by the addition of even a small proportion of *e.g.* fine-particle silica, to a pigment

TABLE 8.4

Particle Diameter Micron	Particle Expansion Factor	*Surface Area and Oil Absorption Value Calculated from Equation* 8.3 ($F = 0{\cdot}64$)							
		$\rho_p = 5{\cdot}2$		$\rho_p = 4{\cdot}0$		$\rho_p = 2{\cdot}6$		$\rho_p = 1{\cdot}8$	
		S $m^2\,g^{-1}$	Y $g/100\,g$	S $m^2\,g^{-1}$	Y $g/100\,g$	S $m^2\,g^{-1}$	Y $g/100\,g$	S $m^2\,g^{-1}$	Y $g/100\,g$
5·0	1·001 86	0·23	10·14	0·3	13·19	0·46	20·29	0·666	29·3
3·0	1·003 1	0·38	10·17	0·5	13·23	0·77	20·35	1·11	29·4
2·0	1·004 7	0·58	10·22	0·75	13·29	1·15	20·44	1·66	29·5
1·0	1·009 3	1·16	10·36	1·5	13·47	2·32	20·72	3·33	29·9
0·7	1·013 3	1·65	10·47	2·14	13·61	3·29	20·93	4·76	30·2
0·5	1·020 5	2·31	10·63	3·0	13·82	4·61	21·27	6·66	30·65
0·3	1·031 2	3·845	10·97	5·0	14·26	7·69	21·93	11·11	31·75
0·2	1·047 0	5·77	11·41	7·5	14·83	11·54	22·82	16·66	33·0
0·1	1·096 0	11·59	12·75	15·0	16·58	23·17	25·51	33·33	36·8
0·07	1·139 0	16·48	13·91	21·43	18·08	32·97	27·81	47·62	40·2
0·05	1·197 8	23·08	15·46	30·0	20·00	46·15	30·92	66·66	44·6
0·03	1·343 0	38·46	19·16	50·0	24·91	76·92	38·32	111·11	55·3
0·02	1·540 9	57·69	23·97	75·0	31·17	115·38	47·95	166·66	169·3
0·01	2·248 1	115·885	39·54	150·0	51·41	231·77	79·08	333·33	114·2
0·007	3·069 5	—	—	—	—	329·67	110·09	476·19	158·75
0·005	4·251 5	—	—	—	—	461·53	153·15	666·66	220·85
0·003	8·406 8	—	—	—	—	769·23	282·01	1 111·11	406·60

suspension. The factors for some particle diameters are given in Table 8.4 together with some oil absorption values calculated for pigment densities corresponding approximately to those of iron oxide, coated rutile, silica and furnace carbon black, assuming that the true packing fraction is 0·64, the value for closest random packing of spheres.

Attempts to use the oil absorption value as a means of assessment of the ease of dispersion of pigments meet with only limited success because of the complexity of the relationships that influence the determined value. Jefferies[67] has pointed out that the surface area/oil absorption (S/Y) ratio used for the assessment of ease of dispersion, or *texture rating* by Carr[68] does not allow for the influence of the packing fraction on the value of Y. Similarly, a method of calculating the thickness of the absorbed layer in the oil absorption paste, proposed by Patton,[69] requires the assumption that two pigments of similar nature can differ in particle size but have the same packing fraction, a rather large assumption for pigments that do not consist of uniform spherical particles.

The absolute value of the packing fraction F, although of considerable theoretical interest, does not need to be known for the purpose of calculating the theoretical relative viscosity of a pigment suspension.

The relative viscosity of pigment suspensions

The addition of particles to a liquid, to form a suspension, causes the suspension to have a viscosity raised above that of the liquid by an amount proportional to the volume fraction of the particles in the suspension. For very dilute suspensions, the increase in viscosity is given by the well-known Einstein equation, but for concentrated suspensions a slight modification of an equation, originally due to Roscoe,[70] is more useful:

$$\frac{\eta}{\eta_0} = \eta_R = \left(1 - \frac{c}{F}\right)^{-2.5} \tag{8.4}$$

where c is the volume fraction of the solid phase and F is the closest possible packing fraction of the particles. At very high concentrations, when the particles are in their close packed array, the theoretical viscosity becomes infinite, at very low concentrations the equation approximates to the Einstein equation.

Ideally, the value of F used in this equation should be that derived from the *critical pigment volume concentration* (c.p.v.c.) determined by the method of Asbeck and van Loo,[71] using the same medium as that in which the suspension is to be prepared.

(Many properties of paint films, including opacity, tensile strength, and permeability, when plotted against the pigment volume concentration (p.v.c.) in the dry film, give curves which have a marked inflexion at or near the same p.v.c. level. This level is called the c.p.v.c., and is considered, with reason, to be the value at which the volume of binder is equal to the void space within the particles, plus sufficient binder to coat each particle. The c.p.v.c. should be determined by measuring the properties of a series of paints; the use of the pigment volume concentration in the oil absorption paste, which gives an adequate approximation, is very common.)

The apparent p.v.c. in the c.p.v.c. or the oil absorption paste is assumed to be that of the pigment particles, each forming the core of a congruently shaped larger particle, due to the adsorbed material on the pigment surface. In a suspension, the effective volume of the disperse solid phase may be increased above the true value by material from the continuous phase immobilised in indentations in the particle surface (as discussed earlier, under Rugosity *etc.*) or inside pigment agglomerates, or by a combination of both effects. Also, even perfectly smooth particles would increase in effective volume if they acquired an adsorbed layer. Reference to Table 8.4 shows that particles of 10 nm diameter, that could have a close-packed solid volume fraction of $F = 0{\cdot}64$, would have an apparent packing fraction of 0·284, if each particle had a monolayer of linseed oil. Use of the apparent volume fraction of pigment in the c.p.v.c. film or oil absorption paste, with the nominal value of c, automatically compensates for any adsorbed layer on the pigment surface and also for occlusion of the medium within particle agglomerates. This latter is particularly important in the case of such pigments as diatomaceous silica and, especially, carbon blacks.

The apparent viscosity of a suspension can be calculated by assuming that the relation between a given shearing stress and its induced rate of shear is Newtonian, so that

$$\eta_{\text{app}} = \frac{\tau}{\varepsilon} \tag{8.5}$$

By this means the apparent viscosity/rate of shear relationship can be plotted and a comparison made with the calculated viscosity. The relationship is not close, because the theoretical viscosity equation does not allow for the effects of particle shape, attractive and/or repulsive forces between particles or for the effect of shearing stresses on agglomerated (not yet dispersed) or flocculated particles. Casson[72] has developed an equation that expresses the shear stress and strain relationship in the form of an equation

$$\tau^{\frac{1}{2}} = k_0 + k_1\varepsilon^{\frac{1}{2}} \tag{8.6}$$

where k_0 and k_1 are constants whose values depend on the properties of the solid and liquid phases of the suspension.

Casson assumed the pigment to be flocculated and that, in conditions of laminar flow, the floccules could be treated mathematically as cylinders of high axial ratio. Weber[73] used Casson's equation and also a generalised equation developed by Heinz.[74] He found that neither equation was completely correct when dispersions of white pigments (principally lithopones and titanium dioxides) in linseed oil were under consideration. Both equations gave curves which fitted his experimental results best at moderate to high rates of shear. This would be consistent with the assumption that at rest or low rates of shear the 'chains' or 'rods' of flocculated particles were curled into a roughly spherical shape. When sheared the floccules would straighten in order to offer the minimum resistance to flow. As a result of his experiments, Weber concluded that the rheological properties of his dispersions were best explained by postulating the existence of layers of adsorbed medium around the pigment particles. He also suggested that multilayer adsorption could occur and that such layers might be 'stripped' stepwise by increasing the rate at which the dispersion was sheared.

Pigment/dispersant/liquid ratio

For economic reasons, the greatest possible proportion of pigment should be in a paste or slurry that is to be milled. (In practice, the absolute maximum pigment content suspension may not be the most convenient, and the formulation actually used may be modified to give a greater margin of stability to the dispersion.) The suspension viscosity rises with increasing pigment content, so the initial viscosity of the liquid should be the minimum possible, that is, the amount of dispersant dissolved in it should be minimal. Agglomerated or flocculated pigment particles occlude a large volume of the liquid phase, so the amount of dispersant should not be less than that needed to deflocculate the pigment. The first simple, practical way of determining this relationship for aqueous surfactant solutions was described by Daniel and Goldman[75] and Daniel[76] subsequently described in detail the method of carrying out the determination with alkyd resin solutions, to obtain the optimum pigment/resin/solvent combination for dispersion by ball-milling. More recently Daniel[77] has shown how the technique can be applied to the formulation of mill bases suitable for dispersion by high-speed impeller milling ('Cowles milling').

The method of determining the optimum pigment/resin/solvent (or pigment/surfactant/water) combination is to make a paste of pigment and dilute resin solution. In the 1946 and 1966 versions of the method, the resin solution was incorporated in the pigment by pasting with a spatula on a glass plate; the technique is simple, but rapid working is essential because of the loss of solvent. In the 1950 version, the resin solution was incorporated in the pigment by stirring with a glass rod in a small jar or beaker; the end-points of the two methods are not quite the same, the 1950 method usually requiring more resin solution. (The analogy with the work effect in the BS method of oil absorption value determination is very close.) In both methods, two end-points were observed, first the *wet-point or ball-point*, at which the pigment just formed a coherent mass which did not (significantly) wet the glass plate or vessel side, and secondly the *flow-point* at which the paste just flowed from the knife (1946) or just flowed from the rod, the paste 'snapping back' to the rod when the stream ruptured (1950). By repeating the determination with a series of resin solutions, curves can be plotted of amount of resin solution per unit weight of pigment required to reach the wet- and flow-points against the resin concentration in solution. Examples are illustrated in Fig. 8.1 for the three systems shown (*a*) strong dilatancy (dilatancy will be discussed again below), (*b*) weak dilatancy, and (*c*) strong flocculation.[77] In strongly dilatant systems the flow-point is usually only 5–15% above the wet-point, for

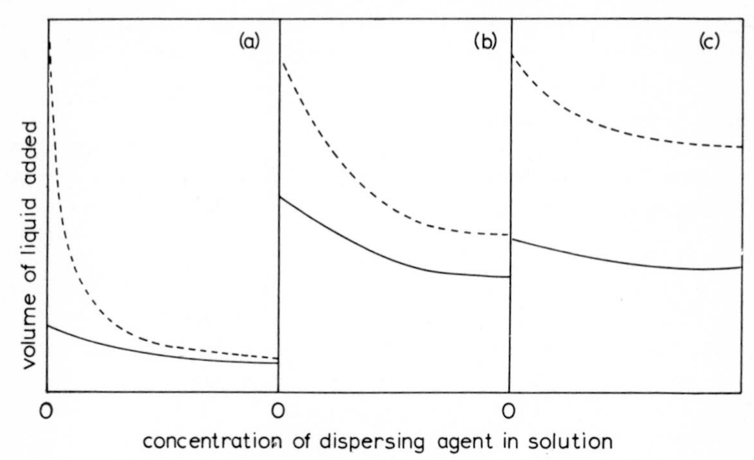

Fig. 8.1. *Daniel flow-point method: examples of systems showing* (a) *strong dilatancy,* (b) *weak dilatancy and* (c) *strong flocculation* (*after Daniel*[77]). *Broken line, flow point; continuous line, wet point.*

weakly or non-dilatant systems 15–30% and when flocculation occurs the gap is considerably wider.[77] A good dispersion requires the smallest further addition of resin solution to convert the wet-point to the flow-point paste.

Attempts to classify dispersions only by the pigment volume concentration in the flow-point paste can be misleading. For example, a conventional grade of coated rutile titanium dioxide may give a p.v.c. at the flow-point of about 49% whereas one of the more recently introduced heavily coated rutiles, consisting of 85% TiO_2 core and 15% w/w coating oxides might give a flow-point paste with only as little as 33% p.v.c. Both are good dispersions, the extra amount of resin solution required to convert the wet-point to the flow-point paste being identical, about 2·5–3·0 ml per 100 g of pigment. The difference between the pigments is in their behaviour approaching the wet-point stage, where the conventional pigment gives a paste having about 52% and the heavily coated grade gives a paste of about 35% p.v.c. The difference in medium requirements is an indication of the difference in the surface structure, the conventional pigment being relatively smooth whereas the heavily coated grade has a very fissured surface, in which a large proportion of the available medium is immobilised. Once the fissures are filled the particles behave as though their surfaces were smooth, the amount of extra medium needed to convert the wet-point to a fluid being similar to that required by the conventional pigment, since the basic particle size and size distributions are similar, being governed by the size of the titanium dioxide core.

Errors in ranking pigments for ease of dispersion can be avoided if the rheological characteristics of the paste during the flow-point determination are observed. The following classification is that of Daniel and Goldman.[75]

Class I: Good dispersions

(i) At the wet-point
   (*a*) shine without or upon only light tapping,
   (*b*) are dry and hard to knead.

(ii) At the intermediate stage (defined as the mixture containing about a tenth to a fifth more liquid than the amount required to make the wet-point paste, but not having reached the flow-point)
   (*a*) lack a distinct intermediate stage.

(iii) At the flow-point
   (*a*) flow without tapping,
   (*b*) offer considerable resistance to sudden pressure, *i.e.* behave as dilatant systems.

(iv) Differ only slightly in composition at the wet- and flow-points.

Class II: Fair dispersions

(i) At the wet-point
   (*a*) shine on sharp tapping.

(ii) At the intermediate stage
   (*a*) flow when tapped or stirred gently,
   (*b*) may offer resistance to sudden pressure.

(iii) At the flow-point
   (*a*) fall, with some elongation at the breaking line,
   (*b*) show no resistance to sudden pressure,
   (*c*) show visible thixotropy.

Class III: Poor dispersions and flocculant systems

(i) At the wet-point

(*a*) remain dull even when tapped.

(ii) At the intermediate stage

(*a*) rise on tapping, so that the surface becomes rigid,

(*b*) show no resistance to sudden pressure,

(*c*) possess marked plasticity (high yield value).

(iii) At the flow-point

(*a*) fall without elongation at the breaking point.

(iv) Require a considerable additional volume of medium to convert the wet-point into the flow-point paste. (In the worst cases, addition of extra medium seems to cause scarcely any reduction in the paste consistency.)

Rheological aspects of dispersion

The rheological characteristics of a typical pigment dispersion can be seen to change markedly as milling proceeds and therefore successive rheological measurements should give a measure of the progress of dispersion. Furthermore, there must be an optimum condition for the operation of each type of mill, and this condition should be definable in rheological terms hence leading to a mill base formulated on a sound scientific basis. However, this concept involves many problems.

The production of pigment dispersion in bulk, as for example in paint manufacture, requires the mixture of pigment and medium to have flow properties which are suited to the type of machine used to make the dispersion. The appropriate mill base consistency may be very high for a heavy duty dough mixer or very low, for a kinetic dispersion machine. In all cases the formulator is faced with the problem that the consistency, and therefore the power required to drive the mixer, follows a standard pattern of a rapidly developed high initial value, which then decreases as the pigment is dispersed. The consistency may also be reduced as a result of the heat developed in the mill base by the mechanical work done on it. The flow of pigment medium mixture, in response to an applied stress, can be measured and expressed graphically either as the rate of shear induced by a shearing stress, or as the viscosity/rate-of-shear relation.

Rigorous mathematical analysis of practical systems is not usually feasible; the variety of particle shapes and sizes, and the fact that the continuous phase of the dispersion is often a colloidal system in itself provide two almost insuperable problems and a third arises from the nature of the dispersion process. Acceptance of the thesis that dispersion is achieved by the acquisition of an adsorbed layer on the pigment surface requires also acceptance of the concept that material is selectively removed from the continuous phase and added to the discontinuous phase, with a resultant change in the proportions of both phases.

The result of these problems is that study of the rheological characteristics of dispersions usually consists of the measurement of those characteristics at intervals during the dispersion process and consideration of the changes in the observed characteristics. These observations are then interpreted by reference to the observed behaviour of simpler systems or to mathematical models. The precise value of studies of the rheology of dispersions, as a means of studying the dispersion process itself, is therefore arguable. Changes in rheological characteristics may mark the progress of the dispersion process, but do not provide information about the processes involved in achieving that stage. The remarks that follow are therefore only postulates, for the most part.

Pigment in the dry state is usually agglomerated, the particles cohering as a result of van der Waals or London forces, and the cohesive energy increases as the size of the particles decreases. When a liquid, in limited amounts, is incorporated into a mass of dry pigment, there is some reinforcement of the point and line contacts between particles by liquid bridges which are held in place by capillary forces. This has been discussed by Rumpf[78] and other workers, including Pietsch.[79] The effect is not very important in pigment dispersion, since the reinforcement of inter-particle adhesion only occurs below a critical ratio of liquid volume to the void volume within the particle mass, and this ratio is quickly exceeded if pigment is added to an agitated liquid, which is the more usual procedure in modern paint and ink manufacture. However, the effect almost certainly helps to increase the initial power requirements above that needed to overcome the cohesive forces between dry pigment particles and inter-particle friction.

Once the air has been displaced and the particles have been brought into contact with the liquid phase, the rheological behaviour of the system will usually fall between two extremes, characteristic of highly dilatant and highly flocculant systems.

A *dilatant* system is one in which the particles are arranged in a way such that the solid volume fraction is at or near a maximum value and any attempt to introduce flow in the system requires the particles to move further apart before they can move. Reference to the dimensions of close-packed arrays given in Table 8.1 indicates that laminar slip-flow can take place in both horizontal and lateral planes in a cubic array, but in only one plane in a rhomboidal array, particle layer separation being necessary for it to occur in the other. In a rhombohedral array, separation in both vertical and horizontal planes is necessary before slip or flow can occur.

The greater the speed with which the force is applied, the greater is the apparent resistance to movement of the system. For this reason a typical dilatant system may flow slowly if its container is tilted but will become rigid and dry in appearance, possibly crumbly, if any attempt is made to stir the paste. Practical mill base formulations are never compounded in such proportions (maximum p.v.c.), but still behave in a manner which is variously described as dilatant or pseudo-dilatant. Much confusion can be avoided by describing their behaviour as *shear-increasing viscosity*.

Tacit in the requirement of the above types of disperse system is that the disperse phase, the pigment particles, must be in a state where minimal particle–particle contact exists, each particle having a surrounding envelope of the continuous, liquid phase, that is, the particles must be *dispersed*. At the other extreme is a system in which each particle is in contact with as many of its surrounding particles as possible, for example a theoretical maximum of twelve for uniform spheres, though an average value of 8–9 contacts is more likely in practice. In the case of pigments, and especially mixtures of pigments having widely different shapes (for example, mixtures of natural calcite, which when ground to make a pigment has roughly cubic particles, and mica, which has a lamellar structure and is ground into platelets, or talc, which has a highly acicular, more or less fibrous structure), it is often found that the pigment paste, at rest, has a very high consistency, which falls quite rapidly when the paste is agitated. This behaviour is produced by the formation in the paste of a more or less rigid skeletal structure of pigment particles having point or line contacts with each other, so that the solid volume fraction of the system may be at or near the minimal value for the system, if a given minimum number of contacts between each particle and its neighbours is specified. In such a system interparticle attraction should exceed the attraction between the particle surface and the medium, but in practice a skeleton of this type may be induced in a system by adding a material, such as water or a suitable surfactant, in sufficient quantity for effective particle–particle contacts to be created by bridges of the additive. Too little additive may have a negligible effect, too much may produce a system in which the particles are dispersed by the action of adsorbed additive, in a continuum of the medium which purports to be the dispersant, but which, in fact, had been unable to perform its task without the further additive.

The formation in a pigment/liquid mixture of a skeletal structure of contacting particles from particles that previously were mostly separate whilst the mixture was agitated, is the preferable definition of the process of flocculation. A high consistency, or structure, induced in a pigment/liquid mixture by the network of particles is best described as *flocculation-induced viscosity*, leaving the term *thixotropic* to be used only for systems which have been tested and found to meet the requirements for such a system, as originally defined by Freundlich.

The term *thixotropy* was coined to describe suspensions in which the solid particles were mutually attractive and in the absence of dispersing forces formed clusters, or flocculates. When a shearing force is applied to the system, the flocculates are dispersed and the system does not display any sign of a residual yield stress. When the shearing force is removed, the structure of the suspension reforms, in a period of time characteristic of the particular system. Plots of rate of shear against shearing stress therefore display a hysteresis loop, the area of which is a measure of the degree of thixotropy of the system.

In the case of pigment suspensions in resin solutions, where the volume concentration of the solid is high, true thixotropic behaviour is rarely exhibited. In most cases the systems are of the kind called *false-bodied*[80] or *thixotropic plastic.*[81] These are suspensions which do not flow until a certain value of shearing stress is exceeded and which retain a finite yield stress even after agitation. In a very few cases, if any, do moderately concentrated suspensions of pigments in resin solutions behave as *Bingham plastic*[82] bodies, having a simple linear relation between rate of shear and shearing stress, once the yield is exceeded.

THE DISPERSION PROCESS IN PRACTICE

The breakdown of aggregates and agglomerates

An aggregate is defined as consisting of particles that are in face-to-face contact, so that the force required for breakdown of an aggregate approximates to that needed for fracture of a crystal. According to Rittinger, the force required must be proportional to the amount of new surface produced, and Rittinger's number is the amount of new surface produced per unit of mechanical energy absorbed by the material. Some values for materials of a pigmentary nature are as follows:

	Specific Gravity	*Rittinger's Number (cm^2 of new surface per kg-cm of energy applied)*	*Hardness (Moh Scale)*
Quartz (SiO_2)	2·66	17·56	7
Sphalerite (ZnS)	4·0	56·2	3·5–6·0
Calcite ($CaCO_3$)	2·71	75·9	3

The Moh scale of hardness is based on the ability of material with a given number to scratch any material with a lesser number. The Moh hardness is *not* a good indicator of the resistance of a material to crushing unless the material is to be crushed very fine, which is the case where pigment comminution is concerned. The Moh hardness *is* a good indicator of abrasivity and so provides a good indication of the probable degree of mill wear.

Assuming that the material is crushed from uniform cubes of 100 μm side length to cubes of 1 μm side length, the new surface created is $5{\cdot}94 \times 10^{-2}$ cm^2 per cube, and the nominal energy needed to crush 1 kg of such cubes of quartz and calcite would be 0·0466 and 0·0106 horse-power hours per kg respectively. To split a cube into halves the energy would be 11.18 ergs and 2·59 ergs respectively for quartz and calcite. For a ball mill using high density ($\rho = 3{\cdot}4$) alumina balls of 2 cm diameter in a mill base such as that in Table 8.5, the force available is that provided by a ball descending under the influence of gravity, $F = mg$, where m may be taken as either the true or the effective mass of the ball. In the latter case, $m = V(\rho_1 - \rho_2)$ where V is the volume of the ball and ρ_1 and ρ_2 the densities of the ball and mill base respectively. The two values of m represent the cases of a ball on the face of the mill charge and one just inside it. The value of m would be 14·3 gm for the free-falling ball and 4·96 gm for the ball immersed in the base, but not influenced by any other part of the charge. The force available would be 14 000 dynes and 4600 dynes respectively, so an impact of 12 ergs would only require the ball to 'fall' approximately 10 μm or 25 μm.

Whilst the breakdown of aggregates in a suspension is within the capability of a ball mill, the task can be carried out more efficiently by the use of a dry-grinding process, preferably as a stage in the manufacture of the pigment and not in the making of a dispersion. If the use of an only relatively coarse pigment is being considered, it should be remembered that the fracturing of an aggregate is only a statistical probability and the probability of fracturing a particle of optimum pigmentary size is rather greater, since these will be present in much larger numbers.

In practically all wet milling processes, the liquid component is to some extent a surface active agent, and the rate of breakdown of aggregates should therefore be accelerated by the Rehbinder effect (*see* Chapter 1). The effect of surfactants on the rate of breakdown is difficult to establish with certainty. There is an optimum type of mill base rheological characteristic at which the highest milling efficiency is attained. The characteristic is transient because the rheology of a suspension must change as the size distribution of the pigment phase alters, and as the degree of interparticle attraction is reduced so that a truly dispersed state is achieved. If the latter state involves selective absorption from the liquid phase there will also be progressive changes in the rheology of the liquid phase itself, an increase in the hydrodynamic volume of the solid phase and a compensating decrease in the effective volume of the continuous phase. A surfactant is usually added to a suspension because it causes a visible change in the rheological characteristics, which also increases the rate of dispersion which changes the rate at which the rheological characteristics of the mill base change. In any typical mill base formulation, deciding whether a surfactant increased the rate of dispersion by causing aggregates to break because of the Rehbinder effect, or because it changed the rheological characteristics of the mill base in a way beneficial to milling efficiency, is a 'chicken and egg' problem.

Agglomerates are held together by weaker forces than those of aggregates, and should be disrupted by shearing forces and not require impaction. In a ball mill breakdown of agglomerates should be swift. In other mills this will only occur if the mill action and the mill base formulation are matched, so that the strongest possible shearing forces are developed. This in its simplest form means that the use of high impeller speeds permits the use of low viscosity suspensions, but low impeller speed necessitates the use of a viscous suspension. Unfortunately suspensions are not Newtonian fluids so a high impeller speed may induce local pseudo-dilatant behaviour, while a low impeller speed may not be great enough for the yield value of a structured suspension to be exceeded. In high-speed impeller mills both types of behaviour may occur simultaneously; the rotor initially impels the mill base which then 'sets-up' so that the rotor is soon spinning in a hole that it has formed in the mill base, the walls of the hole being formed by a nearly rigid 'dilatant' suspension. The small amount of mill base relaxing and flowing back to the impeller is thrown back at the wall with sufficient momentum to maintain the dilatant state. Away from the impeller zone the mill base may be quite stiff due to massive flocculation of the pigment, causing a very high viscosity. This kind of situation can be avoided by using the formulating technique described by Daniel,[77] which has been found to give a better indication of the optimum mill base formulation than the method proposed by Guggenheim,[83] which does not allow for the influence on the dispersion process of the intrinsic pigment dispersing power of the milling medium.

Pigment wetting in practice

Typical formulations used for efficient ball-milling of pigments in alkyd resin media are given in Tables 8.5 and 8.6 for rutile and carbon-black respectively. These pigments have been chosen as representing a typical and a very small particle size pigment, since the wetting of large particles does not present problems (usually) and that of pigments with a wide range of particle size may be considered as the wetting of a series of pigments each of uniform size, covering the range.

In the case of the titanium dioxide the pigment surface accessible to the fluid is about $9{\cdot}5\ m^2\ g^{-1}$, significantly less than the total surface area accessible to nitrogen ($12\ m^2\ g^{-1}$). The thermodynamic work done in the wetting process is therefore quite small, relative to that involved in the wetting of the carbon black. The bulk density of the dry pigment is equivalent to a solid volume fraction of about 0·25, indicating an open structure with less than four interparticle contacts per particle. (Four interparticle contacts per particle, as in diamond, give a solid volume fraction of 0·330; a number of possible open packings of spheres with less than

TABLE 8.5

Pigment: Coated rutile titanium dioxide
Bulk density: Approximately 1 kg/litre
Mean particle size: 220 nm (Electron microscope, graticule size-count)
117 nm (From total BET surface area, 12·5 $m^2 g^{-1}$)
154 nm (From corrected BET area, 9·5 $m^2 g^{-1}$)

Ballmill Base Formula

	Weight %	*Density*	*Volume*	*Volume* %
Rutile	80	4·1	19·55	44·6
Alkyd (63% FA)	4	1·06	3·78	8·6
White spirit	16	0·78	20·51	46·8

Gross volume of dry pigment = 80 units
Void volume in dry pigment = 60·45 units
Volume of liquid = 24·29 units
Pigment volume concentration in oil absorption paste = 55·8%

four contacts per sphere have been described by Heesch and Laves.[84]) The diameter of the capillary passages between the dry particles can be estimated by considering the pigment as an 'expanded' array of uniform spheres. For random close packing ($F = 0{\cdot}64$) the centre-to-centre spacing of the particles would be 1·37 diameters, or using the value for closest packing derived from the oil absorption paste composition ($F = 0{\cdot}56$) it would be about 1·3 diameters. From these values, the mean pore diameter, calculated from the cross-sectional area of the space between four adjacent 220 nm diameter particles, would be about 260 nm or 235 nm respectively.

TABLE 8.6

Pigment: High colour channel black
Bulk density: Approximately 95 g/litre
Mean agglomerate size: 9 nm (Equivalent circle diameter, from electron micrographs)
Mean particle size: ca. 3 nm (Estimated from electron micrographs)
3·5 nm (From BET total surface area)

Ballmill Base Formula

	Weight %	*Density*	*Volume*	*Volume* %
Carbon black	10	1·8	5·55	5·0
Alkyd (63% FA)	30	1·06	28·3	25·5
White spirit	60	0·78	77	69·5

Gross volume of dry pigment = 105 units
Void volume in dry pigment = 99·5 units
Volume of liquid = 105 units
Pigment volume concentration in oil absorption paste = 9·85%

(Using the smaller particle diameters derived from surface area measurements would significantly reduce these values.) The minimum value for the mean pore diameter (calculated in the same way but assuming that the four particles were in contact) would be 102 nm. These values have no experimental foundation and serve only to provide some reasoned estimate of the magnitude of the pore sizes.

Using Washburn's equation (1.14, given earlier) and assuming a mean capillary radius of 80 nm, a liquid viscosity of 5 centipoise and that $\gamma_{L/V} \cos \theta = 30$ dyne/cm, the rate at which the fluid would penetrate the capillaries in the pigment mass can be calculated, and is found to be 0·38 cm/minute.

In practice, adequate wetting of the pigment particles reduces the interparticle friction that maintains the voluminous structure of the dry pigment. As soon as the interparticle attraction exceeds the friction, a collapse takes place. Gravity alone cannot overcome interparticle friction; if the wetting liquid penetrates the mass of pigment from below then, under static conditions, the wetted pigment may collapse leaving an arch of dry pigment above it and out of reach of the liquid. If penetration is spontaneous, the reduction in pore diameter in the collapsed pigment increases the capillary attraction (Jurin's law) and so partially compensates for the reduced rate of penetration, provided the air can escape freely from the mass of pigment.

The agglomeration of pigment particles in the dry state is a result of van der Waals or London forces, and the cohesive energy is inversely proportional to the particle diameter. If only a limited quantity of liquid is incorporated into a mass of dry pigment, then the point and line contacts between particles may be reinforced by liquid bridges that are held in place by capillary forces. The effect is not very important in pigment dispersion, except in such special cases as the over-rapid charging of pigment to a horizontal trough mixer or pug-mill. The very sharp peak in power demand that occurs when pigment is added to medium in a mixer of this type, and that can be demonstrated very readily using a Brabender Plastograph or a similar type of instrument, may be partially contributed by this effect. Fortunately the reinforcement of interparticle adhesion only occurs below a critical ratio of liquid volume to the void volume within the pigment mass, and this ratio is quickly exceeded if pigment is added to an agitated liquid at a sensible rate.

Ideally the pigment should be sifted into the wetting liquid, which should be agitated so that the largest possible proportion of the pigment immediately contacts the liquid. As the pigment falls on to the liquid, penetration of aggregates and agglomerates by liquid can occur easily, the displaced air escaping from the dry upper surface of the pigment.

In less than ideal circumstances, large lumps of pigment may be immersed in the liquid. Particularly in the cases of rutile, chrome yellows or iron oxides it is possible for the outer layers of the lump to be wetted, and to collapse to give a much denser layer of pigment in nearly a close-packed arrangement. The capillary attraction of the liquid into the lump of pigment is then counterbalanced by the pressure required to force air out of the lump, through the dense wetted pigment layer, while the hydrostatic head of the liquid over the lump also opposes the escape of air. Unless the lump is broken by mechanical action little further wetting will occur and the contracted outer layers will exert a considerable compressive force on the dry core.

The formation of partially wetted pellets of pigment is particularly likely to occur during the preparation of pre-mixes for sand mill or bead-mill dispersion, if the long-established correct procedure for mixing pigment and medium is ignored. In one case dumping 1000 kg of titanium dioxide, in 25 kg (bag) lots, into about 600 kg of a 16·7% solid content solution of a short oil alkyd in xylol-butanol, stirred slowly by a simple propeller-type agitator, resulted in the formation of about 30 kg of pellets of diameter 0·5–2·0 cm. The pellets completely blocked the protective grill on the inlet side of the pump feeding the sand-mill, and stopped production. Formation of pellets was prevented by reducing the initial charge of liquid and alternating the addition of pigment and liquid so as to keep the pre-mix paste as stiff as the agitator could move, finally thinning the paste (carefully) to the required composition before milling. The extra time required for pre-mixing was more than compensated for by the increased pumping rate at which an acceptable level of dispersion could still be obtained.

A second type of pigment agglomeration is caused by complete incompatibility between the pigment and wetting fluid. Such behaviour can be demonstrated by exposing a quantity of titanium dioxide to an atmosphere saturated with water vapour. The pigment containing about 1% of adsorbed water, when trundled with about 2 volumes of benzene rapidly forms extremely hard, small (about 2 mm diameter) pellets. As trundling continues the pellets contract slightly until their surfaces become visibly indented, suggesting either that the pellets are formed initially as agglomerates of small pellets which then collect still smaller pellets and/or primary particles, or that water is displaced from within the pellet, causing shrinkage. With toluene or xylene the pellets are usually smaller and of lower density; with lower primary alcohols a stable dispersion is formed.

The converse process in which a titanium dioxide previously coated with, for example, a fatty acid is pelletised by trundling in water has not been observed; perhaps all the coated rutile surface is not available to the acid for adsorption, hence the desired completely hydrophobic surface is not produced. Treatment of the pigment with a methyl silane to replace hydroxyl by methyl groups might produce a surface more analogous to that of the moist pigment.

The wetting of inorganic pigment particles of moderate size (200 nm diameter and above) is a relatively simple process compared to that of material such as channel black (Table 8.6). The black particles look like fused aggregates of spheres and the variety of particle shapes and sizes is much greater than that of titanium dioxide. The electron microscope (EM) particle size is the diameter of the circle of equal area to the projected image of the particle and therefore analogous to the EM size of the titania, as measured using an 'equal areas' graticule. The surface area measured by nitrogen adsorption includes a considerable area, contributed by the walls of pores and fissures, that is probably inaccessible to most if not all of the molecules that constitute the liquid phase. The high (16%) 'volatile' content and low pH value of the black indicate that the oxy-carbon complexes that form the pigment surface contain a fairly high number of carboxylic groups per unit area. Given the opportunity, the pigment could adsorb a considerable amount of atmospheric moisture. (A monolayer of water on the whole of the nitrogen surface area would amount to about 25 g per 100 g of pigment.)

The bulk density of the dry black is very low, 0·096 g/cm^3, equivalent to a solid volume fraction of 0·053, indicating a very open structure with an average number of interparticle contacts that is not very much above the necessary minimum of two per particle. The structure of the dry black may therefore be visualised as random coils of particle chains. An estimate of the mean pore size in the dry black can be made by considering it in terms of a random close packed array of uniform spheres, 'expanded' to reduce the bulk density to the observed value. The necessary centre-to-centre spacing of the spheres would be 2·9 diameters, the cross section of the open space between four adjacent spheres being about 617 nm^2, equivalent to a mean (circular) pore diameter of 28 nm.

The high oil absorption value, 475 g oil/100 g pigment, corresponds to a p.v.c. in the paste of only 9·8%. Calculating, as before, gives a centre-to-centre spacing of 1·87 diameters or if each particle is assumed to have a monolayer of adsorbed oil, of 1·37 diameters. (These figures cannot be used to assess the volume of oil occluded by each particle; the figure obtained can vary from about 1 to 6 particle volumes, depending on the initial assumptions.) For the 1·87 diameters (16·8 nm) separation, the mean pore diameter, calculated as before, would be about 17 nm.

These calculated pore diameters are likely to bear as little relation to reality as do those calculated for titanium dioxide; they merely provide an indication of the order of magnitude.

The rate of penetration of the resin solution into the dry black will be slower than in the case of titanium dioxide; assuming a pore radius of as much as 40 nm (half that assumed for the titania), the additional effect of the higher viscosity, about 15 centipoise, of the more concentrated resin solution will further reduce the rate of penetration, to 0·6 cm/min.

The wetting of the black is likely to take place in two stages, the first consisting of the penetration of the main pores through the mass of pigment and the second of penetration into the interstices of the individual particles. This second stage is likely to be slow, partly because the trapped air will have difficulty in escaping; some form of mechanical milling action is probably essential for this stage to proceed to completion. The expansion of this displaced gas as the ball mill contents are heated by the mechanical energy expended, makes it necessary to vent the mill after the first few hours running. (This practice is advisable with all pigments, but especially so in the case of carbon blacks, Prussian (iron) blues and similar materials.) The practice of loading a ball mill with a premixed slurry has the advantage that almost all the air has been expelled from the mill base and no milling time is lost in 'wetting out' the pigment in the ball mill.

Stabilisation of pigment suspensions

A stable disperse state requires the particles in a suspension to be sufficiently widely separated for the repulsive force (electrical or steric; *see* Chapter 1) to exceed the van der Waals attractive force. Particles in the colloidal size range are subject to Brownian movement, so that a 'buffer' is needed if flocculation is to be avoided. In non-aqueous fluid suspensions it is expected that adsorbed material on the pigment surface will serve this purpose.

Crowl[85] has published data for the attraction potential V_A between various sizes of particle, with and without adsorbed layers. By assuming that the particles are uniform spheres, with a closest random packing solid fraction of 0·64, the value determined experimentally by a number of workers, it is possible to calculate the maximum pigment volume concentration in a suspension that will permit adequate separation between the particles for the suspension to be stable. For this purpose it is assumed that at a value of $V_A = -5kT$, flocculation will not occur. An adsorbed layer of resin on the particle surface will reduce interparticle attraction, but to obtain the requisite separation between the outer surfaces of the adsorbed layers, there must be a corresponding increase in the centre-to-centre separation of the particles. This enforces a corresponding reduction in the maximum possible p.v.c. compatible with this minimum particle separation.

In Table 8.7 are given the maximum permissible volume concentrations of spheres, calculated from data given by Crowl for the attraction potential/particle separation relationship calculated using Vold's[86] equation, and assuming that the closest possible packing of the uncoated spheres has a solid volume fraction of 0·64 (64%). The (sphere + adsorbed layer)/sphere volume ratios are also given. These values cannot be converted to weight ratios without assuming a value for the density of the adsorbed layer, a value that in practice is influenced by the resin solution composition, the configuration of the resin in the adsorbed state, and the quantity of solvent associated with the adsorbed resin.

The effect of an adsorbed layer on the effective volume of particles of less than 200 nm diameter is very marked, and causes a severe limitation on the permissible concentration of such particles if a reasonably low interaction potential is to be achievable. The use of larger particles permits higher volume concentrations, but the effect of gravitational forces on these should be taken into account. Given sufficient time, these particles would settle and form a dense tough sediment, of very dry appearance and very difficult to disperse.

TABLE 8.7

THEORETICAL MAXIMUM VOLUME CONCENTRATION OF UNIFORM SPHERICAL PARTICLES IN DISPERSE SYSTEMS

Particle Diameter nm	*Adsorbed Layer Thickness nm*	*(Particle + Layer)/ Particle Volume Ratio*	*Separation Between Surfaces*		*Maximum p.v.c. in Suspension (Closest Packing = 64% p.v.c.)*	
			at −5kT nm	*at −10kT nm*	*at −5kT*	*at −10kT*
40	5	1·953	0·40	(0·15)	32	(32·4)
100	5	1·331	0·82	0·32	47·1	48
200	5	1·158	1·6	0·63	54·1	54·8
400	5	1·077	4·9	1·6	57·3	58·7
1 000	5	1·030	18·8	7·2	58·8	60·8
2 000	5	1·015	41·6	18·4	59·3	61·3
200	0	1·000	4·8	2·6	59·6	61·6
	2·5	1·076	2·6	1·0	57·3	58·6
	5	1·158	1·5	0·82	54·1	54·7
	7·5	1·242	1·2	0·82	50·7	51·0
	10	1·331	1·0	0·60	47·4	47·7
1 000	0	1·000	25·0	12·2	59·5	61·7
	2·5	1·015	21·0	9·2	59·3	61·3
	5	1·030	18·5	6·8	58·9	60·9
	7·5	1·045	16·0	5·4	58·4	60·3
	10	1·061	13·9	4·4	57·9	59·5

Various workers have reported values for alkyd resin adsorption on titanium dioxide, mostly of the order of 2–3 mg/m^2, but these values are for adsorption from dilute solutions. More recent papers, by Rehacek[87] and by Goldsbrough and Peacock,[88] give results of adsorption from more concentrated resin solutions. These suggest that in the case of the titanium dioxide millbase (Table 8.5) adsorption of resin would be about 2 mg/m^2, but that this would be associated with 3–4 mg/m^2 of associated solvent. This gives a total adsorption of 1·52 g alkyd and 2·28 g solvent on the 80 g of pigment, assuming only the accessible surface (9·5 m^2 g^{-1}) to be coated. In volume terms, this would increase the volume of the disperse phase to 23·87 ml and reduce the continuous phase volume to 19·92 ml, that is, increase the effective p.v.c. of the suspension from 44·6% to 54·5%. In practice, it is likely that the milling action would remove adsorbed material from the pigment almost as soon as it was attached, but the increase in effective p.v.c. would occur once the mill was stopped; this increase in p.v.c. would account for the rapid increase in apparent consistency of the millbase when left undisturbed and provides a more reasonable explanation than 'cooling-out', a very slow process for a mass as great as that of a large pebble-mill and its charge. (Pebble-mills cannot be water-cooled, unlike steel ball-mills.)

The increase in volume of the millbase may also be examined in terms of average particle diameter increase. The enlarged spheres would be 1·07 times the mean diameter of the pigment, equivalent to an adsorbed layer 7·7 nm thick on the mean particle size (220 nm) determined by electron microscopy, and allowing a maximum separation, in terms of uniform spheres, of 12·6 nm between the outer surfaces of the adsorbed layers. From Crowl's data, this would indicate an attraction potential of only just over $-1kT$, so that the dispersion should be quite stable.

In the case of the carbon black dispersion (Table 8.6) deductions about the probable stability of the millbase are much more difficult to make. One reason is that a quantity of the medium, about four times the solid volume of the black, will be occluded by the pigment, with or without any adsorption taking place. Also, the maximum possible separation between the aggregates that are the pigmentary particles, will be correspondingly reduced. The volume concentration in the millbase of aggregates plus occluded medium will be about 25%, but the corresponding hypothetical closest packing (derived from the oil absorption paste composition), would have a solid volume content of 49%, giving a maximum separation of 20 nm between pigment aggregates with an average diameter overall of 80 nm. However, the Vold equations indicate that particles of that order of magnitude would have very small attraction potentials, even without adsorbed resin layers, at separations as small as 5 nm.

Work such as that reported by Hess and Garrett[62] supports the view that a highly oxidised (16% volatile) black of this type would adsorb a large amount of the medium and probably induce a considerable degree of orientation of the polymer in the vicinity of the pigment particles. The very marked structural viscosity of the system at rest and fairly rapid reduction in apparent viscosity with increasing shear rate that characterise dispersion of this type of black are in accordance with these hypotheses.

The influence of an electrical repulsive force on the stability of the dispersions has not been considered. The zeta potential is to some extent affected by the chemical composition of the pigment surface, that affects the kind of material adsorbed. However, if the adsorbed molecules are large, entropic effects are likely to be more influential than the zeta potential of the particles.

The probable stability of a dispersion can be assessed from its composition, or rather, from the relative proportions of its components and taking into account the adsorption characteristics of the pigment, molecular weight distribution of the polymer and the nature of the solvent. However, the intrinsically stable dispersion may, largely because of its high pigment content, develop a very high structural viscosity if left undisturbed, and be very sensitive to small changes in composition, such as solvent loss by evaporation. For these reasons it is usual to dilute the dispersion with medium, to reduce the pigment content and, often, to increase the intrinsic viscosity of the suspension. (This in itself, by reducing pigment mobility hinders pigment flocculation and therefore promotes dispersion stability.)

The stabilisation or 'let-down' stage of dispersion manufacture is apparently simple, but contains pit-falls for the unwary. The failure rate is sufficiently high for the term *colloidal shock* (sometimes called pigment shock or resin shock) to have achieved international status in the jargon of paint and ink technologists. Two cases of 'let-down' will be considered.

'Let-down' of pastes prepared by roll-milling or heavy-duty mixing

Pigment dispersion by heavy-duty mixing or roll-milling requires a paste of high consistency and internal cohesion, so that the powerful shearing forces act throughout the mixture of pigment and medium in the shearing zones, between the inter-meshing blades of the mixer or in the nip of the rolls. A medium that is an effective dispersant for the pigment is an essential for satisfactory results. In the case of roll-milling, the paste fed to the rolls must also have good 'tack', to enable the paste to adhere so strongly to the rolls that it is drawn through the roll-nip. A flocculent paste cannot be roll-milled satisfactorily; its apparent consistency in the feed-hopper is high but the mill through-put is extremely low. The paste that is collected on the mill apron is of thin consistency, either due to temporary dispersion of the pigment or, more often, to a reduction in the pigment content. The material in the feed-hopper, in the latter case, progressively increases in pigment content and if it is eventually drawn through the rolls, is likely to appear as hard flakes of almost dry pigment on the take-off apron. This type of behaviour is not kin to that involved in 'shock'; the most satisfactory cure for it is reformulation of the product, or at least of the millbase, to provide a suitable medium for pigment dispersion.

In some cases, most often heavy-duty mixer pastes, the pigment : medium ratio may be high and the medium barely able to produce a stably dispersed system. If such a paste is thinned with a liquid lacking in pigment dispersing capacity, the pigment may flocculate at once. If, on the other hand, the paste is diluted with a solvent that has intrinsic pigment dispersing capability (*e.g.* a high dielectric constant) then pigment flocculation may not become apparent until the film-forming stage, after the paint or ink has been applied and the solvent evaporates. In this case also, reformulation of the product is the only satisfactory solution.

Any high consistency paste should be thinned slowly, to minimise the risk of the paste breaking up into lumps that are subsequently penetrated by solvent at a rate so slow as to be negligible. Ideally the thinning liquid should be added at such a rate that only a thin film spreads on freshly exposed paste surfaces that are then folded in on themselves, so that the thinner is compressed within an envelope of paste. This reduces the time that a marked difference in composition of thinner and paste exists.

Pastes with a high content of pigment (heavy-duty mixer formulations) preferably should be diluted with more of the medium used in the paste, to avoid the formation of interfaces between liquids of different composition. The rate of addition of the diluent should be slow; in practice the 'let-down' stage for this type of dispersion may take several times as long as the 'dispersion' stage. 'Shock' in systems of this type is an infrequent occurrence and is almost always due to incompetence.

Under the conditions existing in the roll-nips or the zones of maximum shear between the blades of a heavy-duty double-sigma bladed mixer, it is unlikely that any resin is adsorbed on the pigment or if adsorbed, is more than momentarily retained. The greater part of the adsorption process must therefore take place in the paste on standing, after the 'dispersion' stage has ended and usually also after the 'let-down' stage. The dispersion process itself is therefore substantially one in which the pigment particles are mechanically separated and reflocculation prevented only by the viscous hindrance to particle movement. In this situation, the reduction of the paste viscosity on 'letting-down' enables the flocculation process to start at the same moment that adsorption of resin by the pigment can take place easily. If the liquid added merely reduces the paste consistency, then the pigment to resin ratio in the original paste may be too high for a stably dispersed system to exist and flocculation is bound to occur. The time for flocculation to become apparent will be longer, the greater the consistency of the diluted paste; a very viscous fluid may delay flocculation but cannot, of itself, prevent it. If the let-down liquid must be one with no pigment dispersing capability, inclusion in it of an appropriate surfactant may be possible, to provide some insurance against flocculation.

'Let-down' of high-pigment, high-solvent, low-resin content dispersions

The 'slurry-grinding' technique is attractive in that it is the best method of milling if pigment aggregates must be broken down, and also, it gives the largest output of milled pigment per unit volume of millbase. For most purposes, it is necessary to reduce the pigment concentration and increase the consistency of the dispersion. It is in carrying out this stage that difficulty is most often experienced.

A typical millbase and 'let-down' solution for the dispersion of rutile titanium dioxide by ball-milling is given below.

	Mill Base		*'Let-down'*	
	Weight g	*Volume* ml	*Weight* g	*Volume* ml
Rutile	183·0	44·6	—	—
Alkyd	9·1	8·6	67·0	63·2
Solvent	36·6	46·8	28·7	36·8
	228·7	100·0	95·7	100·0

At the end of the milling cycle an equal volume of concentrated alkyd solution is added to the millbase in the ball mill and incorporated by milling for about an hour, according to the size of the mill. At equilibrium the composition is as follows:

	Weight		*Volume*	
	g	%	g	%
Rutile	183·0	56·4	44·6	22·3
Alkyd	76·1	23·5	72·3	35·9
Solvent	65·3	20·1	73·4	41·8

The mill could then be discharged and the millbase pumped into storage tanks pending further processing. Generally no problems due to sedimentation should occur even if the stored dispersion is not agitated for three or four weeks.

The trouble that may occur can be easily demonstrated by measuring into a glass jar 100 ml of the unstabilised dispersion of the above composition. On to this is poured 100 ml of the concentrated resin solution. The resin solution will penetrate the layer of dispersion, but is quickly forced upwards and a well-defined interface should be visible. If the jar is closed and allowed to stand, then after about 20 minutes it will be noticed that the level of the millbase has fallen slightly and a definite 'skin' formed on it, beneath the resin layer. After 24 hours the skin has become thicker and stiffer and, with a little care, can be removed intact. If, instead, the jar is shaken thoroughly, the 'skin' is broken up into small hard nibs that settle rapidly on standing; the nibs do not redisperse if left to stand, nor can they be very easily redispersed by milling.

The decrease in volume of the millbase indicates that solvent has passed from the dilute resin solution to the more concentrated. If solvent passed until the resin solution concentrations were equalised, then for the examples quoted the equilibrium compositions would be as follows:

	Mill Base			*'Let-down'*
	Weight	*Volume*		*Weight*
	g	*ml*	%	*g*
Rutile	183·0	44·6	70·6	—
Alkyd	9·1	8·6	13·6	67·0
Solvent	7·8	10·0	15·8	57·5

These values indicate that the postulated model is over-simplified; the pigment volume concentration markedly exceeds that of the closest packing value that it is reasonable to assign to the titanium dioxide, but the general indications are in agreement with the practical results.

The skin formation is therefore a function of the time that an interface exists between millbase and concentrated resin solution, and of the difference in concentration of the two solutions. The use of a dilute stabilising resin solution is desirable, but most important is the minimising of the time during which an interface can exist. For this reason if a large quantity, say 500 litres, of concentrated resin is to be added to a ball mill, it is best to do it in stages, since the time required to get in such a quantity of liquid, even by pumping, means that skinning is unavoidable. The time taken by the step-wise stabilisation is not as great as is usually necessary to completely redisperse the highly pigmented skin.

Osmotic transfer of solvent can also be reduced by using a solvent with low affinity (little solvent power) for the resin in the millbase and one with high affinity for the stabilising resin solution. But in most cases other restrictions, such as required evaporation rate, prevent the use of solvents sufficiently different in character to have a significant effect, if there is a considerable difference in the resin concentration of the two solutions.

PIGMENT DISPERSION IN THE FINAL PRODUCT

Paint film structure

The preparation of a dispersion is in most cases only a unit operation in the production of, for example, a paint or ink film, a sheet of plastic or a plastic moulding. The most relevant criterion is therefore the state of dispersion of the pigment in this final application, but it is only fairly recently that much interest has been shown in the direct examination of this state. Gordon[89] has discussed the use of permeability measurements on plastic films as an aid to elucidation of polymer structure, and the 'reinforcing' action of carbon black in rubber has been closely studied.[49,50] The work of Funke[90] on the factors affecting the absorption, permeability and diffusion of water by, into and through paint films may eventually permit a clearer understanding of the interaction of the various components of a disperse system in its 'final' (actually always transitory) state.

Most workers have used such properties as paint film gloss and gloss retention as indicators of the probable state of dispersion/flocculation of the pigment. Concurrently, numerous papers have been published on adsorption from alkyd solutions, or from drier solutions, but not from alkyd solutions containing the appropriate mixture of drier soaps (lead, manganese, zirconium and/or cobalt naphthenates or octoates, with or without calcium or zinc derivatives of the same acids). Yet it is well-known that the driers are adsorbed on the pigment surface in most paints, without, according to most workers, apparently affecting their efficiency as accelerators or catalysts of the drying process. Nor is there adequate evidence to decide, one way or the other, about the possibility of ion exchange reactions, such as cobalt displacing aluminium from the surface of a coated titanium dioxide. In any study of such reactions, considerable attention would have to be paid to the amount of free acid in the drier 'soap'; calcium naphthenates in particular vary tremendously in their free naphthenic acid content, which may explain why some brands are more satisfactory than others in preventing the precipitation of lead phthalate from driered alkyd solutions. This last item is a reminder that the interaction of the drier and the alkyd has been a rather neglected field.

The fundamental study of the drying process is in any case much neglected, though the increasing availability of modern analytical apparatus may change this situation. Some observations worthy of further investigation are the following:

(i) An 'air-drying' paint film, based on a typical long oil-modified alkyd, if stored in an inert gas atmosphere for about 24 hours immediately after application dries only extremely slowly on subsequent exposure to air.

(Presumably a free-radical initiated reaction is swiftly terminated by the lack of oxygen; by delaying access to oxygen, the conditions necessary for initiation of the ordinary drying reaction are superseded. Passing a current of inert gas through the storage chamber, to remove volatilising solvent, does not seem to affect the results.)

(ii) If an air-drying alkyd paint is dried in a stream of air and the volatile oxidation products collected in a nitrogen trap, for subsequent analysis, there is a marked difference in composition between those produced by drying in a dry air stream and those from drying in air of normal humidity.

(The effect of the moisture content of the pigment on this effect might repay investigation. The 'chalking' of anatase titanium dioxide pigmented films, that does not occur in the absence of moisture even under intense ultra-violet radiation, seems to have some curious similarities.)

Examples of oddities in paint properties that are more obviously related to adsorption on the pigment surface are provided by the behaviour of rutile titanium dioxide in long oil alkyds:

(iii) Two similar types of coated rutile, each dispersed in two fairly similar long oil alkyds, with identical drier concentrations etc., will give two pairs of paint films, in one of which the colour of that containing one pigment is superior to that containing the other, whilst in the other pair the rankings are reversed.

(iv) Any coated rutile/long oil alkyd based paint will take at least ten days before the initial colours of paint films prepared on consecutive days are similar. During the early part of this 'stabilising' or 'homogenising' period the films tend to become slightly less bright and bluer in shade, towards the end brightness may recover a little and the hue become a little yellower. The 'stable' colour is apparently always a little less bright and bluer than that of a film prepared from the paint 24 hours after its manufacture.

(v) The colour of a white gloss paint exposed to diffuse daylight indoors has a diurnal fluctuation superimposed on a seasonal variation and, in the case of media drying by oxidation, considerable progressive yellowing. The seasonal fluctuation is shown most clearly by weekly measurements of the colour of a film of a non-drying (*e.g.* lauric acid modified) alkyd/butylated melamine resin based baking enamel. The diurnal fluctuation is of constant pattern, overnight yellowing, followed by bleaching during the hours of daylight to a degree proportional to the brightness of the day.

(Street lighting, especially by mercury vapour lamps, may largely prevent the 'reversible' overnight yellowing, and this must be taken into account when choosing a site for the exposure of the panels.)

(vi) If two sets of films are prepared from a white baking enamel (alkyd/amino resin based), one set being given a longer time, lower temperature bake and the other a shorter time, higher temperature bake, it is usually found that the longer time, lower temperature bake gives superior colour. On interior exposure of the panels it will be found that one set will yellow and the other bleach until after a few months the influence of baking schedule has disappeared. The effect is observed with both short oil oxidising alkyds and non-oxidising alkyds, combined with either urea- or melamine-formaldehyde resins.

The various items noted above indicate that a paint film is a far from inert substance at the molecular level, even if it is not subject to the direct physico-chemical attack that causes film degradation on exterior exposure. One implication of this film characteristic is that both the ingredients of the paint, the film-formation conditions, and the history of the film should all be recorded in much more detail than is usual, if it is desired to draw valid general conclusions from a study of paint film properties. This is especially the case if film properties are used as a foundation for speculation about the state of pigment dispersion.

Certain observations can be made about the film-forming process that may seem trite, but are a necessary foundation to further discussion:

Consider a suspension of uniform spheres whose volume concentration in the suspension is 9·14%. The suspension has a volatile content of 47·8%, so that the volume concentration of spheres in the solid residue is 17·5%. If the closest possible packing of the spheres has only 54% solid content, and the spheres are 220 nm in diameter, then the mean distance between adjacent spheres will be 0·81 diameters, or 178 nm. If a film of suspension is spread on a flat rigid surface and allowed to dry, then the decrease in area covered can be considered negligible and the reduction in film volume can be considered as due entirely to a reduction in the height of the film normal to the substrate.

In this case the final mean separation between the spheres in the dry film *cannot* be 0·45 diameters (99 nm), as would be the case for uniform shrinkage in three dimensions. A close packing fraction of 0·54 (54%) is roughly equal to the 0·524 solid fraction of cubic packing of uniform spheres. Assuming uniform distribution of the spheres throughout the film of suspension, 40 000 nm thick when first applied, then the spheres will be arranged in vertical stacks, each containing 100 spheres with a separation of 178 nm between the spheres. If there is no lateral motion, then the spheres would move together to form a stack 22 000 nm high. But the total shrinkage, 47·8 % of the volume of the film of suspension, would give a mean dry film thickness of 20 880 nm. The surface of the dry film, in theory, would therefore be slightly indented between the heads of the columns of stacked spheres.

In practice, surface tension would cause lateral movement of the upper spheres, but the initial gap between adjacent stacks is less than one sphere diameter, so that spheres in each plane of stacks would tend to move laterally as well as normally to the plane. In an orderly displacement this could produce a rhomboidal packing array, with a solid volume fraction of 0·605. Random displacement would permit even closer packing, up to 0·64 solid volume fraction.

The most important conclusion from this consideration of a model structure is that a section through its dry film would inevitably give the impression that all the spheres were flocculated.

The values quoted, of 220 nm spheres, at 9·14 % volume content in the bulk suspension, 17·5 % in the dry film, with a closest packing fraction of 0·54, were deliberately chosen as approximating to rutile titanium in a typical long oil alkyd medium. The volume shrinkage is also of the correct order for an ordinary alkyd-based paint. A study of particle size distribution in sections of rutile pigmented alkyd paint films has been reported by Murley and Smith[91] who found that a 100–150 nm thick layer of medium, almost totally deficient in pigment, was formed at the upper face (paint/air interface) of the paint film, with a layer compensatingly rich in pigment just below the clear layer. They suggested that pigment near the paint/air interface was repelled from the paint surface as a result of the electrostatic 'image force' and deduced an equation for the distance travelled in a given time. This required, for simplicity, the assumption that the viscosity of the suspension did not change with time, an assumption greatly at odds with reality. The final form of the equation was

$$a^3 - a_0{}^3 = \frac{K_0 K_1 d\zeta^2 t}{4v} \frac{K_1 - K_2}{K_1 + K_2} \tag{8.7}$$

where

a = particle distance (metre) from boundary after time t

a_0 = particle distance (metre) from boundary when $t = 0$

K_1 = dielectric constant of medium surrounding particle

K_2 = dielectric constant of second medium (*i.e.* air)

K_0 = permittivity of free space, $8{\cdot}85 \times 10^{-12}$ farad/metre

ζ = zeta potential on particle (volt)

d = particle diameter (metre)

t = time for which particle moves (second)

v = suspension viscosity (Newtons per sq metre).

Using values of $K_1 = 3$, $K_2 = 1$, $d = 250$ nm, $\zeta = 30$ mv, $t = 600$ seconds (the time for 15% of the total solvent to evaporate) and a constant viscosity of 1 Nm^{-2} (10 poise) the equation gave a value of 760 nm for the thickness of the clear layer, about 5–7 times too great. Substitution of a value of 200 nm for the particle diameter, in accordance with particle size data given in their paper, only reduced the calculated clear layer thickness to 710 nm. Calculating back indicates that the viscosity assumed by Murley and Smith was much too small, a viscosity of 106 Nm^{-2} being necessary to give a calculated clear layer thickness of 150 nm for 200 nm particles.

This apparently high film viscosity is not too great to be acceptable. Solvent loss by the film requires a steady diffusion of molecules to the surface but, as shown by the demonstration of the mechanism of 'let-down shock', the rate of solvent diffusion from a dilute to a concentrated resin solution is rapid, so there is not likely to be a significant solvent concentration gradient in the paint film from slightly below the film/air interface to the substrate. After 10 minutes air-drying time, a normal gloss paint is just about 'set-to-touch' (though still capable of being lapped-in in a 'wet-edge time' test when the solvent-rich fresh paint dilutes the original coat). A tactile comparison with a heavy lithographic stand oil of 1000 poise viscosity suggests that a value of about 1000 poise for the upper layers of a paint in the set-to-touch state is quite reasonable.

An additional factor that should be considered also is the influence of the disperse phase on the viscosity of a suspension; as solvent is lost the disperse phase volume fraction in the suspension increases. Also, Murley and Smith found that at the advancing front of the clear layer the pigment concentration was about 2·5 times the average concentration in the film. Using eqn. (8.4), the relative viscosity of the suspension can be calculated for suspensions giving final dry film pigment contents of 10, 17·5 and 30% by volume. (The value of η_0, the viscosity of the continuous phase, also increases as solvent is lost.)

INFLUENCE OF PIGMENT VOLUME FRACTION ON SUSPENSION VISCOSITY DURING CLEAR LAYER FORMATION

Initial State			*15% of Solvent Gone*			*All of Solvent Gone*		
Resin in Soln. %	*p.v.c.* %	η/η_0	*Resin in Soln* %	*p.v.c.* %	η/η_0	*Resin in Soln* %	*p.v.c.* %	η/η_0
55	5·29	1·29	59	5·68	1·32	100	10·0	1·66
55	9·14	1·59	59	9·80	1·65	100	17·5	2·67
55	16·9	2·56	59	18·0	2·75	100	30·0	7·56

EFFECT OF PIGMENT CONCENTRATION AT CLEAR LAYER ADVANCING FRONT ON SUSPENSION VISCOSITY (INCREASE = 2·5 × MEAN P.V.C.)

Resin in Soln %	*p.v.c.* %	η/η_0
59	13·22	2·14
59	24·5	4·54
59	45·0	88·7

The paint used by Murley and Smith had a dry film p.v.c. of about 10%[92] so that relative clear layer thicknesses can be calculated for the above suspensions by assuming that pigment movement ceases after 10 minutes drying and that the estimated value of v for the suspension giving 10% p.v.c. in the dry film is either $(80{\cdot}4 \times 1{\cdot}32) = 106\ \text{Nm}^{-2}$, or $(49{\cdot}6 \times 2{\cdot}14) = 106\ \text{Nm}^{-2}$:

Dry Film p.v.c. %	*Clear Layer Thickness*			
	Using Mean p.v.c. Viscosity Factor		*Using Increased p.v.c. Viscosity Factor*	
	v	*nm*	*v*	*nm*
10	106	150	106	150
17·5	132·5	139	225	118
30·0	221	117	4 400	43

The lower value, 43 nm for the clear layer thickness of a 30% p.v.c. film compares fairly well with the value of 60 nm, determined by an interference method, for a 30% p.v.c. film of rutile in an alkyd medium, reported by Kawabata.[93] Wilska[94] has reported clear layers of 540–1000 nm in films of rutile pigment at 14% p.v.c. in both alkyd and acrylic media, but the thicknesses were calculated from the weight lost by etched films, assuming that only unpigmented medium was removed.

Paint film gloss

Subjective rating of the relative gloss of a series of paint films presents difficulties that arise from the fact that observers assess gloss by personal techniques that are often quite dissimilar, so that one observer may rate best a film rated worst by others, and that the eye tires, limiting the number of panels that can be reliably rated in a session. In addition, the human observer, often subconsciously, integrates a variety of factors in making his assessment, of which not all are necessarily to do with the gloss, or lack of it, of the film under examination. To be set against these defects is the fact that on occasion the subjective assessment and the instrumental rating of gloss are grossly at variance.

Paint film 'gloss' may be defined as the extent to which the surface of a paint film functions as a specular reflector, or conversely as the extent to which it does not act as a diffuser of incident radiation. The latter definition is a more helpful aid to an understanding of the problem. A perfect diffuse reflector would scatter the incident light ray equally in all directions, so that the locus of equal intensity of reflected radiation would form a hemisphere around the point of incidence. In practice, this hemisphere is slightly distorted at its lower edge and markedly distorted by a protrusion from its surface at the angle of reflectance of the incident ray. The cross-section of the locus can be conveniently plotted two-dimensionally using a gonio-reflectometer, for any given angle of incidence. For a perfect mirror, the emergent ray should have the same dimensions and intensity as the incident ray, with no dispersion of light by diffuse reflection. This is scarcely achieved by the finest metallic mirrors. For a surface such as a paint film, that is quite likely to have a thin layer of clear medium over the pigmented bulk, true 'mirror image' gloss is unattainable. Even if the paint/air interface were optically flat, the pigmented layer below the clear layer is bound to be uneven, to a degree dependent on the intrinsic particle shape and the degree of dispersion (aggregate and/or agglomerate shape), and also non-uniform in refractive index, depending on the concentration of pigment particles in the medium. Pigments that have refractive indices greater than that of the medium, especially the white 'hiding' pigments, will scatter the incident light and thus inevitably cause some diffusion. In addition, the relationship of the clear layer thickness to its refractive index and the wavelength of the incident light may be such that an interference effect is set up; such an effect, especially in the case of white pigments, can cause a variation in the apparent hue of the film.[95]

If the paint film surface is not smooth, its effect on the gloss of the film will depend on the size of the irregularities and the wavelength of the incident light. This has been described by Dunderdale *et al.*[96] who found that large irregularities caused angular divergence of the specular beam of light, whereas small irregularities, of the same order of size as the incident light wavelength, caused scattering (diffusion) of the light. The relationship between the state of dispersion of the pigment and the gloss of the paint film is a fairly simple one, but not always that which might be expected; it is affected by the type of medium and the pigment concentration in the film.

In the case of alkyd/amino resin based baking enamels, shrinkage during film formation arises not only from the loss of solvent but also from the further condensation of the butylated amino-formaldehyde resin. This latter effect can be considerable but is a function of the resin composition, and therefore specific to the resin used. The effect of whether the pigment was dispersed in the alkyd or the amino resin, the kind of alkyd and amino resin used, the alkyd/amino resin ratio and the grade of titanium dioxide have been shown to be very marked.[97] McCausland[98] showed that there was a marked correlation between the size of pigment agglomerates in the paint and the major irregularities of the surface of the baked film of a rutile pigmented alkyd/amino paint. The effect of improved dispersion (increased ball-milling time) on the gloss, and the gloss/p.v.c. relationship of alkyd/amino resin based films is quite different to those for the same pigments in air-dried long oil alkyd based paints.[97] This difference might be attributed to the extra shrinkage of the baked film due to condensation of the amino resin, but if the state of pigment dispersion is very good, there is a marked correlation between gloss loss on overbaking and the presence or absence of coating oxides on titania pigments, both anatase and rutile. This may be due to a greater amount of adsorbed resin, giving an increased 'clear layer' thickness and therefore less surface roughening on added shrinkage due to overbaking of the film.

In all cases, there is a steady fall in gloss values as the pigment concentration in the film increases. Improved dispersion reduces the gradient of the gloss/p.v.c. curve and the gloss value at any given p.v.c. increases. The improvement in gloss is more marked when measured using 20°/20° incident/reflected light beams than when 45°/45° beams are used, indicating that the chief cause of the improvement is a reduction in diffusion of the incident light, or, in other words, a reduction in the haziness of the film.

In the case of long oil alkyd based paints, the influence of pigment content on gloss is not marked until a pigment content of 15% v/v is exceeded. Even then, the effect is more marked in the case of a more readily dispersed grade of rutile than in that of one less easily dispersed. The gloss of the films containing the less easily dispersed pigment falls slightly with increasing p.v.c.; longer milling to give improved dispersion results in slightly lower gloss and also causes the reduction in gloss with increasing p.v.c. to take place a little more rapidly. The easily dispersed pigment, in marked contrast, gives films with much reduced gloss, whether measured at 20°/20° or 45°/45°, when the films have more than about 15% p.v.c. At 20% p.v.c. and above, the pigment content/gloss relationship is to some extent dependent on the medium; in a long linseed alkyd the 20° gloss levelled out at about 78% at 20 and 25% p.v.c. (compared with a 15% p.v.c. gloss of 90%). Increased milling to improve the dispersion only resulted in very slightly reduced gloss of the linseed alkyd paint films containing less than 15% p.v.c. and a more marked reduction, to about 73%, of the gloss values at 20 and 25% p.v.c. In the long soya alkyd based paints the more easily dispersed pigment gave films with markedly superior 20° and very slightly superior 45° gloss up to 15% p.v.c., above which the gloss of the films dropped markedly, the 20° gloss being equal to that of the paint containing the less well dispersed pigment at 20% p.v.c. The gloss of the films containing the less well dispersed pigment was substantially unaffected by the p.v.c., up to 20% p.v.c., but sharply declined at higher p.v.c. values, reaching a minimum (20° gloss = 16%) at about 50% p.v.c., whereas the better dispersed pigment had a 20° gloss of 40% at that level. At 60% p.v.c., the 20° gloss value was independent of the pigment, both films giving readings of 80%.

In all cases, the fineness-of-grind was better than could be measured with a Hegman gauge, so that the effect was primarily due to the difference in the size distributions of the pigments below a maximum agglomerate size of about 1000 nm. From electron micrographs of paint film sections, one apparent difference was that the less easily and the easily dispersed pigments contained respectively 6·6% and 1·5% by weight of aggregates each formed from 10–12 primary particles, the larger of these being about 600 nm in diameter.

The gloss/p.v.c./dispersion relationship may be explained on the basis that in the lower p.v.c. films the clear layer thickness is similar for both pigments. At higher pigment contents, the larger total number of particles and the larger number of small particles, that will be less strongly repelled from the paint/air interface, combine to turn the statistical probability that there will be some particles in the clear layer into a certainty, so that the gloss is reduced. At more than 30% p.v.c. the effective volume of the particles, assuming a closest possible packing solid fraction of about 0·54, exceeds 55% and the pigment particles start to form a significant proportion of the paint film surface. The clear layer is likely to be reduced to the adsorbed resin layer on the pigment and ceases to make a significant positive contribution to the specular reflection from the surface. The combination of unpigmented medium and pigment particles forming the surface will increase the proportion of the incident light that is scattered.

As the pigment particles form an increasing proportion of the paint surface at still higher p.v.c. values diffusion will increase, but the more dispersed pigment, with fewer large aggregates, will cause less distortion of the residual specularly reflected light beams, so that its measured gloss will be higher. At 54% p.v.c. the void volume between the particles will exceed the total volume of the medium and the paint surface will consist almost entirely of pigment particles. The number average particle diameter, about 180 nm (less than the weight average of about 220 nm) is less than the half-wavelength value for the radiation forming the visible spectrum (400–720 nm), so that only those surface irregularities due to aggregates will cause light scattering, though some of the incident light will be scattered after it has entered the pigment layer. (The number of aggregates of 10–12 particles is about $17 \times 10^6/\text{cm}^2$, assuming that distribution of the various sizes of aggregate is uniform throughout the film (20 microns thick). The total number of primary particles is about $24 \times 10^8/\text{cm}^2$.) The proportion of incident light that is scattered is relatively small, however, so the measured specular reflectance becomes quite high (60% at 20°/20°) for both pigments.

Flooding and floating

If a paint film contains two or more pigments, it is possible that, during film formation, these pigments may be substantially separated from each other. Pigment segregation into strata parallel to the paint film surface is described as *flooding* (the film colour being uniform, but the hues of the upper and lower surfaces of a stripped film differing considerably). Pigment segregation into 'columns' perpendicular to the surface is described as *floating* and is characterised by the appearance on the paint film surface of more or less hexagonal Bénard cells. These are the more visible the more marked the contrast in hue of the two pigments.

Assuming the paint to be thoroughly mixed before application, segregation can only occur in the period during which the film is fluid enough to allow pigment particle movement; likelihood of the defect therefore increases with film thickness and with reduction in the rate of solvent loss from the film.

A vast number of papers have been published describing these defects and suggesting methods for their prevention. All the latter can be summed up as devices for reducing the tendency to separate and/or achieving conditions in the film such that pigment separation will be hindered. The latter condition is most readily achieved by inducing an acceptable degree of flocculation; the immobilisation of a large volume of the liquid phase within the flocs causes a marked increase in the viscosity of the suspension and in addition the relatively large size of the flocs makes them much less mobile than the individual pigment particles. A corollary to this is that if one pigment is flocculated and the other dispersed, then flooding or floating will be accentuated. Also, if both pigments are fully dispersed but of considerably different mean particle size, flooding or floating is very likely.

One of the chief mechanisms causing separation is thermal (convection) currents, set up as a result of the cooling action of solvent evaporation on the paint film surface. Convection in the film may be increased by application of the paint to a warm substrate, or reduced by application to a cold one, especially if the substrate is a good conductor of heat. In this latter case, the reduced rate of solvent loss and extended period of paint film fluidity are more likely to assist flooding or floating than the increase in the viscosity, due to the lower film temperature, is to hinder them.

A second mechanism is electrical in nature. The dispersed particles are likely to have some amount of electrical charge,[85,99] and if both pigments have charges of the same sign, their mutual repulsion will cause separation; this in itself could not cause flooding or floating, but an electrical charge is usually built up at the paint/air interface, through solvent evaporation, and this could cause horizontal stratification, that is, flooding. Unlike charges would cause association, pigment flocculation, the most reliable preventive of flooding or floating.

The Bénard cells that are associated with floating are almost certainly present in all paint films, but without pigments of highly contrasted hue to mark their boundaries, they are not visible. Exposure of the paint film to ultra-violet radiation and a humid atmosphere will sometimes make the Bénard cell structure apparent in the very early stages of paint film erosion. A problem of interpretation can arise here, because a very similar Bénard cell pattern may be produced by a stress-relieving action in a paint film. This is especially important in the case of alkyd/amino resin blends, where the shrinkage of the amino resin (as it is further condensed during baking) may set up quite considerable stresses in the film.

It is perhaps worth remarking that the popular association of flooding and/or floating with particular pigment combinations is not entirely justified. Floating, in particular, is the more visible, the greater the contrast in hue of the pigments. White and black or blue offer maximum contrast and therefore are more noticeable. Careful examination of films, under a good light, and especially using a low-power wide-field microscope, will often reveal the characteristic cell pattern of floating, even though the contrast in hue of the two pigments is too small to be apparent to an ordinary observer.

Pigment dispersion and hiding power

Early in this chapter it was pointed out that there is no one optimum pigment particle size, for either white, black or coloured pigments. For white pigments, for example, optimum size varies with the pigment concentration in, and the thickness of the film to be opacified. The optimum particle size *for a given application* must be determined by experiment, since there is no theory of opacification able to do more than indicate the probable size range as being large, medium or small. However, before making a positive statement concerning the best particle size, it is best to know the particle size distribution of the pigment in the test-piece. This may or may not correlate with the particle size distribution determined using a pigment dispersion prepared for that specific purpose.

There is no absolutely satisfactory method of determining the pigment particle size distribution in its intended end-use. Paint-film sectioning, though often used, has the disadvantage discussed earlier, that the film-forming process can give a flocculated appearance to the particles. The higher the pigment volume concentration, the worse is this effect and the more difficult the sizing and counting of the particles. Methods such as those described by Hess and Garret[62,100] have the disadvantage, in the one case of being very time consuming and, in the other, that relatively low pigment concentrations (3% p.v.c. of carbon black) must be used, or else films pressed-out very thin. The latter procedure could easily disperse agglomerates and so give an incorrect result.

It is difficulties such as these that make especial care necessary to ensure that any report dealing with opacity is not of the *ipse dixit* or *ipso facto* type. 'A set of dispersions were prepared and property 'X' was measured; the ranking order of the dispersions in this test was paralleled by the ranking order in respect of opacity of films made from the dispersions; therefore it is concluded that an increased value of property X is productive of higher opacity.' The conclusion may be valid, but some causality should be demonstrated.

Franklin[99] showed that hiding power of titanium dioxide dispersions in alkyd increased with the zeta-potential (calculated from electrophoretic mobility) but that the effect of zeta-potential became less important as the molecular weight of the resin increased. The problem of the adsorbed layer thickness/adsorbed resin weight per unit area was discussed by Doorgeest.[101] Thus, whilst it is reasonable to consider that a thicker adsorbed layer on the surface of pigment particles will reduce the likelihood of pigment flocculation, and that a high molecular weight resin should give a thicker adsorbed layer, it is necessary to take into account the influence of the pigment surface on resin configuration in the adsorbed layer. Goldsbrough and Peacock[88] have suggested that the configuration of alkyd molecules adsorbed on alumina and on silica coatings on rutile is markedly different and that a negligible amount of solvent is associated with the resin adsorbed on silica. As the two surfaces are respectively basic and acidic in character, it seems possible that on alumina adsorption is through carboxylic groups, that are by definition end-groups for alkyd molecules, whilst adsorption on silica is through the residual hydroxyl groups, that are distributed along the polymer chain. The alkyd molecules could thus be envisaged as being 'end-on' to the alumina-coated and laying flat on the silica-coated pigment surface. The alumina-coated pigment would be expected to have superior opacifying power, because the thicker layer should help to prevent excessive crowding and consequent reduced effectiveness of the particles as light scatterers. Evidence in support of this hypothesis has been presented by Franklin *et al.*[102]

The simplistic view of the pigment dispersion and paint film opacity relationship outlined above does not deal adequately with the problems of pigment/medium interactions in commercial paint formulations. Some of the effects on paint properties of various additives have been described by Blakey.[103] Quite often two or more such additives are used in one paint; a study of their interaction and effects on the properties of the suspension might provide a useful insight into some frequently encountered problems.

BIBLIOGRAPHY

Inorganic pigments

J. J. Mattiello (Ed.) *Protective and Decorative Coatings*, Vol. 2, John Wiley, New York (1947).

H. Kittel (Ed.), *Die pigmente*, Wissenschaftliche Verlag, Stuttgart, (1960).

W. A. Glozer, *Kirk–Othmer Encyclopedia of Chemical Technology*, 2nd Edition, Vol. 15, Interscience, New York (1968).

L. A. Tysall, *Industrial Paints*, Pergamon, Oxford (1964).

C. J. A. Taylor (Ed.), *OCCA. Paint Technology Manuals*, Vol. 6, *Pigments Dyestuffs and Lakes,* Chapman & Hall Ltd, London (1966).

REFERENCES

1. A. Brockes, *Optik*, **21** (1964) 550.
2. S. R. Orchard, *J. Oil Colour Chem. Assoc.*, **51** (1968) 44.
3. P. Kubelka and F. Munk, *Z. Tech. Phys.*, **12** (1931) 593.
4. *e.g.* G. Schenkel, *Plastics Extrusion Technology and Theory*, Iliffe, London, 1966.
   The Plastics Institute publications (Iliffe, London):
   P. D. Ritchie (Ed.), *Physics of Plastics*.
   E. G. Fischer, *Extrusion of Plastics* (2nd Ed.).
   W. R. Groves and M. G. Munns, *Plastics Moulding Plant*, Vols. 1 & 2.
5. W. H. Stockmayer, *J. Polymer Sci.*, **9** (1952) 69.
6. R. Wilson and A. H. Robson, *Off. Digest*, **27** (1955) 111.
7. A. R. H. Tawn, *J. Oil Colour Chem. Assoc.*, **39** (1956) 223.
8. A. J. Seavell, *J. Oil Colour Chem. Assoc.*, **42** (1959) 319.
9. R. Brett, *J. Oil Colour Chem. Assoc.*, **41** (1958) 428.
10. J. H. Hildebrand and R. Scott, *The Solubility of Nonelectrolytes*, 3rd Ed., Reinhold, New York, 1949.
11. H. Burrell, *Off. Digest*, **27** (1955) 726 and 1069.

12. J. H. Hildebrand, *J. Amer. Chem. Soc.*, **38** (1916) 1452.
13. G. Scatchard, *Chem. Rev.*, **8** (1931) 321.
14. P. Flory, *J. Chem. Phys.*, **10** (1942) 51.
15. M. L. Huggins, *J. Phys. Chem.*, **46** (1942) 151; *Ann. NY Acad. Sci.*, **43** (1942) 1; *J. Amer. Chem. Soc.*, **64** (1942) 1712.
See also:
P. J. Flory, *Principles of Polymer Chemistry*, Chapter 12, Cornell Univ. Press, New York, 1953.
H. Tompa, *Polymer Solutions*, Chapter 2, Butterworths, London, 1956.
16. R. C. Nelson, R. H. Hemwall and G. D. Edwards, *J. Paint Tech.*, **42** (1970) 636.
17. H. Burrell, Proc. Sixth Fatipec Congress (1962) 21.
18. H. Burrell, *J. Paint Tech.*, **40** (1968) 197.
19. J. H. Hildebrand, *Chem. Rev.*, **44** (1949) 37.
20. H. Tompa, *Polymer Solutions*, Butterworths, London, 1956.
21. T. G. Fox and P. J. Flory, *J. Phys. and Colloid Chem.*, **53** (1949) 197.
22. Los Angeles Society Technical Committee 10, *Off. Digest*, **33** (1961) 1329.
23. G. Kimball, *A Treatise on Physical Chemistry*, Vol. 2, Chapter 3, H. S. Taylor and S. Glasstone (Eds.), Van Nostrand, New York, 1951.
24. C. M. Shaw and J. F. Johnson, *Off. Digest*, **25** (1953) 339.
25. E. G. Bobalek *et al., Ind. Eng. Chem.*, **48** (1956) 1956.
26. W. W. Reynolds and H. B. Gebhart, Jr., *Off. Digest*, **29** (1957) 1174.
27. H. E. Weisberg, *Off. Digest*, **34** (1962) 1154.
28. L. Dintenfass, *J. Oil Colour Chem. Assoc.*, **41** (1958) 125.
29. C. Hansen, *J. Paint Tech.*, **39** (1967) 505.
30. Toronto Society, *Off. Digest*, **35** (1963) 1211.
31. P. Sorensen, *J. Oil Colour Chem. Assoc.*, **50** (1967) 226.
32. A. I. Medalia, *J. Coll. Interface Sci.*, **24** (1967) 393 and **32** (1970) 115.
33. E. K. Fischer, *Colloidal Dispersions*, John Wiley, New York, 1950.
R. D. Cadle, *Particle Size Determination*, Interscience, New York, 1955.
34. Proceedings of Particle Size Analysis Conference, Loughborough 1966, Soc. Analyt. Chem., London, 1967.
35. K. S. W. Sing, *J. Oil Colour Chem. Assoc.*, **54** (1971) 731.
36. B. C. Lippens and J. H. de Boer, *J. Catalysis*, **4** (1965) 319.
37. K. S. W. Sing, *Chem. and Ind.* (1967) 829.
38. K. S. W. Sing, *Surface Area Determination* (D. H. Everett and F. S. Stone, Eds.), p. 25, Butterworths, London, 1970.
39. R. E. Day and G. D. Parfitt, *Trans. Faraday Soc.*, **63** (1967) 708.
40. D. Urwin, *J. Oil Colour Chem. Assoc.*, **52** (1969) 697.
41. P. C. Carman, *Flow of Gases Through Porous Media*, Butterworths, London, 1956.
42. B. H. Kaye and M. R. Jackson, Ref. 35, above, p. 313.
43. P. C. Carman, Symposium on Particle Size Analysis, 1947, *Trans. Inst. Chem. Engrs.*, **25** Supplement, p. 118.
44. R. H. S. Robertson and B. S. Emödi, *Nature*, **152** (1943) 539.
45. F. M. Lea and R. W. Nurse, *J. Soc. Chem. Ind.*, **58** (1939) 277.
46. R. B. Anderson and P. H. Emmett, *J. App. Phys.*, **19** (1948) 367.
47. S. G. Ward and R. L. Whitmore, *Brit. J. App. Phys.*, **1** (1950) 325.
48. H. D. Jefferies, *Verfkroniek*, **37** (1964) 436.

49. B. B. Boonstra and A. I. Medalia, *Rubber Age*, **92** (1963) 802 and **93** (1963) 82.
50. E. Micek, F. Lyon and W. M. Hess, *Rubber Chem. & Tech.*, **41** (1968) 1271.
51. L. L. Ban and W. M. Hess, Paper given at Ninth Biennial Conf. on Carbon, Boston, 1969.
52. D. F. Harling and F. A. Heckman, *Materie Plastiche ed Elastomeri*, **35** (1969) 80; Cabot Corp. Research Paper 37–72.
53. C. Fisher and M. Cole, *The Microscope*, **16** (1968) 81.
54. A. I. Medalia, *J. Coll. Interface Sci.*, **32** (1970) 115.
55. H. F. Clay, *J. Oil Colour Chem. Assoc.*, **40** (1957) 935.
56. A. W. Evans and R. D. Murley, Proc. Sixth Fatipec Congress (1962), 125.
57. G. Kämp and H. G. Völz, *Farbe u. Lack*, **74** (1968) 37.
58. G. D. Parfitt, *J. Oil Colour Chem. Assoc.*, **54** (1971) 717.
59. D. Rivin, Proc. Fourth Rubber Techn. Conf. (1962), 261.
60. N. Scott, *J. Oil Colour Chem. Assoc.*, **49** (1966) 559.
61. D. Dollimore, *J. Oil Colour Chem. Assoc.*, **54** (1971) 616.
62. W. M. Hess and M. D. Garret, *J. Oil Colour Chem. Assoc.*, **54** (1971) 24.
63. J. G. Oldroyd, *Rheology of Disperse Systems*, Ed. C. C. Mill, Pergamon, London, 1959, p. 1.
64. R. Roscoe, *Suspensions, Flow Properties of Disperse Systems*, Ed. J. J. Hermans, North Holland, Amsterdam, 1953, p. 1.
65. E. Marsden, *J. Oil Colour Chem. Assoc.*, **32** (1949) 183.
66. E. Marsden, *J. Oil Colour Chem. Assoc.*, **42** (1959) 119.
67. H. D. Jefferies, *J. Oil Colour Chem. Assoc.*, **52** (1969) 635.
68. W. Carr, *J. Oil Colour Chem. Assoc.*, **49** (1966) 831 and **50** (1967) 1115.
69. T. C. Patton, *J. Paint Techn.*, **42** (1970) 665.
70. R. Roscoe, *Brit. J. App. Physics*, **3** (1952) 267.
71. W. K. Asbeck and M. van Loo, *Ind. Eng. Chem.*, **41** (1949) 1470.
W. K. Asbeck, D. D. Laiderman and M. M. van Loo, *Off. Digest*, **30** (1952) 326.
72. N. Casson, *Rheology of Disperse Systems*, Ed. C. C. Mill, Pergamon, London, 1959, p. 83.
73. H. H. Weber, *Deutsche Farb.-Zeit.*, **14** (1960) 312.
74. W. Heinz, *Material Prüfung*, **1** (1959) 311.
75. F. K. Daniel and P. Goldman, *Ind. Eng. Chem. Anal.*, Ed., **18** (1946) 26.
76. F. K. Daniel, NPVLA Sci. Sect. Circ. No. 744, 1950.
77. F. K. Daniel, *J. Paint Tech.*, **38** (1966) 534.
78. H. Rumpf, *Agglomeration*, Ed. W. A. Knepper, Interscience, New York, 1962, p. 379.
79. W. P. Pietsch, *Nature*, **217** (1968) 736.
80. J. Pryce-Jones, *J. Oil Colour Chem. Assoc.*, **19** (1936) 295.
81. J. M. Burgess and G. W. Scott-Blair, *Report on Nomenclature*, North Holland, Amsterdam, 1949.
82. E. C. Bingham, *Fluidity and Plasticity*, McGraw-Hill, New York, 1922.
83. S. Guggenheim, *Off. Digest*, **30** (1958) 729.
84. H. Heesch and F. Laves, *Z. Krist.*, **85** (1933) 443.
85. V. T. Crowl, *J. Oil Colour Chem. Assoc.*, **50** (1967) 1042.
86. M. J. Vold, *J. Colloid Sci.*, **6** (1951) 492.
87. K. Rehacek, *Farbe u. Lack*, **76** (1970) 656.
88. K. Goldsbrough and J. Peacock, *J. Oil Colour Chem. Assoc.*, **54** (1971) 506.

89. M. Gordon, *The Structure and Physical Properties of High Polymers*, Plastics Monograph No. C. 10, The Plastics Institute, London, 1957.
90. W. Funke, *J. Oil Colour Chem. Assoc.*, **50** (1967) 942.
W. Funke, U. Zorll and B. G. K. Murthy, *J. Paint Tech.*, **41** (1969) 210.
W. Funke and R. Brandt, *J. Oil Colour Chem. Assoc.*, **45** (1971) 230.
91. R. D. Murley and H. Smith, *J. Oil Colour Chem. Assoc.*, **53** (1970) 292.
92. R. D. Murley, Private communication.
93. A. Kawabata, *Shikizai Kyokaishi*, **42** (1969) 98.
94. S. Wilska, *J. Paint Tech.*, **43** (1971) 65.
95. W. G. Armstrong and W. D. Ross, *J. Paint Tech.*, **38** (1966) 462.
96. J. H. Colling, W. E. Craker and J. Dunderdale, *J. Oil Colour Chem. Assoc.*, **51** (1968) 526.
97. H. D. Jefferies, *J. Oil Colour Chem. Assoc.*, **45** (1962) 681.
98. R. J. McCausland, *Double-Liaison*, **91** (1963) 53.
99. M. J. B. Franklin, *J. Oil Colour Chem. Assoc.*, **51** (1968) 499.
100. M. D. Garret and W. M. Hess, *J. Paint Tech.*, **40** (1968) 367.
101. T. Doorgeest, *J. Oil Colour Chem. Assoc.*, **50** (1967) 1079 and discussion, pp. 1106–14.
102. M. J. B. Franklin, K. Goldsbrough, G. D. Parfitt and J. Peacock, *J. Paint Tech.*, **42** (1970) 740.
103. R. R. Blakey, *J. Paint Tech.*, **43** (1971) 65.

CHAPTER 9

DISPERSION OF ORGANIC PIGMENTS

J. BERESFORD and F. M. SMITH

INTRODUCTION

A pigment is defined as a coloured, black or white substance which is insoluble in the medium in which it is applied and imparts colour or opacity to this medium, or, to use the definition of the Society of Dyers & Colourists 'a substance in particulate form which is applied to bodies by mechanical incorporation, by chemical precipitation or by coating to modify their colour and light scattering properties'.

Whereas pigments in their widest sense are used for a variety of purposes such as decoration, identification, obliteration, corrosion resistance or radiation resistance, organic pigments are almost invariably used for decoration or identification and in the case of certain printing inks for rheology modification. Thus they constitute only a very small proportion by weight of total pigment production and usage. When compared with coloured or black inorganic pigments, organic pigments tend *in general* to be tinctorially stronger, brighter, chemically less reactive, less stable to solvents and media, less stable to light and heat, and higher in specific surface area.

The performance of a pigment can be divided into *fastness properties* and *application properties*. It is, however, not possible to define easily which properties are dependent entirely on chemical constitution and which properties are dependent on physical properties such as particle surface, size and shape. In fact, there is evidence to suggest that most pigment properties are both chemically and physically dependent. This can be seen from the following discussion of pigment properties.

Hue

The hue of a pigment depends on its chemical structure, *i.e.* the number and type of chromophore and auxochrome groups present and the degree of chemical purity. In the case of dyes a maximum absorption at a particular wavelength in a particular solvent can be used to assist in identifying the constitution of a soluble colouring matter; in pigment technology, however, differences in crystal structure and particle size *with the same chemical constitution* can produce two or more pigments of different hue. For example, copper phthalocyanine blue (CI Pigment Blue 15) can be prepared in α (reddish blue shade) and β (greenish blue shade) forms, and α, β and γ linear quinacridone (CI Pigment Violet 9) provides three red pigments varying from yellow shade red to blue shade red. In other cases, notably with dinitraniline orange (CI Pigment Orange 5), reduction in particle size by grinding, either dry or in a medium, can result in marked changes in hue and tinctorial strength, *i.e.* from a yellow shade red to a strong pure orange. Electron micrographs indicate that this hue change is due to particle reduction by the shattering of single pigment crystals, and *not* by the alteration of crystal habit as in the case of copper phthalocyanine or linear quinacridone.

Tinctorial strength

Each chemical type has a maximum obtainable tinctorial strength. Within specific chemical types, notably arylamide yellows and toluidine reds, wide variations in pigment strength are apparent between commercially available pigments due to differing crystal sizes within the same crystal habit. It is also clear, however, that all anilide bisarylamide yellows (CI Pigment Yellow 12) are tinctorially stronger than all arylamide yellow G's (CI Pigment Yellow 1), due entirely to their chemical nature and crystal form.

Fastness

Fastness to solvent bleed, chemicals and heat is almost entirely dependent on chemical properties. Lightfastness is largely dependent on chemical properties but Birrell[1] has drawn attention to the large lightfastness difference between amorphous and crystalline forms of dianisidine blue (CI Pigment Blue 20). This is, however, an unusual phenomenon and may indeed be associated with changes in the chemical nature of the pigment surface.

Rheological properties

The rheological properties imparted by a pigment to a system are often characteristic of a chemical type but the large variations possible within chemical types are due entirely to different physical properties, probably a combination of particle size, specific surface area, particle shape and pigment vehicle interactions.

CHEMICAL CLASSIFICATION OF ORGANIC PIGMENTS

In any discussion on the dispersion of organic pigments some knowledge of pigment chemistry and of general application performance, is desirable.

For a more detailed discussion of pigment chemistry it is necessary to consult standard text books and papers.[2–5] It is possible to classify all organic pigments in the following groups; lakes, metal salts, metal chelates, metal free azo pigments, cationic dye complexes, phthalocyanines, vat pigments and polycyclic pigments.

It is sufficient for present purposes to briefly discuss these accepted classifications of organic pigments.

Lakes

Lakes are metal salts of readily soluble anionic dyes. These are dyes of azo, triphenylmethane or anthraquinone structure containing one or more sulphonic or carboxylic acid groupings. They are usually precipitated by a calcium, barium or aluminium salt onto a substrate of alumina, blanc fixe or a mixture of the two. Typical examples are peacock blue lake (CI Pigment Blue 24) and Persian orange (lake of CI Acid Orange 7). Normally lakes are cheap, bright and tinctorially weak, have good solvent fastness and poor resistance to light and chemicals.

Metal salts

These are salts (Ca^{++}, Ba^{++}, Mn^{++}) of sparingly soluble anionic azo dyes. Usually these dyes contain one sulphonic acid grouping in their structure and, often, one carboxylic acid grouping in addition. They are not precipitated on to a base and are usually marketed as full strength pigments. Examples are calcium red 4B (CI Pigment Red 57), barium lake red C (CI Pigment Red 53) and calcium, barium and manganese red 2B's (CI Pigment Red 48). In general, these salts have good resistance to solvents and heat, are tinctorially strong and bright, have poor chemical resistance particularly to alkalis and have moderate lightfastness.

Metal chelates

These are organic compounds insolubilised by a chelating metal. Whilst a large number of compounds have been reported there are only a limited number of commercial pigments in this class. These compounds have good solvent resistance and better chemical resistance than metal salts. Examples are pigment green B (CI Pigment Green 8), nickel azo yellow (CI Pigment Green 10) and some more recent azo-methine chelates (CI Pigment Yellow 117).

Metal free azo pigments

This group is composed of an extremely wide range of yellow, orange, red and some blue pigments, all of which are pure organic azo compounds without aqueous solubilising groups and without any metal or inorganic component. As a class they tend to bleed in resin and solvent systems.

Their heat stability is good although some of them sublime or crystallise from supersaturated solution. Lightfastness may vary from moderate to excellent across the class. Examples of earlier azo pigments include arylamide yellow 10G (CI Pigment Yellow 3), anilide diarylide yellow (CI Pigment Yellow 12), toluidine red (CI Pigment Red 3) and trichloraniline red (CI Pigment Red 112).

In order to improve solvent and heat resistance higher molecular weight compounds have been introduced, in particular the azo condensation series as represented by CI Pigment Yellow 95 and CI Pigment Red 144.

Cationic dye complexes

These are manufactured by insolubilising cationic (basic) dyes with complex inorganic acids such as phosphomolybdotungstic acid or ferrocyanide ions. These pigments, usually in the shade range red-violet-blue-green, are tinctorially strong and brilliant but have poor resistance to light, heat, solvents and chemicals. They are used almost exclusively in printing inks. Typical examples are PTMA Rose Red (CI Pigment Red 81) and PMA Violet (CI Pigment Violet 3).

Phthalocyanines

Phthalocyanine pigments are metal free or copper chelated tetrabenzoporphyrazine. They give blue and green pigments of outstanding strength, brilliance and all-round fastness. The most common forms are α and β copper phthalocyanine blue (CI Pigment Blue 15) and polychloro copper phthalocyanine green (CI Pigment Green 7). Copper phthalocyanine blue was the first pigment compound to be marketed in two distinct polymorphs.

Vat pigments

This group consists of pigments derived from insoluble textile vat dyes. Their high price demands very high all-round fastness properties (although many suffer from poor tinctorial properties in terms of brightness). Their introduction was prompted by demands made by the automotive industry for high lightfastness and by the plastics industry for heat fastness. Now they are being superseded by the newer polycyclic pigments. Important vat pigments include indanthrone (CI Vat Blue 6) and anthrapyrimidine yellow (CI Vat Yellow 20).

Polycyclic pigments

Into this last group has been placed a very wide range of complex chemical types introduced in recent years. They are expensive, high quality pigments designed for such demanding applications as automotive finishes and melt spun synthetic fibres. Obviously they must have excellent all-round fastness properties. Examples which have achieved commercial significance are quinacridone red (CI Pigment Violet 9), carbazole dioxazine violet (CI Pigment Violet 23) and the isoindolinones (CI Pigments, Red 180 and Yellow 109).

From this discussion it is clear that most of the fundamental properties are dictated by chemical class. The choice of pigment for any particular system must be based on chemical considerations. Having chosen the correct chemical type to suit fastness, hue and economic considerations, one is left with a quite wide range of application properties which are substantially independent of the organic chemistry of the molecule. It will become obvious that many of these properties are functions of the fine state of division of the pigment in the application vehicle.

THE PHYSICAL PROPERTIES OF PIGMENTS

Pigments as powders

Organic pigments are finely divided solids (surface area approximately 10–120 m^2/g) of relatively low density (about 1·4–2·5 g/ml) and are usually crystalline. It is not possible to generalise any other properties of organic pigments because of the very wide range of chemical species. So, for example, one cannot talk of a general lyophobicity; some pigments are lyophobic in certain applications and not in others.

At formation organic pigments are in a state of fine division but under the influence of ionic salts in the vat, for example, and during filtration and drying, they agglomerate or aggregate to levels which are unrepresentative of the original state. Honigmann and Stabenow[6] have formalised a nomenclature (*see* Chapter 7) to standardise description of the various levels of co-existence of particles. Furthermore, it has been recognised that in application it is frequently advantageous to retain or regain the manufactured state of fine division and consequently such things as dispersed pastes, flushed pastes and chips have been used.

Nevertheless, a significant proportion of organic pigment is sold as powder and the powder properties are of considerable interest since both manufacturers and users must handle these finely divided, low density solids. Very little has been published about the flow and handling properties of organic pigment powders although manufacturers are obviously aware of problems of this type because in recent years some organic pigments have been offered in the form of granules. Normally it is standard practice to quote bulk densities, solid densities and surface areas by nitrogen adsorption.

It is difficult to relate the powder size to application properties since the relationship is not obvious. Cartwright *et al.*[7] and Carr[8] have, respectively, attempted to relate specific surface area to flow properties of a lithographic ink and to texture in relation to ease of dispersion.

It is surprising that the lead given by the white and black pigment field in surface chemical specification has not been followed by organic pigment studies. Zettlemoyer[9] and Doorgeest[10] have reported on the adsorption of water vapour. Beresford, Carr and Lombard[11] have used nitrogen adsorption to derive average particle sizes but further gas adsorption experiments have not been reported. This may be due to the difficulty in achieving organic pigments with extremely pure surfaces, often an essential precursor to basic surface studies.

Organic pigment research, until very recently, has been concerned with the production of new chromophores in attempts to improve the basic fastness and coloristic properties. It is unlikely, however, that synthesis of new compounds in itself will improve significantly the surface chemistry of the pigment.

Earlier in this book (Chapter 1) it has been suggested that the dispersion process can be considered conveniently as three distinct processes occurring simultaneously, *i.e.* wetting, separation and stabilisation. If one uses such a model then properties of the pigment powder influence all three processes.

Wetting

Obviously any study of wetting must begin from the powder. Bikermann[12] employed a sessile drop method to determine the contact angle between surfaces and liquids. Although this method is ideally suited to flat extensive surfaces such as metals, an experimental technique can also be developed for finely divided powders. Recently, Johnson and Dettre[13] published a theoretical treatment for idealised rough surfaces. Although measurements of this type can be made, the repeatability is poor and the results are only of limited value; with powders the method is not able to differentiate better than $\theta \simeq 0°, <90°, >90°, \simeq 180°$. This may be sufficient for some purposes, but, in general, with liquids of viscosities in the poise range, *e.g.* paints, inks, the method breaks down.

An alternative approach is to obtain the 'work of adhesion' by methods such as that suggested by Bartell[14] and improved more recently by Bouvet.[15] Basically one measures the pressure, and hence the force, required to just prevent the liquid rising through a plug of the powder. A similar technique was adopted by Heertjes and Kossen.[16] Bouvet quotes an equation

$$\frac{l^2}{t} = \frac{W\varepsilon}{\eta S}$$

where l is the length of liquid penetrating a plug of the powder in time t, and W is the work of adhesion and is equal to $\gamma \cos \theta$. γ is the solid–liquid interfacial tension, θ the contact angle, ε the plug porosity, S the powder surface area and η the liquid viscosity.

Obviously from measured ε, S, η values W can be determined by plotting l^2 against t.

There are drawbacks to this method. First, there is the practical difficulty of packing the plug evenly so that liquid advances through the plug in a plane face. Secondly, S is strictly the surface area of tortuous capillaries through the plug and not that of the uncompressed powder; consequently S should be determined by a suitable technique such as the permeability method of Carmen and Malherbe[17] and then is no longer constant but depends upon ε and the packing pressure used to form the plug. Similar criticisms apply to other work of this type.

Finally it is possible to measure energies of wetting in highly sensitive calorimeters. This method is direct and simple and thermodynamically sound but, of course, with organic pigments one is faced with practical problems created by the highly aggregated state of the powder. The heat change measured when a powder is rapidly immersed in a liquid is proportional to the surface area wetted but there is no simple way of measuring the area presented to the liquid. In addition, the wetting of external surface and pores will proceed at different and indeterminate rates and consequently heat changes may occur slowly over extended periods, thus reducing the accuracy.

Separation

Separation is the reduction in size of agglomerates and aggregates. The extent of the size reduction achieved is a balance between the effective external force applied and the forces of particle/particle interaction, assuming that the acts of separation and stabilisation can be treated separately. The externally applied force is independent of the pigment but the interaction forces between agglomerates, aggregates and particles depend strongly upon the size and type of particle and the previous history of the powder.

Whereas several authors have concentrated their efforts upon surface areas by nitrogen adsorption or permeability without reference to application, Carr[8] has related surface area and oil absorption. From his results it is obvious that to a first order correlation the dispersion size achieved (as estimated by the oil absorption figure) depends upon the size of particles in the powder. The authors are not aware of any more rigorous treatment of this subject.

Again, no attempt has been made to measure directly the strengths of particle/particle interactions of organic pigments which would seem to provide the key to separation. It might be argued that these interactions vary with the vehicle but this is not true; the particle/particle interactions remain constant but particle/vehicle and vehicle/vehicle interactions, which also influence the total dispersion process, vary. Casson[18] has suggested that 'yield value' is a description of these particle/particle interactions;

however, the authors' opinion is that it describes the total interactions of the system and not any one in isolation. Such a measurement would be most useful if it truly represented the total system interactions, but 'yield values' as measured generally within the paint and ink industries are dependent on the measuring instrument and are not very repeatable.

Stabilisation

Any stabilisation must be created by forces of repulsion between particles. Whether such forces are created by electrical or entropic effects, they are most likely to arise from interaction between the pigment surface and the vehicle. The extent of the interaction depends as much upon the pigment surface as the vehicle composition. Therefore, knowing the specific chemical nature of the surface and the vehicle it should be possible to predict the interaction level, assuming that bulk chemical/physical reaction knowledge can be applied at the surface. One can go further and design the surface so that it does indeed provide favourable sites for reaction with a particular vehicle. There is a vast weight of patent literature on the surface treatment of organic pigments to improve their compatibility with specific systems, such treatments often being arrived at empirically. The characterisation of surfaces even in simple liquid systems is extremely difficult, let alone the highly complex industrial systems. Vapour phase studies, however, have proved remarkably instructive with other powders and must be equally valuable with organic pigments.

The crystallography of organic pigments

Until recently little attention had been paid to this topic and yet the vast majority of organic pigments are crystalline and many are found to possess more than one crystal form. Outstanding examples are the commercially available α and β forms of copper phthalocyanine, although other crystal modifications have been claimed. In fact crystal studies of phthalocyanine pigments have proved far more popular than studies on any other organic pigment, probably because of the clear commercially significant differences between the α and β forms.

In the late 1930s Robertson and his co-workers carried out most of the fundamental single crystal work on copper and several other metal phthalocyanines; the reader is referred to the book of Moser and Thomas[19] for an excellent summary of their findings. Although in the 1950s and 1960s one or two authors suggested small amendments to Robertson's figures, in general his results still stand. Recent workers have been much more concerned with mechanisms of polymorphic change and growth of phthalocyanines; in particular Suito and Uyeda[20] and Honigmann *et al.*[21] have published extensively. Honigmann has studied the grinding of copper phthalocyanine which results in $\beta \rightarrow \alpha$ changes and also crystallite disintegration. He proposes, from powder photography studies, that the various polymorphic forms of copper phthalocyanine reported (*e.g.* α, β, γ, δ, ε) are really only various physical forms of α and β. Honigmann argues that the reported different polymorphs have arisen because of interpretation difficulties in powder photographs. If crystallite sizes are below 0·1 μ then lines broaden, adjacent lines superimpose and appear as broad single lines, and line displacement may occur because of lattice imperfections. One can only regard pigment specimens as having different crystal phases if true differences exist in the unit cell; those with the same unit cell but different crystallite diameter or lattice imperfection are merely different physical forms. Even though their colouristic properties differ this is no indication of polymorphism as claimed in some patent literature. Honigmann says the claimed γ copper phthalocyanine is really the α form in very fine crystals. The differences between β, αI and αII (Honigmann nomenclature) exist chiefly in the stacking structure. β has the central metal atom octahedrally co-ordinated whereas in α there is a tendency towards closest packing. Within a stack the forces are strong while the arrangement towards three dimensional structure proceeds without strong or specific bonds between stacks. Thus needle shaped crystals should cleave easily along their major axis and highly elastic bending should be possible. α stacks permit several possible arrangements, distinguished chiefly by movement in this stack axis direction, αI and αII being two such possibilities. In the stacking of these molecules two opposing forces act, (1) tendency towards closest packing, and (2) tendency to octahedrally co-ordinated central metal atoms. Whichever is stronger determines α or β stacking. Chlorine substitution opposes the octahedral co-ordination and hence chlorinated copper phthalocyanine is only found in the α modification, although again very different physical forms can be differentiated. This last conclusion by Honigmann disagrees with the findings of Shigemitzu[22] who claims to have demonstrated the existence of polymorphs.

Honigmann's arguments are well developed from sound observation and therefore his hypotheses are convincing. It is difficult, however, to believe in the elastic bending of phthalocyanine needles especially when the salt grinding process for crude copper phthalocyanine is known to produce α crystals of very low acicularity. The salt grinding of crude copper phthalocyanine causes immediate crystallite disintegration with rapid reduction of crystal size. Again, although the authors accept Honigmann's results qualitatively the actual results quoted are functions of the mechanical design, additive ratios and the starting material. Honigmann's results on crystallising effects of organic solvents confirm that $\alpha \rightarrow \beta$ conversion and Ostwald ripening occur. The final results of size and crystal shape are functions of the solvent, the original crystal size and any additives (such as surfactants).

Suito and his co-workers studied the conversion and growth of copper phthalocyanine both *in vacuo* and in solvent, drawing similar conclusions from both. Significantly they concluded that $\alpha \rightarrow \beta$ transformations are preceded by preliminary growth of α, and this growth is ascribed to recrystallisation effects caused by solubility differences between large and small crystals. Rearrangement to β is postulated to occur at any corner of an individual crystal where the solvent, even given very small solubility, allows the lattice to relax, thus triggering formation of β nuclei by giving enough spaces for individual molecules to rotate to new positions. One would expect this triggered nucleation to occur with the original small crystallites but at that stage it must compete energetically with crystal growth. Suito quotes the unit cells of β ($a = 19{\cdot}6$ Å, $b = 4{\cdot}79$ Å, $c = 14{\cdot}6$ Å, $\beta = 120{\cdot}6°$ and molecular plane at 45·8° to b-axis) and α ($a = 29{\cdot}52$ Å, $b = 3{\cdot}79$ Å, $c = 23{\cdot}92$ Å, $\beta = 90{\cdot}4°$ and molecular plane at 26·5° to b-axis). Both polymorphs tend to grow in the direction of the b-axis leading to the formation of needles, a tendency which seems to be common to many polycyclic aromatic compounds. When crystals of α are acicular along the b-axis they convert readily with little energy to the β, thus allowing further growth of the β to take place along the b-axis. On the other hand, crystallites of α which form obliquely arranged raft structures, require greater molecular rearrangement, *i.e.* greater energy to convert to β and therefore there is less energy available to promote growth of the β.

Zhdanov and Vorona[23] detail the appearance and structure changes in copper phthalocyanine as a function of temperature between $-10°$ to 330°C, the pigment being formed by vacuum condensation. Their extensive work supports the findings of Honigmann and Suito.

Although the salt milling and solvent conversion processes seem to have aroused a lot of interest no one has offered an explanation of the $\beta \rightarrow \alpha$ conversion in salt milling and only one worker (Raidt[24]) has reported studies on the acid solution–water precipitation process. He suggests that several 'acids of crystallisation' ($CuPc.xH_2.SO_4$) can be formed, especially by the use of organic solvents as drown-out media or by varying the ratio of copper phthalocyanine sulphuric acid solution and drown-out water. Structure determinations on these acid crystals were not possible due to decomposition by water uptake although this stepwise decomposition can be followed by X-ray analysis.

When one considers other organic pigments the published information is scant by comparison. Mez[25] has reported the crystal structure of CI Pigment Yellow 1 (Arylamide Yellow G) and its chlorine analogue. More recently Chapman and Whitaker[26] have reported data for CI Pigment Yellows 1, 3 and 4, CI Pigment Red 2 and the mono and dibromo analogues of CI Pigment Yellow 3. Honigmann[27] has reported briefly on pigments other than phthalocyanine as have Daubach[28] and Herbst and Merkle,[29] but there is nothing on crystal change, growth or processing of pigments such as quinacridone, metal salt or azo pigments which can compare with that of Honigmann, Suito and Raidt for phthalocyanine.

Whilst there are quite large numbers of publications dealing with the crystallography of phthalocyanines, it would seem there are several unsolved mysteries. Nevertheless, significant progress has been made already and will ultimately contribute to a better understanding of the pigment surface.

The optical properties of pigments

Although the optical properties of pigments only truly become significant when related to their surrounding application medium it is not possible to determine them in such circumstances and therefore all reported work has been carried out on pigment powder or carefully grown single crystals. In 1948 Cooper[30] reported a very simple method, based on the Brewster angle determination, for measuring the refractive index of polished, compressed discs of organic pigments. Whilst aware of the birefringent nature, the light absorptive capacity and optical dispersion character of organic pigments he was not able to make allowance for any of them. Patterson and Hannam[31] reported briefly on investigations into the birefringence of azo pigments. They measured the lower refractive index by liquid immersion, Becke line techniques, and were able to estimate the higher refractive index by using the optical rotation caused by large crystals of known thickness. Account was taken of the wavelength dependence but it was not possible to allow for the high absorption of organic pigments. In a theoretical treatment based on the Mie theory of light scattering Brockes[32] calculated the relation between the optical properties (refractive index and absorptivity) and the particle size of pigments. Brockes drew the relationship between refractive index and adsorptivity and the absorption and scattering coefficients of the Kubelka-Munk[33] theory. In a later paper Felder,[34] using a more elegant adaption of the Mie theory to the case for absorbing particles, proved the validity of his derived equations by measurements on model spherical particles of dyed polymers. In a second paper Felder[35] extended his work to include the effect of particle shape, as well as size; in particular he included the effect of elliptical particles. Again by using model polymer particles of given shape Felder was able to substantiate his findings which were that a particle of minor axis b, major axis a behaved as a/b spherical particles of radius b. To elaborate his theories Felder used the complex refractive index $m = n - ik$ where n = the real refractive index and k = the absorption per unit length. The problem remained to measure the refractive indices of real pigment particles as opposed to model particles. Nassenstein[36] investigated this problem and came up with a solution based on the early apparatus of Cooper but using the Fresnel laws of reflection at many angles for a medium with a complex refractive index, and utilising the polarisation of light effects at dielectric boundaries. Like Cooper, Nassenstein used a compressed disc of pigment powder which was polished, and consequently paid no attention to the known birefringence. It could be argued that this omission was serious but the authors support the view that in most practical applications of organic pigments the particles are randomly orientated and therefore the Nassenstein measurements made, as they were, on a packed bed of randomly orientated particles will yield the most significant refractive index. If one were to make any criticism of Nassenstein's results it would be that no evidence was presented to show that the act of compression did not alter the particles. With soft organic crystals it would not be difficult to imagine pressure causing a polymorphic change.

Organic pigments in dispersion

In general, organic pigments comprise very fine crystals in the 0·01 μ to 0·4 μ size range although their powder surface areas rarely exceed 120 m^2/g. Pigment powders may contain lumps of very large dimensions—millimetres even—and it is obvious, therefore, that work must be applied to achieve the incorporation of these highly agglomerated powders into specific vehicles. The methods used depend upon the wide range of vehicles which can vary from very thin water emulsions, such as emulsion paints, through viscous letterpress inks to highly viscous molten plastics. Examples of these techniques will be described in a later section; at the moment it is sufficient to realise that the dispersions achieved can be a function of many factors.

Given that such things as cost, hue and fastness requirements are satisfied by the chosen pigment, its level of dispersion, or particle size, can determine many of the important application properties. A list of properties which depend upon size includes

(1) tinctorial strength
(2) resistance to flocculation and settling
(3) transparency
(4) opacity
(5) gloss
(6) flow, or rheology of the pigment-vehicle system.

It is worthwhile, therefore, to begin this section by discussing briefly the sizing methods that have been applied to pigmented systems.

Particle size measurement in liquids and films

Beresford, Carr and Lombard[37] reported on the use of the Coulter Counter, a sizing instrument dependent on the alteration of electrical properties of an electrolyte by the presence of particles. They concluded that for organic pigments the Coulter Counter was (*a*) insufficiently sensitive to submicron particles, (*b*) inaccurate in that dispersions tend to flocculate when added to the electrolyte carrier, and (*c*) inaccurate due to the possibility of flocculates being broken down when passing the measurement orifice. But the most significant reason why they rejected this method was that they found the practical lower size limit to be 0·8 μ, *i.e.* too high for organic pigment dispersions with mean diameters at 0·2–0·4 μ.

Atherton, Cooper and Fox[38] introduced another technique, the disc centrifuge, applied to disperse dyes in water. This technique has been well described elsewhere and is based on the well-known sedimentation behaviour of particles dispersed in a liquid when subjected to a centrifugal field. Beresford[39] reported on a critical examination of the instrument when applied to aqueous dispersions of pigments and found satisfactory reproducibility, minimum disturbance of the dispersion, and a size range extending down to 0·05 μ so long as the 'buffered line start' technique described by Jones[40] was used. There are limitations to the use of this technique. Whatever the starting concentration of pigment dispersion, one is forced to use no more than about 2% pigment weight concentration. Thus it is necessary to dilute the original suspension and, whilst Beresford did not report trouble with the aqueous media in which he worked, dilution flocculation can be obtained in non-aqueous media. In fact, obtaining a satisfactory dilution can be the single most difficult step. Streaming problems, that is non-Stokesian sedimentation, have been observed by several workers but again the authors found the 'buffered line start' usually overcame such problems, although they would recommend observing the sedimentation every time and rejecting anything suspicious. Fractions are cut from the spinning liquid to a fixed depth and concentrations determined separate from the machine. By selecting a series of disc speeds, start radii and sedimentation times one is able to cover the size range of interest. Obviously this size range also depends upon the liquid viscosity and the density difference between solid and liquid but normally with organic pigments the instrument has a practical size range of about 2 to 0·05 μ. Beresford found no evidence of breakdown of flocculates during sedimentation even when examining presscake material. The particle diameter determined has a real physical significance; it is the diameter of a sphere of the same density which would sediment at the same rate under the same conditions, often referred to as the Stokes diameter. There is no sound way of relating this diameter to those often quoted as equivalent spheres for non-spherical particles obtained by other techniques.

Amongst others, Crowl[41] has reported on the use of the electron microscope to determine the size distribution of pigment dispersions. The size range of most dispersions and the magnification and resolving power of the electron microscope seem to make it ideally suited for this application. There are disadvantages, however. The electron microscope examines dry specimens of low particle population and drying and dilution are both liable to create changes in the dispersion state. In general, organic pigment particles are not spherical and dispersion does not break down aggregates to the crystallite level; the electron microscope, however, has the power to resolve many of these crystallites and as a consequence one always has the dilemma 'is the observed group of crystallites a single particle in the dispersion or several particles which have agglomerated as an artifact of the microscope slide preparation technique?' Again, the non-sphericity leads one to quote an equivalent spherical diameter. There are several equivalent diameters that may be used and it is claimed that useful information can be obtained by comparing and contrasting them. Automatic film scanning methods such as are employed in the 'Quantimet'[42] greatly facilitate this counting and sizing.

Whilst there are drawbacks to both centrifugal sedimentation and electron microscopic sizing methods, they are the only ones to have been used to determine the influence of state of dispersion on application properties. Crowl[43] determined the size by electron microscopy when studying flotation, flocculation and flooding of phthalocyanine pigments. In this excellent work he demonstrated that charge effects could not explain the observations on flocculation in alkyd paint media by measuring the several factors of size of particle, electrophoretic mobility, dielectric constants and Hamaker constants, thus allowing theoretical calculation of stability from the DLVO theory. These calculations showed criteria for stability were not met by such a system. However, adsorption of resin could be shown to relate to observed stability and therefore Crowl concludes that entropic repulsion mechanisms must dominate. Carr,[44] on the other hand, used centrifugal sedimentation to determine the particle size and demonstrated that the effect of particle size on tinting strength is very much a function of the particular organic pigment. With phthalocyanine blue in alkyd paints a markedly non-linear dependence is observed where there is little effect of size on tinting strength until $0{\cdot}4\ \mu$ then the strength increases almost exponentially until $0{\cdot}1\ \mu$, the lowest size Carr recorded. However, pigment green B in the same paint system showed only very slight dependence of tinting strength on particle size. It is interesting to note that both these results agree qualitatively with the theoretical results of Brockes[32] if one assumes a low refractive index for both blue and green and low absorption for the green but high absorption for the blue. Brockes predicts a maximum in the case of high absorption but Carr was not able to demonstrate this, although his results cannot be said to deny the existence of such a maximum. By demonstrating the difference between two organic pigments, Carr highlights a commercially significant dilemma—should one seek ultimate tinting strength by continued grinding or should one compromise? Obviously for phthalocyanine blue the dilemma is a real one, but it does not exist for pigment green B—fine sizes only give marginally increased strength. Because of his results Carr also questions whether the traditionally notorious phthalocyanine blue flocculates more readily than other pigments. Is it merely that any very small change in size for the phthalocyanine blue is readily seen as a loss of strength whereas changes in size of pigment green B will go largely unnoticed?

Significantly, however, both Crowl and Carr concluded that dispersions with the highest proportion of fine particles (less than 0·2 μ) produced flocculation-free paints even though their sizing methods and flocculation assessments were both different. In a similar exercise to that of Carr, Hauser, *et al.*[45] showed a qualitative agreement between size, strength results and Brockes calculations. Carr used one commercial pigment and several different dispersion times in a long oil alkyd system whereas Hauser, used several different phthalocyanine preparations, most of which were dispersed under identical conditions in a linseed oil ink varnish. Nevertheless, the qualitative agreement between these two works is encouraging.

Whilst there have been many papers over the years describing how opacity, transparency, gloss and flow vary with dispersion they have largely avoided the question of size measurement by quoting dispersion assessed by fineness of grind gauges, by colour strength or even by inference from the properties they are measuring and therefore cannot be considered fundamental. However, there are a few isolated results. For example, Carr[44] found the opacity of organic pigment paint stainers to be largely independent of the size of particles within the size range he examined. This rather startling result can be explained as the combination of two factors and the authors believe that whilst Carr's results cannot be disputed, his conclusion is only valid over the size range examined. If one combines the contribution of light scattering (which passes through a maximum around 0·6 μ to 0·2 μ) and light absorption (which is only beginning to rise steeply at 0·3–0·4 μ) then in the region 0·7–0·1 μ it is possible to imagine a roughly constant opacity. This analysis requires that the pigment be of the strongly absorbing type and therefore would not apply to whites such as titanium dioxide.

Beresford[46] reported briefly on the effect of particle size on the high gloss of paint stainers. He demonstrated that down to a certain particle size (about 0·4 μ) the surface roughness of paint films falls, thus allowing the specular gloss to increase. Below this size specular gloss is constant but the haze reduces as size reduces and therefore the total visual gloss improves as size is progressively reduced.

Hauser *et al.*[45] measured plastic viscosity and yield value of his phthalocyanine ink pastes by applying the Casson[18] equation. Hauser concluded that plastic viscosity at infinite shear was independent of particle size but that yield value depended most strongly on size below a critical size of

$0{\cdot}12\ \mu$, indicating that the very low shear flow of these pastes would deteriorate very rapidly below this size limit. Unfortunately, these results formed only a small part of Hauser's report and thixotropic and concentration effects were deliberately excluded. This exclusion of two very important factors limits the value of these results. Cartwright,[7] on the other hand, used published surface area values and related these to Casson yield values and also to a thixotropic index and again similar results were obtained in that both yield value and thixotropic index depended most strongly upon high surface areas but only weakly upon low surface areas. It might be argued that relating surface areas of dry powders to rheological characteristics is not describing the effect of dispersion, but the qualitative value of these conclusions is increased if one bears in mind the remarkable agreement between surface area and oil absorption demonstrated by Carr.[8]

To summarise, it has been demonstrated that the dispersion state of organic pigments, as defined by their particle size, dramatically affects application properties. We have seen that organic pigments are non-ideal particles—non-spherical, aggregated and crystalline—and that practical measurements will not agree quantitatively with theories developed for ideal particles. Nevertheless, the qualitative agreement between practical results and the theories of Brockes and Felder for the optical behaviour of pigments shows that these theories offer valuable tools to the pigment chemist. Dispersion state is extremely important in determining the colour strength of any organic pigment. Although these results have been obtained for only a limited number of systems the authors believe the qualitative results are sufficiently well established to permit wider application.

The importance of dispersion size to dispersion stability has been demonstrated by Crowl and Carr for two systems, but one cannot draw general conclusions for all systems from their work. It would appear that existing theories of colloid stability could provide the basic understanding of these systems although more evidence is needed covering a wider range of systems than has so far been studied. The authors would make one reservation to this statement in that they feel that dispersion stability is perhaps the wrong concept to apply to highly pigmented pastes, such as lithographic and letterpress inks where it is difficult to imagine the 'non-interacting particles' so popular in theoretical colloid chemistry. At the concentration levels found in industry it is impossible for one pigment particle to exist outside the influence of its neighbours. It is felt that such systems will require another theoretical approach.

The importance of dispersion to the flow properties of pigmented systems has been demonstrated, although again much more information is needed before general conclusions will be possible. There is also a need for a satisfactory theoretical model against which one can compare any practical results.

THE APPLICATION OF PIGMENTS

It is possible to classify most industrial applications of organic pigments by method of dispersion and viscosity of vehicle at the dispersion stage rather than by the most usual classification into end uses, such as lithographic ink, automotive lacquer, injection moulding or paper coating, etc.

The majority of pigment users use dry pigment powder, granule or lump as starting product but of course other physical forms or physical states are marketed. In *idealised* form Fig. 9.1 gives a production flow chart for

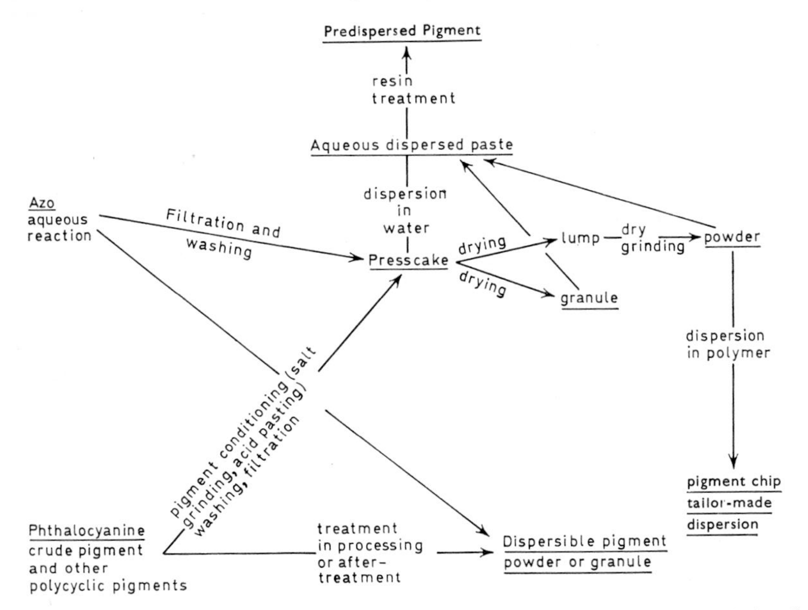

Fig. 9.1. *Production flow chart for organic pigments (marketed forms are underlined).*

organic pigments indicating at what stages the products may be marketed. The processes of pigment conditioning referred to as acid pasting and salt grinding have been described by Smith and Easton.[47] However, whether the pigment dispersion for application is carried out in the pigment factory or the user factory, the starting product for dispersion is either pigment presscake or more commonly dried powder, lump or granule.

Extremely high viscosity systems

The solid dispersions of pigment in solid, friable resins such as nitrocellulose, cellulose acetate, various polyamide resins, etc., are usually referred to as pigment chips. Under conditions of heat and extremely high rates of shear, pigments are dispersed in these resins at round about the melting points of the resins to obtain extremely fine dispersions often not achievable by other methods. The equipment used for this type of dispersion is usually a heavy duty two-roll mill with heated rollers. This efficient but expensive dispersion method gives products which can be dissolved in solvents to produce low viscosity gravure or flexographic inks. Such inks are printed on transparent films and on aluminium foil or other substrates where maximum transparency and gloss are required. It is not normally possible to achieve the state of dispersion of a *chip* ink by more conventional fluid-dispersion techniques. The pigment content of these resin chips can be as high as 60–70%. There is no theoretical lower limit but obviously the higher the grinding pigmentation the lower the relative cost.

Very high viscosity systems

The pigmentation of most thermoplastic and thermosetting resins commonly in use in the plastics industry falls into this category. The polymers most commonly pigmented with organic pigments include flexible and rigid PVC, polyolefines, polystyrene, acrylonitrile-butadiene styrene, cellulose polymers and melamine and phenol based thermosetting resins. It is also usual to include natural and synthetic rubbers in this group. The pigmentation level in these systems does not often exceed 1% of organic pigment and 5% of white pigment, such as titanium dioxide. In most cases the organic pigment level is more likely to be in the region of 0·1%.

Pigment dispersion in these systems is usually carried out at about the melting point of the polymer under conditions of high shear on a two-roll mill or in a heated Z-arm mixer of the Banbury type. The temperature of dispersion is of extreme importance; if the melting range of the polymer is exceeded this results in a rapid drop in viscosity of the mix and serious loss of shear in the dispersion equipment. Particular care is necessary in the case of polyolefines which have very narrow melting ranges.

Under large scale production conditions in injection moulding and extrusion, thermoplastic polymer chips or granules are often coloured by tumbling with pigment powder. Dispersion is expected to take place under conditions of low shear in the injection or extrusion screw. In these cases the use of a pigment pre-dispersed or coated with polymer is usually preferred.

High viscosity systems

This category includes most letterpress and lithographic ink systems based on oils or synthetic varnishes. This type of ink can contain up to 30–35% of organic pigment, the necessary high degree of structure and high viscosity of the system being imparted by the presence of the pigment. During printing the rheological properties, ink transfer properties and gloss of ink systems depend to a large extent on the surface properties and degree of dispersion of the pigment in the system. It is in this field where

Fig. 9.2. *Molteni mill.*

very slight apparent differences in the surface and dispersion properties of organic pigments can have a marked effect on the application properties of the end product.

Pigment dispersion in these systems is usually carried out either entirely on a high speed water cooled three-roll mill or by initially pre-mixing and dispersing in a heavy duty Z-arm mixer at very high viscosity followed by refining on a high speed three-roll mill.

It is becoming apparent, however, that in the future, pigment dispersion in these high viscosity systems may be carried out in high speed cavitation mixers fitted with scraper blades and revolving dispersion pans to enable the highly viscous mill base to be moved into the sphere of activity of the dispersion head. One example of this latter type of machine, made by Molteni is shown in close up in Fig. 9.2.

Medium viscosity systems

The majority of non-aqueous paints come into this category during the dispersion operation. Paints are very rarely manufactured at their application viscosity but are often prepared as viscous white paint bases and viscous coloured pigment stainers which are blended and reduced with further resin and solvent to produce the finished paint. Alternatively, in stronger shades all constituents may be dispersed together in a so-called *co-mix* formulation to produce a coloured paint base which is likewise reduced to application viscosity with further resin and solvent.

Organic pigments constitute only a small proportion of most paints; exceptions are the high tinctorial strength, bright shade paints such as post-office red which may contain up to 10% of toluidine red pigment (CI Pigment Red 3). Pastel and medium shades may contain up to 2% of organic pigment, the rest of the dispersed phase consisting of inorganic white pigments and extenders.

Pigmentation problems associated with paint systems are usually related to economics, colouristic requirements and pigment fastness properties required during paint manufacture, application and end use. Problems associated with rheological properties, whilst occurring frequently in paint manufacture, are almost always cured by well tried empirical techniques. Perhaps the commonest pigment-vehicle problem in non-aqueous paint systems is the so-called flocculation or loss of colour strength and flooding–flotation in paints based on mixtures of titanium dioxide and phthalocyanine blue (Chapter 7). Whilst again most cures for these problems are empirical, involving blue pigment modification by the pigment maker or paint modification by the paint maker, these problems have been the subject of many practical and theoretical examinations. In Crowl's view[43] the degree of flocculation in alkyd and alkyd amino paints shows a high degree of correlation with the differences in particle size distribution of the pigments in the system. Those with the highest distribution of very fine particles show the least flocculation. Conversely, flotation effects are more pronounced in pigment dispersions containing a high proportion of fine particles except in certain cases (desirable) where physico-chemical co-flocculation of the blue and white pigments occur.

Dispersion equipment used for these medium viscosity systems on the technological scale is currently undergoing great changes. The traditional machines are the simple ball mill and the two-roll or bar mill (Fig. 9.3) often preceded by a simple pre-mix on low energy machines such as a vertical Pug. However, both these machines are extremely slow in production rate and they are gradually being superseded by more rapid methods.

There are two basic types of new equipment coming into use both relying on extremely high power input into dispersion for a relatively short time. The first is the cavitation mixer as represented by the Cowles Dissolver and the Torrance mixer. Dispersion in this type of machine depends entirely on developing high rates of shear in the pigment vehicle mix with a disc impeller. Dispersion takes place without the aid of grinding media such as balls, pebbles, ballotini, etc. The other type, as represented by the sand grinder, the attritor and the Perl mill, again depends on high energy input transmitted through an impeller system but in these cases the impeller is used to drive a dispersion medium such as sand, glass ballotini or small

Fig. 9.3. *Two-roll mill.*

steel or porcelain balls. The cavitation-type mixer is usually a batchwise process whereas the latter types often operate *continuously*, the degree of dispersion depending to a large extent on the dwell time in the equipment.

Low viscosity systems

Gravure and flexographic printing for periodicals, coloured newspapers, paper packaging, foil and film packaging, require printing inks of very low viscosity in the region of 500 centipoise. The organic pigmentation level in such systems can vary between 3% and 10% in media such as nitrocellulose, metal resinates, cellulose derivatives, alcohol soluble polyamides, shellac in solvent and aqueous solutions and a variety of water soluble or water thinnable resins. Pigment dispersion is often carried out at application viscosity or slightly above application viscosity. The dispersion of pigments in water for use in resin emulsion paints, paper dyeing and coating, and viscose rayon mass dyeing falls into this category of viscosity during dispersion. For these low viscosity systems the commonest dispersion machine used is the ball mill, but again this type of machine whilst it is cheap to use is extremely slow in production rate and is being replaced by high-energy methods as is the case in the paint industry. In addition to the cavitation mixer type, the sand grinder type and

Fig. 9.4. *Kady mill.*

the Perl mill type which are often used for these systems, machines are also available which have been specifically developed for low viscosity systems.

A typical example is the Kady mill (Fig. 9.4) which depends on extremely high shear being developed on the pigment particle as the pigment vehicle mixture is forced through the slots in the dispersion head by a high speed impeller.

It should also be mentioned that low viscosity inks, particularly for aluminium foil printing where high transparency is required, are often made by *dissolving*, in simple mixing equipment, pigment chips which have been made by dispersing pigment in solid resin at high temperature and viscosity.

THE DISPERSION OF ORGANIC PIGMENTS

It may be recognised from the foregoing that the dispersion of a pigment in a medium is a difficult process and can be very expensive in terms of capital equipment and in the experimental development of techniques of dispersion. The user, particularly if he manufactures coloured plastics and to a lesser extent paint, is expending this effort on a relatively minor portion of his total system and in the case of organic pigments a portion used only for decoration or identification. In most printing inks, of course, the organic pigment portion is a major part of the system.

However, due to these factors there has always been pressure from the pigment user for pigments which are more easily dispersible and give less unpleasant side effects, such as flocculation, in use. In recent years this pressure has become more intense due to rationalisation in production, more informative production costing information and perhaps the major factor the development and introduction of new types of dispersion equipment. Many organic pigments are much inferior in ease of dispersion in high speed or high energy equipment when compared with inorganic pigments and extenders. These latter pigments often represent a higher proportion of a system by weight if not by cost and this leads again to an obvious demand for organic pigments showing better dispersion characteristics pushing more and more the responsibility for pigment dispersion into the lap of the pigment manufacturer.

For many years some pigment users have been used to buying aqueous pigment dispersions for use in viscose rayon, resin emulsion paints, paper and some other applications. This was a development which merely put the responsibility for pigment dispersion into the hands of the pigment manufacturer who used and still does use traditional *hard work* dispersion methods for making these products. It is, of course, worth noting that the pigment manufacturer can withdraw his product at the most suitable stage of manufacture relevant to his dispersion method rather than having to rely upon marketed forms of pigment powder, granule or presscake.

Of more interest are recent and continuing developments in pre-dispersed pigments and more easily dispersible pigments. It is necessary to deal with these two groups separately in order to cover adequately the principles involved.

Pre-dispersed pigments

Strictly speaking, the aqueous dispersions mentioned above come into this category in that they are pigments pre-dispersed in water by the pigment manufacturers. Much more recent are developments in pigments pre-dispersed in polymer system. It is possible, of course, for the manufacturer to pre-disperse pigments by traditional chipping methods in certain polymers such as nitrocellulose, polyolefines, polyamides, etc., but this is merely a transfer of dispersion responsibility. The dispersion of pigment in water, however, is a relatively simple operation and with modern equipment, surface active agents and pigment manufacturing techniques it is possible to prepare very fine aqueous dispersions with particle sizes below 0·5–1 μ. In the preparation of pre-dispersed pigments such dispersions can be co-flocculated with a relatively large proportion of polymer emulsion or polymer solution to produce a powdered pigment–resin composition in which the pigment particles are either coated with resin or are associated with polymer particles in such a way that they remain in a state of dispersion and can be subsequently used for colouring a polymer mass or polymer solution by simple mixing. Examples of the types of resins in which pigments are commercially available in this form include maleic resins, zinc/calcium resinates, low MW polyethylene, PVC, PVC/PVA copolymer, polyamide and acrylics.

This type of product is, however, necessarily expensive as it involves a two stage manufacturing process, however simple, involving high manufacturing and handling costs.

Dispersible pigments

More easily dispersible pigments are not strictly speaking pre-dispersed but are manufactured by more simple techniques and although they are technically not as good as the pre-dispersed types, tend to be not much more expensive than traditional pigments and are quite adequate in dispersibility in modern high speed equipment. Manufacturing methods for this type of product are still shrouded in commercial *know-how* and in obscure patent literature but certain trends in methods are becoming clear. Examples are:

(*i*) chemical modification of arylamide yellows and diarylide yellows during manufacture to include in the finished product a proportion of a long chain fatty amine adduct which can assist in separating pigment agglomerates;

(*ii*) addition of small proportions of resins during manufacture which will assist in separating agglomerates, the choice of resin to maintain multipurpose application being obviously important, and

(*iii*) strict control of manufacture at all stages, particularly in precipitation, filtration and drying, to minimise undue crystal growth, aggregation and agglomeration.

From the nature of this type of trend and the extremely small proportion of any additives used these *dispersible* pigments tend to be far more multipurpose in application than the strictly pre-dispersed type.

It is the authors' view that these forms of pigment will be progressively improved and will eventually replace most organic pigments in use today. In this way responsibility for pigment dispersion or the provision of pigments which will disperse easily will shift more and more to the pigment manufacturer. This will enable the user, the ink, paint or plastics technologist, to employ much more useful effort in improving the applicational performance of their mass and surface coating systems rather than expending much wasted effort in redispersing a coloured pigment to a form which existed early in its scheme of manufacture.

CONCLUSIONS

Despite the fact that the dispersion problems appear to move from the pigment user to the pigment maker as a result of the above trends, the general situation is still far from satisfactory. Let us re-examine the main physical factors and see how they are affected.

Particle size reduction is probably more readily achieved if the pigment is pre-dispersed during manufacture, particularly if it is dispersed into its final vehicle by a standardised technique. But particle size is not necessarily the complete answer. Certainly the main concern of the pigment user is to avoid oversized particles which result in adverse performance characteristics such as brushing-up, poor gloss, 'tramlines' in printing, poor colour value and poor storage properties, but in the course of providing a pigment which can be easily incorporated without oversized particles, very little control is exercised over the *smallness* of particles in the 0·025 to 1 μ range. It is within this range that the pigment can have a big effect on transparency and rheological properties thus affecting flow out properties of paint and ink for example. Means of measuring particle size distribution in this area in a final pigmented system are very meagre; the disc centrifuge is very helpful but even this must examine a *prepared* sample. The next objective of controlling particle size in the 0·025 to 1 μ range *during* manufacture and incorporation is still to come. Only then can we finally identify what sizes and size distributions are desirable for different performance properties.

That pre-dispersed pigments are usually made for specified applications, is recognition of the importance of the pigment-vehicle relationship. This normally recognises the principle that like materials are compatible with each other. Consequently the use of a prepared pigment in a somewhat offbeat vehicle can result in incompatibility. Supplier and user are not often both prepared to put all their cards on the table. If they did, would we be any wiser? It is doubtful, because there is far too little information available on what is adsorbed on pigments in relation to desirable performance properties. Current academic work on *type* compounds and *clean* surfaces is necessary to make measurements in identifiable situations but ultimately simple methods will need to be evolved so that pigment surfaces can be identified on a routine basis. Resins could similarly be characterised by their effect on known pigment surfaces. In this way pigmented resin systems could be made with a good degree of assurance that their performance would be satisfactory. Heats of wetting techniques probably represent the best approach.

REFERENCES

1. P. Birrell, *J. Oil Colour Chem. Assoc.*, **47** (1964) 878.
2. L. S. Pratt, *The Chemistry and Physics of Organic Pigments*, Chapman & Hall, London, 1947.
3. H. Kittel, *Pigmente Wissenschaftliche Verlagsgesellschaft m.b.H.*, 1960.
4. *Paint Technology Manual Vol.* 6. *Pigments, Dyestuffs and Lakes*, Oil & Colour Chem. Assoc., Chapman & Hall, London, 1965.
5. F. M. Smith, *J. Soc. Dyers Colourists*, **78** (1962) 222.
6. B. Honigmann and J. Stabenow, VIth Fatipec Congress Report (1962) 89.
7. P. F. S. Cartwright, C. H. Smith and W. Carr, *Printing Technology*, **2** (1967) 81.
8. W. Carr, *J. Oil Colour Chem. Assoc.*, **50** (1967) 1115.
9. A. C. Zettlemoyer, *Off. Digest*, **395** (1957) 1238.
10. T. Doorgeest, Thesis, Delft Tech. High School, 1965.
11. J. Beresford, W. Carr and G. A. Lombard, *J. Oil Colour Chem. Assoc.*, **48** (1965) 293.
12. J. J. Bikermann, *Trans. Faraday Soc.*, **34** (1938) 634.
13. R. E. Johnson and R. H. Dettre, *Am. Chem. Soc. Advances in Chemistry Series*, **43,** Chapter 7, p. 112.
14. H. J. Osterhof and F. E. J. Bartell, *J. Phys. Chem.*, **34** (1930) 1399.
15. P. Y. Bouvet, VIIth Fatipec Congress Report (1964) 224.
16. P. M. Heertjes and N. W. F. Kossen, *Powder Technology*, **1** (1967) 33.
17. P. C. Carmen and P. le R. Malherbe, *J. Soc. Chem. Ind.*, **69** (1950) 134.
18. N. Casson, *Rheology of Disperse Systems*, C. C. Mill, Ed., Pergamon Press, New York, 1959.
19. F. H. Moser and A. L. Thomas, *Phthalocyanine Compounds*, Reinhold, New York, 1963.
20. E. Suito and N. Uyeda, *Kolloid Zeitschrift*, **193** (1963) 97.
21. B. Honigmann, H. Lenne and R. Schrödel, *Zeitschrift Krist.*, **122** (1965) 185.
22. M. Shigemitzu, *Bull. Chem. Soc. Jap.*, **32** (1959) 607.

23. G. S. Zhdanov and Yu. M. Vorona, *Izv. Akad. Nauk., SSSR, Ser. Fiz.*, **27**(9) (1963) 1232.
24. H. Raidt, *Krist. Tech.*, **1**(3) (1966) 511.
25. H. C. Mez, *Ber. Bunsenges. Phys. Chem.*, **72** (1968) 389.
26. S. J. Chapman and A. Whitaker, *J. Soc. Dyers Colourists*, **87** (1971) 120.
27. B. Honigmann, *J. Paint Technol.*, **38** (1966) 493.
28. E. Daubach, *Farbe u. Lack*, **70** (1964) 687.
29. W. Herbst and K. Merkle, *Deutsche Farben Zeit*, **8** (1970) 365.
30. A. C. Cooper, *J. Oil Colour Chem. Assoc.*, **31** (1948) 343.
31. A. R. Hannam and D. Patterson, *J. Soc. Dyers Colourists*, **79** (1963) 192.
32. A. Brockes, *Optik*, **21** (1964) 550.
33. P. Kubelka and F. Munk, *Zeits. für tech. Physik*, **12** (1931) 593.
34. B. Felder, *Helv. Chim. Acta*, **47** (1964) 688.
35. B. Felder, *Helv. Chim. Acta*, **49** (1966) 440.
36. H. Nassenstein, *Ber. Bunsenges. Phys. Chem.*, **71**(3) (1967) 303.
37. J. Beresford, W. Carr and G. A. Lombard, *J. Soc. Dyers Colourists*, **81** (1965) 615.
38. E. Atherton, A. C. Cooper and M. R. Fox, *J. Soc. Dyers Colourists*, **80** (1964) 521.
39. J. Beresford, *J. Oil Colour Chem. Assoc.*, **50** (1967) 594.
40. M. Jones, *Soc. Anal. Chem., Proceedings*, **3** (1966) 116.
41. V. T. Crowl, Proceedings Particle Size Analysis Conference, Loughborough, *Soc. for Anal. Chem.* (1966) 36.
42. Metals Research Ltd., Hertfordshire, England.
43. V. T. Crowl, *J. Oil Colour Chem. Assoc.*, **50** (1967) 1023.
44. W. Carr, *J. Paint Tech.*, **42** (1970) 696.
45. P. Hauser, M. Herrmann and B. Honigmann, *Farbe u. Lack*, **76** (1970) 545.
46. J. Beresford, Advances in the Chemistry and Technology of Pigments. Leeds University, July 1971.
47. F. M. Smith and J. D. Easton, *J. Oil Colour Chem. Assoc.*, **49** (1966) 614.

INDEX

Absorption oil, 164, 263, 341
Adhesion, 68, 71, 74
  force of, 74
  work of, 3, 68, 388
Adhesional wetting, 3
Adhesives, dispersion equipment for, 245
Adsorption, 58, 77, 87, 98, 134
  electrolytes, of, 80
  isotherm,
    for gases, 58
    for solutions, 77
  non-electrolytes, of, 78
  polymers, of, 80
  solution, from, 63, 77
Ageing, 211
Agglomerate, 2, 270, 357, 387
  breakdown of, 353
  pigment-medium, 272
Aggregate, 2, 270, 387
  breakdown of, 353
  cemented, 271
  sintered, 271
Alkyd resins, 320
  solubility of, 322
  structure of, 321
Amonton's law, 76
Amphipathic behaviour, 133
Andreason pipette, 285
Anti-skinning agents, 310
Attraction, forces of, 14, 44
Attractive potential energy, 14, 44
Attritor mills, 240, 402
Azo pigments, 385

BET,
  equation, 58, 79
  method, 58, 60, 62
Ball mills, 237, 402
  horizontal, 237
  planetary, 239
  vibratory, 238
Ball-point, Daniel, 347
Beilby layer, 52
Bénard cells, 298, 377
Bingham flow, 279
Bingham plastic, 353
Boltzmann distribution, 102
Booth equation, electrophoresis, 128
Brabender plastograph, 262
Brownian motion, 10, 272

Capacity of double layer, 104
Capillary condensation, 66
Cationic dye complexes, 386
Ceramics, dispersion equipment for, 246
Charge,
  discreteness of, 99
  zero point of, 96
Chemicals, dispersion equipment for, 246
Cluster, 2
  formation and stoichiometry of, 177
Coagulation, 2
Coalescence, 213
Coating of pigments, 336
Cohesion, 68
Cohesive energy density, 323
Colloid mill, 237
Colloidal shock, 363
Colour of inorganic pigments, 313
Colour strength, 313
Composite isotherm, 78

Concentrated dispersions, 39
Contact angle, 4, 68, 138
  hysteresis of, 70
  measurement of, 6, 70, 139
Coulombic forces, 14, 17, 45
Coulter Counter, 284, 288, 394
Critical cluster size, nucleation, 194
Critical pigment volume concentration, 345
Critical supersaturation, 176
  ratio, 184
Crystallisation maximum, 196
Crystallites, 270

Daniel Flow Point, 224, 347
Debye–Huckel linear approximation, 104
  parameter, 96
Defects, Shottky and Frenkel, 64
Dendritic crystal, 208
Dielectric properties of dispersions, 302
Diffuse double layer, 80, 96, 102
  free energy of, 109
Diffusion, surface, 52
Dilatant flow, 279
  system, 351
Disc cavitation mixers, 234, 402
Discrete ion effect, 99
Discreteness of charge effect, 99
Dispersers, dry disc, 236
Dispersibility, 3
Dispersible pigments, 166, 406
Dispersing agent, 310
Dispersion, 1, 267
  colloidal, 1
  degree of, 267
    assessment of, 282
  equipment for,
    adhesives, 245
    ceramics, 246
    chemicals, 246
    dyestuffs, 249
    paint, 247
    paper, 248
    pharmaceuticals, 249
    pigments, 249

Dispersion—*Contd.*
  equipment for—*Contd.*
    plastics, 250
    printing inks, 251
    refractories, 246
    rubber, 252
  formulation, 255
  inorganic pigments, of, 339, 353
  media, 318
  operations, economics of, 253
  optimisation, 255
  organic pigments, of, 405
  process, three stages in, 1, 272, 340
  spontaneous, 5
  stability, 3, 10, 267
    assessment of, 282
  state of, 268
  technical aspects of, 221
  technological properties of, 273
Distortion, surface, 55
DLVO theory, 13
  application to,
    aqueous systems, 23
    non-aqueous systems, 25
Dorn effect, 111
Double layer, 17, 86
  capacity of, 104
  electric (*see* Electric double layer)
  free energy of, 109
  Gouy–Chapman diffuse, 80, 96
  solid side of interface, on, 107
  Stern, 23, 80, 94
Driers, 310
Dupré equation, 3, 4, 68, 73
Dyestuffs, dispersion equipment for, 249

Economics of dispersion operations, 253
Edge energy, 57, 183
Einstein equation, viscosity, 296
Elasticity, surface, 153
Electric double layer, 17, 76, 86, 140
  diffuse part, 102
  inner part, 94
  origin of, 86
Electric potential, 87

Electrical conductivity of dispersions, 301
Electrochemical potential difference, 88
Electrokinetic potential, 97, 109, 112
zero point of, 97
Electrokinetic properties, 111
Electron diffraction, 52
Electron micrographs of paint films, 300, 304
Electron microscopy, 62
for size analysis, 287
Electro-osmosis, 111
Electrophoresis, 111
Electrophoretic effect, 127
Electrophoretic mobility,
accuracy of, 124
measurement of, 27, 114
zero point of, 97
Electrostatic forces, 46
Energy,
edge, 57, 183
potential,
attraction, of, 44
repulsion, of, 44
specific surface free, 51
surface, 49, 182
free 53, 71, 204
total, 49
total,
ionic crystals, for, 56
rare gas crystals, for, 55
Energy barrier,
flocculation, to, 13
nuclei formation, to, 181
Entropic repulsion, 33
Entropy surface, 49
Epitaxy, 185, 190
Esin–Markov effect, 99

False-bodied systems, 353
Fastness of organic pigments, 384
Ferranti–Shirley viscometer, 296 224, 227, 261
Field emission microscope, 52
Flexographic inks, 251
Flocculate, air, 271
Flocculating action, 39
Flocculation, 2, 213, 267, 340
concentration, 24
induced viscosity, 353
maximum rate of, 10
ortho-kinetic, 13
slow, rate of, 22
Flooding, 30, 153, 160, 298, 376
Flotation,
froth, 71
paints, 30, 160, 298, 376
Flow,
Bingham, 279
dilatant, 279
Newtonian, 279
plastic, 74, 279
pseudoplastic, 279
thixotropic, 280
viscous, 74
Flow-point, Daniel, 224, 347
Fluid-energy milling, 331
Flushing, 168, 170, 269
Foaming, 152
Forces,
adhesive, 74
attractive, 14, 44
between,
ions, 45
molecules, 44
permanent dipole and ion, 46
two permanent dipoles, 46
coulombic, 14, 17, 45
electrostatic, 46
repulsive, 14, 17, 44
Formulation of mill base, 263, 347
Free energy,
specific surface, 51
surface, 53, 71, 204
Frenkel defects, 64
Friction,
rolling, 77
sliding, 76
Fuchs' theory, flocculation, 22

Galvani potential difference, 90
Gibbs equation, 87, 138
Gibbs free energy, 35, 49, 138
of critical nucleus, 177

Gibbs–Kelvin equation, 181, 184, 214
Gibbs–Wulff condition, 182, 204
Gloss, 373
Gouy–Chapman diffuse layer, 24, 81, 96, 102
Gravure inks, 251
Grinding, 296
Growth,
  diffusion controlled, 200
  interface controlled, 200

Hamaker constant, 14, 23, 34
  evaluation of, 14
Hamaker theory, 14
Hegman gauge, 282
Henry equation, 127
Heteroflocculation, 30
Heterogeneity of solid surfaces, 63
Heterogeneous nucleation, 183
Hiding power, 377
High shear rate mixers, 234
Homogeneous nucleation, 181
  rate of, 182
Homotattic surface, 53
Horizontal ball mills, 237
Hückel equation, 126
Hue, 383
Hydrophilic–lipophilic balance (HLB), 137, 143
Hydrophilic sol, 1
Hydrophobic sol, 1
Hynetic (Kady) mills, 236, 404
Hysteresis,
  adsorption isotherms, in, 66
  contact angle, of, 70

Immersional wetting, 3
Inner Helmholtz plane, 91
Inorganic pigments,
  classification of, 308
  colour of, 313
  dispersion of, 339
  manufacture of, 312
  milling (micronising) of, 331
  opacity of, 313
  particle size of, 313
Inorganic pigments—*Contd.*
  physical and chemical aspects of, 327
  surface chemistry of, 335
  wetting of, 355
Interface,
  gas–liquid, 50
  liquid–liquid, 50
  solid–liquid, 44
Interfacial energy,
  nucleus size, and, 194
  precipitate morphology, and, 215
Interfacial tension, 49, 138, 388
Isoelectric point, 180
Isotherm,
  adsorption, 58, 77
  composite, 78
  individual, 79
  standard, 67

Jet milling, 331

Kady (Hynetic) mill, 236, 404
Kelvin equation, 64
Kubelka–Munk theory, 313, 393

Lakes, 385
Langmuir equation, 77, 79
Laplace equation, 53, 65, 75
Lattice match, nucleation, 189
Let-down process, 363
Levelling of paint films, 296
Lifshitz theory, 15
Light absorption coefficient, 313, 393
Light scattering coefficient, 313, 393
Linseed oil, 319
Lithographic inks, 251
London constant, 14
London–van der Waals forces, 14
Low shear rate equipment, 230
Lubrication, 76
Lyophilic sol, 1
Lyophobic,
  sol, 1, 113
  solid surface, 112

Macropore, 65
Mechanical breakdown of clusters, 8, 272
Mercury porosimeter, 65
Metal chelates, 385
Metastable,
  cluster, critical, 181
  limit, 176
  phase, 177
Micelle formation, 137
Microelectrophoresis, 115
  cells,
    choice of, 117
    circular cross section, of, 115, 123
    rectangular cross section, of, 116
    stationary levels in, 115
Microflow mills, 241
Micronising of pigments, 330
Micropore, 65
Micro-potential, 99
Microscopic counting for particle size, 286
Mill base formulation, 280, 347
Milling, jet-, 331
Mills,
  attritor, 240, 402
  ball, 237, 402
  colloid, 237
  horizontal ball, 237
  hynetic (Kady), 236, 404
  microflow, 241
  multiple-roll, 244
  perl, 241, 402
  planetary ball, 239
  pug, 230, 402
  roll, 242
  sand, 241, 402
  single-roll, 243
  triple-roll, 244, 401
  two-roll, 244, 402
  vibratory ball, 238
Mixers, 229
  disc cavitation, 234, 402
  high shear rate, 234
  low shear rate, 230
  planetary, 231
  stator rotor, 236
Mixers—*Contd.*
  trough, 230
  Z blade, 231
Mobility, surface, 51
Monolayer capacity, 59, 63
Morphology and supersaturation, 214
Moving boundary method, electrophoresis, 115
Multiple-roll mills, 244

Nernst equation, 90, 98
Newtonian,
  flow, 279, 346
  liquid, 279
Nibs, 272
Nucleation,
  electric field, effect of, 216
  heterogeneous, 183
  homogeneous, 181
    rate of, 182
  mechanism, effect on precipitate characteristics, 195
  polymers, of, 184
  secondary or ancillary, 185
  theory, 181
  theory, criticisms of, 184
    experimental tests of, 190
Nucleators, 194
Nucleus, 177
  critical, 177
  stoichiometry of, 178

Oil absorption, 164, 263, 341
  value, 341
Opacifying power, 314
Opacity, 393
Optical corrections, microelectrophoresis, 123
Optical microscopy for size analysis, 286
Optical properties,
  dispersions, of, 297
  inorganic pigments, of, 313
Organic pigments,
  classification of, 384

Organic pigments—*Contd.*
dispersible, 406
dispersion of, 388, 394, 405
particle characteristics of, 387
pre-dispersed, 406
surface area of, 394
Ortho-kinetic flocculation, 13
Outer Helmholtz plane, 94
Overbeek equation, electrophoresis, 128

Paint,
dispersion equipment for, 247
emulsion, 147
film, gloss, 373
film, structure, 367
films, electron micrographs of, 300, 304
water-based stoving, 154
Paper, dispersion equipment for, 248
Particle concentration effect, flocculation, 29, 39
Particle diameter,
arithmetic mean, 328
length mean, 328
median, 328
volume–surface mean, 328
Particle, primary, 270
Particle shape, 327
Particle size analysis,
Coulter Counter, 288
methods, 283
microscopic counting, 286
sedimentation methods, 284
sieving, 284
Particle size distribution, 328, 395
Pellets, pigment, 358
Penetration of liquid into capillaries, 6
rate of, 7
Perl mills, 241, 402
Pharmaceuticals, dispersion equipment for, 249
Phthalocyanine pigments, 386
Pigment dispersion,
aqueous media, in, 145
equipment for, 249
non-aqueous media, in, 156
Pigmented film structure, 315
Planetary ball mills, 239
Planetary mixers, 231
Plastic flow, 74, 279
Plastic viscosity, 279
Plastics, dispersion equipment for, 250
Point B method, 59
Poisson–Boltzmann equation, 24, 102
Poisson equation, 102
Polarisability, 47
Polycyclic pigments, 386
Pore,
macro-, 65
micro-, 65
radius of, 65
transitional, 65
volume, 65
Porosimeter, mercury, 65
Porosity, 65
Porous solids, 65
Potential,
determining ions, 90
difference,
electrochemical, 88
Galvani, 90
Volta (contact), 90
electric, 87
electrokinetic (zeta), 97, 109, 112
energy,
attraction, of, 14, 44
energy, curve, 20, 45
repulsion, of, 19, 44
total, 19, 44
inner (Lange), 89
ki, 89
micro-, 99
outer (Lange), 89
self-atmosphere, 99
Stern, 23, 38, 76, 101
streaming, 111
surface (wall), 89, 95
Powders, nature of, 268
Precipitates, morphology of, 184, 197, 203, 214
Precipitation, 175
curves, 202
electric field, effect of, 216
homogeneous solution, from, 198

Precipitation—*Contd.*
  kinetics, 197
  rate of, 200
Pre-dispersed pigments, 406
Printing inks, dispersion equipment for, 251
Protective action, 38
Pseudoplastic flow, 279
Pug mills, 230, 402

Recrystallisation, 211
Refractories, dispersion equipment for, 246
Rehbinder effect, 9, 354
Relaxation effect, 127
Repulsive forces, 14, 17, 44
Rheogoniometer, Weissenberg, 296
Rheological behaviour, 261, 277, 340, 350
Rheological properties, 295, 384
Rheopexy, 281
Rittinger's number, 353
Roll mills, 242
  multiple, 244
  single, 243
  triple, 244, 401
  two, 244, 402
Rolling friction, 77
Roughness,
  factor, 329
  surface, 69
Rubber, dispersion equipment for, 252
Rugosity, 329

Sagging of paint films, 296
Sand mills, 241, 402
Schulze–Hardy rule, 23
Secondary nucleation, 185
Sedimentation behaviour, 275, 291
  methods for particle size analysis, 284, 394
  potential, 111
  rate, 291
  volume, 292
Seed crystals, 201
Self-atmosphere potential, 99
Shear increasing viscosity, 352
Shottky defect, 64
Sieve measurements, 284
Single-roll mills, 243
Sintering, 51
Slater–Kirkwood expression, 15
Sliding friction, 76
Smoluchowski,
  equation (electrophoresis), 126
  theory (flocculation), 12
Smoothness,
  factor, 329
  ratio, 330
Sol,
  hydrophilic, 1
  hydrophobic, 1
  lyophilic, 1
  lyophobic, 1
Solid–liquid interface, 44
Solubilisation, 137
Solubility parameter, 323, 324
Specific surface area, 58, 62, 79
Spreading, 70
  coefficient, 70, 138
  spontaneous, 5
  wetting, 3
Stability, 3, 10, 272, 360
  adsorbed layers, due to, 31, 379
  concentrated dispersions, of, 39
  DLVO theory of, 13
  ratio, 22
Staining power, 313, 318
Standard isotherm, 67
Stator rotor mixers, 236
Steric factors in stability, 33, 140
Stern,
  adsorption isotherm, 99
  double layer, 80, 94
  plane, 81
  potential, 23, 38, 76, 101
Stickiness of particles, 75
Stokes law, 275, 284
Streaming potential, 111
Supercooling, 176
Supersaturation, 176
  critical, 176
    ratio, 184
  precipitate morphology, and, 214

Surface,
  active agents,
    aqueous media, in, 142
    classification of, 134
    non-aqueous media, in, 143
    properties of, 136
    selection of, 141
  activity, 133
  area, 58, 62
  coatings, 336
  composition of inorganic pigments, 335
  diffusion, 52
  distortion, 56
  elasticity, 153
  energy, 49, 182
    calculation of, 54
    ionic crystals, for, 56
    rare gas crystals, for, 55
  entropy, 49
  excess, 138
  free energy, 53, 71, 204
    specific, 51
  homotattic, 54
  mobility, 51
  phase, 48
  potential, 89
  roughness, 69
  solids, of, 50
    conditioning, 52
  tension, 48, 138
    liquids, of, 50
    solids, of, 53
    solubility, and, 64
  tension, surface free energy, and, 53

t-plot, 67
Technological properties of dispersions, 273
Textile printing, 169
Texture, 164, 345
Thixotropic,
  flow, 280
  plastic, 353
Thixotropy, 140, 272, 280, 353
Tinctorial strength, 384
Tinters, universal, 167
Tinting strength, 313, 318
Traffic jam mechanism, 209
Transitional pore, 65
Triple-roll mills, 244, 401
Trough mixers, 230
Two-roll mills, 244, 402

Undertone, 317
Universal tinters, 167

Vat pigments, 386
Vibratory ball mills, 238
Viscose pigmentation, 146
Viscosity, 278
  apparent, 295
  Einstein equation of, 296
  flocculation induced, 352
  kinematic, 278
  shear increasing, 352
Viscous flow, 74
Volta potential difference, 90

Wall potential, 89, 95
Washburn equation, 7, 357
Wedge gauge, 282
Weissenberg rheogoniometer, 296
Wet-point, Daniel, 347
Wetting, 2, 3, 68, 73, 138, 272, 340, 388
  adhesional, 3
  agent, 73, 310
  immersional, 3
  pigments, of, 355
  spreading, 3
Wiegner tube, 285
Work of adhesion, 3, 68, 388

Yield point, 228, 279, 341
Young's equation, 4, 68, 73, 138

Z blade mixers, 231
Zero point,
  charge, of, 96
  electrokinetic potential, of, 97
  electrophoretic mobility, of, 97
Zeta potential, 97, 109, 112, 378
  electrophoretic mobility, from, 125
  stability, and, 27